HYBRID ROCKET PROPULSION DESIGN HANDBOOK

HYBRID ROCKET PROPULSION DESIGN HANDBOOK

ASHLEY CHANDLER KARP

ELIZABETH THERESE JENS

ACADEMIC PRESS
An imprint of Elsevier

ELSEVIER

Academic Press is an imprint of Elsevier
125 London Wall, London EC2Y 5AS, United Kingdom
525 B Street, Suite 1650, San Diego, CA 92101, United States
50 Hampshire Street, 5th Floor, Cambridge, MA 02139, United States
The Boulevard, Langford Lane, Kidlington, Oxford OX5 1GB, United Kingdom

Notices

Knowledge and best practice in this field are constantly changing. As new research and experience broaden our understanding, changes in research methods, professional practices, or medical treatment may become necessary.

Practitioners and researchers must always rely on their own experience and knowledge in evaluating and using any information, methods, compounds, or experiments described herein. In using such information or methods they should be mindful of their own safety and the safety of others, including parties for whom they have a professional responsibility.

To the fullest extent of the law, neither the Publisher nor the authors, contributors, or editors, assume any liability for any injury and/or damage to persons or property as a matter of products liability, negligence or otherwise, or from any use or operation of any methods, products, instructions, or ideas contained in the material herein.

ISBN: 978-0-12-816199-9

For information on all Academic Press publications
visit our website at https://www.elsevier.com/books-and-journals

Publisher: Matthew Deans
Acquisitions Editor: Chiara Giglio
Editorial Project Manager: Emily Thomson
Production Project Manager: Fizza Fathima
Cover Designer: Matthew Limbert

Typeset by VTeX

Dedication

Dedicated to Lachlan, Kira and Chloe.

The love and support of our families, especially Ross Allen and Jonathan Karp, enabled us to write this book.

We appreciate reviews, input, and support for this book given to us by Brian Cantwell, Arturo Casillas, Marty Chiaverini, Jason Rabinovitch, George Story, and Greg Zilliac. A special thank you goes to Adrien Boiron and Martina Faenza for developing the regression rate chapter material with us. We would also like to acknowledge Brian Cantwell, Arif Karabeyoglu, and Greg Zilliac for allowing us to use some material from their courses. Finally, to the wonderful propulsion engineers who have mentored us on our journey, we hope you see the benefit of your time and insights throughout these pages.

This work was done as a private venture and not in the authors' capacity as employees of the Jet Propulsion Laboratory, California Institute of Technology.

Contents

8. Hybrid design challenges

9. Hardware design

10. Design examples

11. Testing

List of tables

List of figures

Nomenclature

Symbol	Description
n_{stages}	Number of stages in a turbo pump [-]
V_{SS}	Suction specific speed for pumps [rad m$^{0.75}$/s$^{1.5}$]
α	Angle of attack [°]
α_L	Coefficient of linear expansion [K^{-1}]
α_{pump}	Empirical pump mass scaling factor [S.I.]
$\mathcal{R}$	Nozzle area ratio [-]
$\bar{g}_e$	Gibbs Free Energy per mol [J/mol]
$\bar{h}_f$	Specific enthalpy of formation [J/mol]
$\bar{h}_{fusion,i}$	Latent heat of fusion [J/mol]
β_{pump}	Empirical pump mass exponent [-]
Δf	Frequency resolution [Hz]
ΔP	Change in pressure [Pa]
ΔP_{pump}	Pressure change required to be generated by the pump [Pa]
ΔV	Change in velocity [m/s]
δ	Momentum boundary layer thickness [m]
δ_l	Characteristic thermal thickness in the liquid phase [m]
$\dot{m}_O$	Oxidizer mass flow rate [kg/s]
$\dot{m}$	Mass flow rate [kg/s]
$\dot{m}_F$	Fuel mass flow rate [kg/s]
$\dot{m}_O$	Oxidizer mass flow rate [kg/s]
$\dot{m}_p$	Propellant mass flow rate [kg/s]
$\dot{m}_{ent}$	Entrained mass flow rate per unit area [kg/m^{2}s]
$\dot{m}_{out}$	Mass flow rate out of the combustion chamber [kg/s]
$\dot{Q}_c$	Convective heat transfer at the fuel surface [J/m^{2}s]
$\dot{Q}_r$	Radiative heat transfer at the fuel surface [J/m^{2}s]
$\dot{r}$	Regression rate [m/s]
$\dot{r}_{ent}$	Regression rate of fuel from entrainment mechanism [m/s]
$\dot{r}_{ins}$	Insulator ablation rate [m/s]
$\dot{r}_v$	Regression rate of fuel from vaporization [m/s]
ϵ	Strain [-]
ϵ	Surface roughness [m]
ϵ_j	Dry mass fraction of stage j [-]
η_{C^*}	Combustion efficiency [-]
η_n	Nozzle efficiency [-]
η_{pump}	Pump efficiency [-]
Γ	Payload mass fraction of rocket [-]
γ	Ratio of specific heats [-]
γ_{fpa}	Flight path angle [°]
κ	Thermal diffusivity [m^2/s]
λ	Thermal conductivity [W/(m.K)]
λ_j	Payload mass ratio of stage j [-]
λ_{ij}	Eigenvalue of the i-j acoustic mode [$-$]
$\mathfrak{P}_{pump}$	Pump power [W]
μ	Dynamic viscosity [kg/(m.s)]
ν	Poisson's ratio [-]
ν_j	Propellant mass fraction of stage j [-]
$\overline{P_c}$	Average combustion chamber pressure [Pa]
Φ	Equivalence ratio [-]
ρ	Density [kg/m^3]
ρ_F	Fuel density [kg/m^3]
ρ_g	Density of the gas [kg/m^3]
ρ_{ins}	Insulation density [kg/m^3]
$\rho_{press,ullage}$	Density of the pressurant gas in the oxidizer tank ullage [kg/m^3]
ρ_{press}	Density of the pressurant gas in the pressurant tank [kg/m^3]
ρ_{tank}	Tank material density [kg/m^3]
σ_r	Radial Stress [N/m^2]
σ_t	Surface tension [N/m]
σ_θ	Hoop Stress [N/m^2]
σ_B	Stefan-Boltzmann constant
σ_y	Yield stress [Pa]
τ_w	Wall shear stress [Pa]
τ_{pump}	Pump shaft torque [N.m]
θ	Angle between thrust vector and horizontal reference [deg]
θ_e	Nozzle half-angle (conical) or exit angle (bell) [deg]
θ_f	Dimensionless variable to account for the velocity profile distortion caused by the flame [-]
V	Volume [m^3]
V_C	Combustion Chamber Volume [m^3]
V_{free}	Free Volume [m^3 or in^3]
V_{ins}	Insulation Volume [m^3]
V_{post}	Volume of the post combustion chamber [m^3]
V_{press}	Total volume of the pressurant tanks [m^3]
V_{pre}	Volume of the pre combustion chamber [m^3]
V_p	Propellant Volume [m^3]
V_{ullage}	Total volume of the oxidizer tank ullage [m^3]
ξ_j	Stage optimization variable [-]
ζ	Temperature recovery factor [-]
A	Cross sectional area [m^2]
a	Empirical regression rate coefficient for regression rate in terms of total mass flux [SI]

a	Speed of sound [m/s]
A_b	Area of the burning surface [m^2]
a_H	Semimajor axis of the Hohmann transfer ellipse [m]
a_{ent}	Regression rate coefficient of the entrained fuel [SI]
A_e	Nozzle exit area [m^2]
A_e	Nozzle exit cross sectional area [m^2]
a_g	Absorptivity of the gas
a_{HL}	Hard liquid speed of sound [m/s]
A_{inj}	Area of a single injector element (hole) [m^2]
a_l	Absorption coefficient of the liquid [m^{-1}]
a_{orbit}	Semimajor axis of the ellipse [m]
a_o	Empirical regression rate coefficient in terms of only oxidizer mass flux [SI]
A_{th}	Nozzle throat cross sectional area [m^2]
B	Blowing parameter [-]
C	Effective exhaust velocity [m/s]
C^*	Characteristic velocity [m/s]
C_D	Drag coefficient [-]
C_d	Discharge coefficient [-]
C_F	Thrust coefficient [-]
C_f	Skin friction coefficient [-]
C_L	Lift coefficient [-]
c_l	Specific heat of a liquid per unit mass [J/kgK]
c_s	Specific heat of a solid per unit mass [J/kgK]
$C_{calorimeter}$	Specific heat of a bomb calorimeter [J/K]
C_f	Coefficient of friction [-]
C_{H0}	Stanton number without blowing [-]
C_H	Stanton number with blowing [-]
c_p	Specific heat capacity at constant pressure [J/(kg K)]
$C_{V,valve}$	Valve flow coefficient [-]
c_v	Specific heat capacity at constant volume [J/(kg K)]
D	Diameter [m]
D_h	Hydraulic diameter [m]
D_m	Mass diffusion coefficient [m^2/s]
D_o	Fuel grain outer diameter accounting for sliver fraction [m]
D_f	Final diameter of the fuel grain port (at the end of the burn) [m]
D_{inj}	Hydraulic diameter of the injector element (hole) [m]
D_i	Initial diameter of the fuel grain port [m]
D_p	Fuel grain port diameter [m]
D_{th}	Nozzle throat diameter [m]
Da	Damköhler number [-]
E	Elastic modulus [Pa]
E	Total energy [J]
e	Specific total energy [J/kg]
e_g	Emissivity of the gas [-]
e_w	Emissivity of the wall [-]
F	Thrust [N]
f	Friction factor [-]
F_D	Aerodynamic drag force [N]
F_G	Gravitational force [N]
F_L	Aerodynamic lift force [N]
f_{ijk}	Acoustic mode frequency where i, j, k are the mode number for the tangential, radial and longitudinal mode, respectively [Hz]
f_{ILFI}	Primary frequency of the intrinsic low frequency instability [Hz]
f_s	Sampling frequency [Hz]
Fr	Correction factor for surface roughness [-]
G	Gravitational constant $= 6.674 \times 10^{-11}$ [m^3/(kg.s^2)]
G	Total (fuel and oxidizer) mass flux [kg/m^{2}s]
g	Gravitational acceleration [m/s^2]
g_0	Earth gravity $= 9.81$ [m/s^2]
G_e	Gibbs Free Energy [J]
G_o	Oxidizer mass flux [kg/m^{2}s]
H	Enthalpy [J]
h	Specific enthalpy [J/kg]
h_c	Enthalpy of the gas at the flame [J/kg]
h_{ent}	Total heat of entrainment [J/kg]
h_{fg}	Latent heat of vaporization [J/kg]
h_{fusion}	Latent heat of fusion [J/kg]
h_f	Enthalpy of formation [J/kg]
h_m	Total heat of melting [J/kg]
H_p	Head rise across a pump [m]
h_t	Stagnation enthalpy [J/kg]
h_v	Total effective heat of gasification [J/kg]
h_w	Enthalpy of the gas at the wall [J/kg]
I	Total impulse [Ns]
I_{sp}	Specific impulse [s]
K	Minor loss coefficient [$-$]
K_P	Pressure Based Equilibrium Constant [-]
K_{prop}	Bulk modulus of elasticity of the propellant [Pa]
K_{sc}	Stress concentration factor [-]
k_s	Safety factor
L	Length [m]
l	Line length [m]
L^*	Characteristic length of the combustion chamber [m]
L_F	Fuel grain length [m]
L_n	Nozzle length [m]
L_{pcc}	Post combustion chamber length [m]
$L_{tank,cyl}$	Tank cylindrical section length [m]
Le	Lewis number [-]
M	Mach number [-]
m_d	Dry or non-propellant mass [kg]
M_e	Mach number at the nozzle exit [-]
m_F	Fuel mass [kg]
m_f	Final mass [kg]
m_g	Mass of gas in the combustion chamber [kg]
m_i	Initial mass [kg]
m_L	Payload mass [kg]

m_n	Mass of the nozzle [kg]	T	Temperature [K]
m_O	Oxidizer mass [kg]	t_0	Initial time [s]
m_p	Usable propellant mass [kg]	t_b	Burn time [s]
m_t	Tank mass [kg]	t_d	Characteristic fluid diffusion time [s]
M_w	Molecular weight [kg/mol]	t_k	Characteristic kinetic reaction time [s]
$m_{d,j}$	Dry mass of stage j [kg]	T_a	Initial fuel temperature [K]
$m_{f,j}$	Final mass of stage j [kg]	t_{close}	Time required for a valve to close [s]
$m_{i,j}$	Initial mass of stage j [kg]	$T_{e,x}$	Freestream temperature at location x [K]
m_{inert}	Inert mass ablated/combusted during a test [kg]	T_g	Average gas phase temperature [K]
m_{ins}	Insulation mass [kg]	t_{ins}	Time that the rocket insulation is designed for [s]
m_O	Total mass of oxidizer [kg]	T_m	Melting temperature [K]
$m_{p,j}$	Propellant mass of stage j [kg]	t_m	Burn time margin [s]
m_{planet}	Planet mass [kg]	$T_{t,c}$	Total temperature in the combustion chamber [K]
m_{press}	Total mass of pressurant gas [kg]	$T_{t,th}$	Total temperature in the nozzle throat [K]
m_{pump}	Mass of a turbopump [kg]	T_v	Vaporization temperature [K]
m_{tank}	Tank mass [kg]	$T_{w,x}$	Wall temperature at location x [K]
MR	Mixture ratio, the oxidizer to fuel mass ratio [-]	th_w	Tubing wall thickness [m]
N	Number of samples	th_{cyl}	Thickness of cylindrical tank [m]
n	Empirical regression rate exponent [-]	th_{ell}	Thickness of an ellipsoidal end cap [m]
n	Number of moles [-]	th_l	Melt layer thickness [m]
N_A	Molecules per mole. Avogadro number $= 6.022141 \times 10^{23}$ [mol^{-1}]	th_{sphere}	Thickness of spherical tank [m]
		$tol_{\dot{m}_F}$	Convergence tolerance for the fuel mass flow rate (η_{C*} method) [kg/s]
N_s	Pump stage-specific speed [$\mathrm{rad\ m^{0.75}/s^{1.5}}$]	U	Internal energy [J]
N_{inj}	Number of injector elements (holes) [-]	u	Specific internal energy [J/kg]
N_{pump}	Pump rotation rate [rad/s]	V	Velocity [m/s]
$NPSH$	Net Positive Suction Head for pumps [m]	v	Specific volume [$\mathrm{m^3/kg}$]
O/F	Oxidizer to fuel mass ratio [-]	V_e	Velocity at the nozzle exit [m/s]
$O/F\vert_{stoic}$	Stoichiometric oxidizer to fuel mass ratio [-]	V_y	Velocity in the y-direction [m/s]
P	Pressure [Pa]	V_{esc}	Escape velocity [m/s]
P_a	Ambient pressure [Pa]	V_{orbit}	Velocity of target orbit [m/s]
P_e	Pressure at the nozzle exit [Pa]	We	Weber number [-]
P_c	Combustion chamber pressure [Pa]	x	Position coordinate, distance [m]
P_d	Dynamic pressure [Pa]	X_i	Mole fraction of species i [-]
P_{inlet}	Inlet pressure [Pa]	y_c	location of the flame [m]
$P_{t,c}$	Total pressure in the combustion chamber [Pa]	Y_i	Mass fraction of species i [-]
P_{vapor}	Vapor pressure [Pa]	Y_{exp}	Expansibility factor for compressible flow through a venturi [-]
Pr	Prandtl number [-]		
R	Specific gas constant [J/(kg.K)]	Z	Compressibility factor [-]
r	Radius [m]		
r_f	Final fuel port radius [m]		
r_i	Initial fuel port radius [m]	**Superscripts**	
r_{cc}	Combustion chamber radius [m]	$*$	Sonic condition
r_{cyl}	Radius of cylindrical tank [m]		
R_{orbit}	Instantaneous radius from the center of the planet [m]	**Subscripts**	
r_{sphere}	Radius of spherical tank [m]	$\oplus$	Earth
r_{tank}	Tank radius [m]	$\mars$	Mars
R_u	Universal gas constant = 8.314 [J/mol K]	$\leftmoon$	Moon
Re_D	Reynolds number based on pipe diameter [-]	$\odot$	Sun
Re_x	Reynolds number based on distance x [-]	BOL	Beginning of life
S	Entropy [J/K]	c	Combustion
Sc	Schmidt number [-]	e	Boundary Layer Edge
SN	Swirl number [-]	ent	Entrainment
SN_g	Geometric swirl number [-]	EOL	End of life

F	Fuel	O	Oxidizer
f	Final	s	Solid
g	Gas	t	Total condition
i	Initial	v	Vaporization
l	Liquid	w	Wall

CHAPTER

1

Introduction

1.1 Chemical propulsion overview

Chemical propulsion systems are used to launch from the Earth, maintain or transfer orbit(s), and navigate the solar system. They turn chemical potential energy stored in the propellant combination into kinetic energy by accelerating the combustion products through a supersonic nozzle. Chemical propulsion systems are classified by the nature and the phase of the propellant. Thus the common types of chemical propulsion systems are: cold gas, monopropellant, solid, liquid bipropellant, and hybrid.

Fig. 1.1 shows cartoons of these common systems. A cold gas system simply accelerates the flow of pressurized gas through a nozzle. Helium and Nitrogen are often used for cold gas systems. Liquid monopropellants utilize a catalyst bed to decompose the propellant, and then the hot byproducts are accelerated through the nozzle to create thrust. Hydrazine is the most common monopropellant used today. However, hydrogen peroxide and nitrous oxide are also interesting options. Bipropellant liquid systems store the fuel and the oxidizer separately in the liquid phase. The propellants are combined in the combustion chamber via an injector, which is the crux of the design. Then products are accelerated through a nozzle. Storable bipropellants have more modest performance but can be stored under typical atmospheric conditions on

Earth and space (e.g., Monomethyl Hydrazine (MMH) and Nitrogen Tetroxide (NTO)). Cryogenic bipropellants have particularly high performance but must be stored at low temperatures (e.g., Hydrogen and Oxygen). In a solid rocket, the fuel and oxidizer are mixed in the correct proportion within a solid grain. Once combustion is initiated, it burns to completion at a rate determined by the propellant's chemical composition, operating pressure and temperature, as well as the geometry of the design. The most common propellant combination for solid rockets is Aluminum in a Hydroxyl Terminated Polybutadiene (HTPB) binder reacting with Ammonium Perchlorate (AP). Hybrid rockets, which typically utilize a solid fuel and liquid or gaseous oxidizer, are the focus of this book and will be described in more detail in the following chapters.

1.2 What is a hybrid rocket?

In the context of chemical propulsion systems, the term "hybrid" refers to the fact that the fuel and the oxidizer are stored in different phases. Since hybrid rockets use both solid and liquid (or gaseous) propellants, they generally combine propellants used for liquid and solid rockets. To this end, they are often believed to have the benefits or the challenges of both more

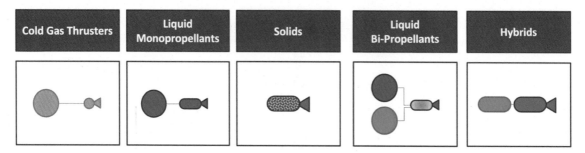

FIGURE 1.1 Cartoons of the most common chemical propulsion systems. Fuel is shown as blue, and oxidizer is depicted as green. In the case of the cold gas system, any gas is possible, including inerts, fuels, or oxidizers.

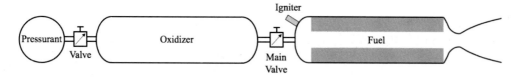

FIGURE 1.2 Standard Hybrid Rocket, including a pressurization system.

conventional configurations. It is the experience of these authors that either or both of them can be true. The diversity of propellants available to a hybrid configuration makes their suitability highly dependent on the application. This text endeavors to equip propulsion engineers with the tools necessary: both in theory and practice, to leverage the benefits of these hybrid rocket systems as much as possible while designing to mitigate the challenges.

Hybrid rockets are generally made up of a liquid (or gaseous) oxidizer and a solid fuel; see Fig. 1.2 for a simplified, standard, axial configuration. The reverse configuration, using a solid oxidizer and liquid or gaseous fuel is also possible but is rarely used due to a lack of high performance and storable (under Earth or space conditions) solid oxidizers [11]. The reader will find further discussion of this in Chapter 5. Gaseous oxidizers do not require a pressurant. Self-pressurizing oxidizers (e.g., N_2O) are generally valued for the simplicity that does not require a pressurant but may require additional pressurization to overcome feed system losses. Operation begins by opening a valve (or valves)

connecting the oxidizer tank to the combustion chamber, where the fuel grain is stored. Oxidizer flows from its storage tank, atomizes or vaporizes across an injector and passes over the fuel grain. An igniter is used to evaporate some of the fuel and provide the activation energy to the system to initiate combustion. Combustion develops within a turbulent boundary layer. The flame is situated above the solid fuel where the vaporized oxidizer and fuel exist in a combustible mixture ratio (Fig. 1.3). This process is self-sustaining.

Classical hybrid rocket combustion is depicted in Fig. 1.3. The oxidizer is shown moving from left to right over a solid fuel grain (gray). It shows four zones that describe the physical regions, heat transfer, and gas properties, including temperature (T), velocity (V), species mass fractions (Y), and movement of materials within the turbulent boundary layer. Combustion is diffusion limited in the classical case. This limited regression rate has been a challenge for hybrid rocket development.

One of the fundamental distinctions in hybrid rockets is based on how they combust. Classi-

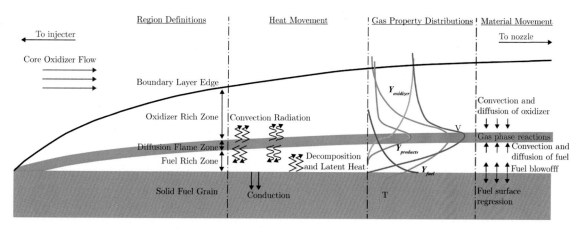

FIGURE 1.3 Classical Hybrid Rocket Combustion Process. Adapted from [12] and [13], also see [14].

cal hybrids follow the diffusion-limited combustion process described above. Others utilize liquefying or high regression rate fuels. These fuels form a liquid layer as they burn, just like a burning candle. When the oxidizer passes over the melt layer, the shear force between them breaks off liquid droplets and entrains them into the flow, essentially creating a fuel injection system (Fig. 1.4). Liquefying fuels burn much faster (3–5x) than classical fuels.

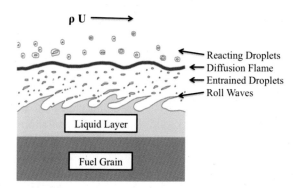

FIGURE 1.4 Liquefying Hybrid Rocket Combustion Process, adapted from [15].

Hybrids are also classified by how they burn. Axial hybrids are the most common and are defined by an oxidizer that flows from the fore end to the aft end of the rocket along a single or multiple ports that run the entire length of the fuel grain. Examples of axial hybrids include center perforated grains and wagon-wheel grains (Fig. 1.5 a and b, respectively). Swirl hybrids have either the oxidizer injected radially, with a swirl component or follow an (often 3D printed) swirling port down the length of the grain (Fig. 1.5, c and d). End-burning hybrids either have the oxidizer injected at the aft end of the grain or axially, through many tiny ports (e.g., Fig. 1.5). The latter introduces a pressure dependence on burn rate [16]. Finally, counter flow hybrids, where the oxidizer is injected at the aft end of the grain, circles to the fore-end, and then exits again at the aft, have also been tested (Fig. 1.5, f). Note that in multi-port fuel grains, the hydraulic diameters of each port must be equal, and the oxidizer distribution in the head end must be uniform to attempt to achieve even burning. (Subsonic flows like this are notoriously subject to instability that could be triggered by something as innocuous as separation from the entrance to one or more of the ports.)

Hybrid rockets are often referred to by the propellants being used, e.g., an HTPB/N_2O hybrid uses hydroxyl-terminated polybutadiene

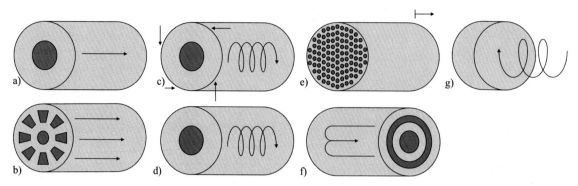

FIGURE 1.5 Types of Hybrids: a) Single Port Axial, b) Multiport Axial (Wagon Wheel), c) Swirl Injection, d) Swirling Port, e) Axial Injection End Burner (Pressure Dependent), f) Counter Flow End Burner, g) Swirl Injection End Burner (can also have two fuel disks). In all cases, the resulting thrust is from left to right.

fuel with nitrous oxide oxidizer. Occasionally additives to the fuel, such as metals, hypergols, strength additives or performance enhancers, can be included. However, while these additives can have a substantial impact on performance, they are not always clearly stated in the naming convention.

Hybrid rocket design requires knowledge of the regression rate of the fuel. The theoretical derivation of the regression rate for both classical and liquefying fuels will be presented in Chapter 3. The regression rate depends nearly exclusively on the oxidizer mass flux for typical hybrid operating conditions. Only a very weak dependence on the chamber pressure exists over most of the oxidizer flux range, allowing the chamber pressure to be optimized in the chamber design process (except in the case of axial end burners.) The hybrid design is dependent on its geometry, including the length-to-diameter ratio, port design and precombustion and post combustion chamber designs. Effects due to injection [17] and scaling up from small to large motors [18], [19] have also been demonstrated and must be considered in the design process. Hybrid rocket design will be discussed further in Chapter 7.

1.3 Comparison of chemical propulsion systems

Table 1.1 gives a comparison of chemical propulsion systems. The advantages and disadvantages of monopropellant, liquid bipropellant, and solid propulsion systems are compared to hybrid propulsion systems. Monopropellant systems are simple, capable of many restarts, and can be throttled but have low performance. They excel in small applications, e.g., Reaction Control Systems (RCS), but have also been used successfully in medium thrust class missions, e.g., the Main Lander Engines for the Skycrane, which landed the Perseverance and Curiosity rovers on Mars [20,21]. Solid rockets have low system complexity but modest performance. They are not tolerant to propellant cracks or debonding, which could lead to an explosion and a catastrophic system failure. Liquid systems are more complex because the dual feed system is required to deliver propellants from their storage tanks to the engine. However, like hybrids, they can be throttled, shutdown and restarted, and can perform non-destructive mission abort modes. Each propulsion system has a unique set of reasons it should be considered for a given mission, and the list for hybrid rockets has been lengthening steadily.

TABLE 1.1 Comparison of chemical propulsion systems. Notes: * NO can be added to NTO to depress the freezing point and reduce the storage temperature. * Performance with O_2 would be Medium-High. [†] Cryogenic liquids are typically stored at their normal boiling point. [‡] Actual temperatures vary by propellant. Typical range for Star Motors (Northrop Grumman) are given here.

Parameter	Monopropellant	Storable Bipropellant Liquid (e.g. NTO/MMH)	Cryogenic Bipropellant Liquid (e.g. H_2/O_2)	Solid (HTPB/Al/AP)	Classical Hybrid (e.g. multiport HTPB with N_2O)	Liquefying Hybrid (e.g. paraffin with N_2O)
Performance	Low	Medium-High	High	Medium	Medium*	Medium*
Thrust	Low-Medium	Medium-High	Medium-High	High	Low-Medium	Medium-High
Propellant density	Medium	Medium	Low	High	Medium	Medium
Shutdown/ restart	Yes	Yes	Yes	No	Yes	Yes
Throttling Complexity	Low (single liquid)	High (two liquids)	High (two liquids)	Very High	Medium	Medium
Utilization of propellants	High [Residual (prop) and hold up in lines]	Medium-High [Residuals (fuel, ox and mixture ratio) and hold up in lines]	Medium (fuel, ox and mixture ratio) and hold up in lines and propellant boil off	Very High	Low [Residuals (fuel, ox and mixture ratio) and hold up in lines]	Medium [Residual (ox, fuel and mixture ratio) and hold up in lines]
Freezing for Liquids/Low Temp Storage Only for Solids	2 °C [22]	−11 °C* [23]	−259.2 °C for H_2 and −218.4 °C for O_2 [24]	−20 °C to −55 °C[‡] [25]	−55 °C for N_2O [26] > −95 °C for HTPB [27]	−55 °C for N_2O [26] and > −100 °C [28] for paraffin
Boiling for Liquids/High Temp Storage for Solids	112 °C [22]	21.2 °C [23]	−252.8 °C for H_2 and −183.0 °C for $O_2^{[†]}$ [24]	−88 °C for N_2O [26] and 40 °C[‡] [25]	40 °C [TBC based on aging]	−88 °C for N_2O [26] and 40 °C for paraffin [29]
System complexity	Low	High	High	Low	Medium	Medium
Applications	Trajectory Correction Maneuvers, Reaction Control Systems, Station Keeping	Main Propulsion, Orbit Insertion	Launch Vehicles	Boosters, Main Propulsion, Spin/Despin	CubeSat, Main Propulsion (small)	Main Propulsion

It should be noted that the temperatures listed in Table 1.1 are not operational limits. Freezing and boiling temperatures are quoted for the liquids. The solid temperatures are harder to quantify (for example, there is a large spread in glass transition temperatures depending on the material composition), but estimates for storage conditions are provided. It is assumed that the high-temperature limit for HTPB is similar in the hybrid and solid configurations.

1.4 Benefits and challenges of hybrid propulsion systems

Interest in hybrid rockets was originally generated by the hope that they could embody the advantages of both their parent systems: liquid and solids. As more research is completed in hybrid rocket propulsion, benefits are continuously being realized. This same research is also identifying (and attempting to mitigate) challenges. The following two sections will introduce the most common benefits and challenges. Specific propellant combinations can introduce others, which will be discussed in Chapter 5.

1.4.1 Benefits

Hybrids can enjoy advantages over more conventional propulsion systems, most of which revolve around the inert nature of the fuel and the separation of the propellants. However, there are several other factors that contribute to their desirability. The following list enumerates the main benefits enjoyed by hybrid rockets. Similar lists can be found in [6] and [30].

1. Safety: Hybrid propulsion systems are inherently safer than liquid or solid systems. The fuel and oxidizer are separated by both phase and physical location. Therefore an accidental, intimate mixture is more difficult to obtain. Even in the event of an explosive situation, the amount of fuel and oxidizer that can react is dramatically less than in a solid or liquid system [31]. Most hybrid propellants are inert on their own, and many are non-toxic (notable exceptions include N_2O_4). However, as for any propulsion system, oxidizers for hybrid rockets require safety and compatibility precautions. Additionally, as in all rocket applications, transient events are cause for concern.

2. Performance: Hybrids typically enjoy a relatively high specific impulse - between that of solid and storable liquid systems. The solid fuel grain also makes it easy to add performance-enhancing/stabilizing materials, enabling a small boost in specific impulse and/or increased density.

3. Throttling: The solid regression depends on the oxidizer flow rate; therefore it can be throttled by opening or closing a single valve. Throttle ratios of 10:1 [32] or more [33] have been demonstrated in hybrids. Momentum matching of fuel and oxidizer is not required in hybrids as it is for throttling in liquid systems.

4. Decreased Complexity (compared to liquids): Hybrids require half the propellant delivery system of a liquid bipropellant rocket. Simple operation requires as little as a single valve. However, mission requirements for range safety or fault tolerance may increase the system's complexity.

5. Temperature Tolerance: The regression rate (and therefore thrust) of hybrid fuels is not first-order dependent on the temperature as it is in solids. However, low temperature will still drive the chemical kinetics, oxidizer velocity, and thermal expansion effects in the motor. Launching hybrids over a wider range of temperatures and with minimal to no thermal conditioning for most applications should be possible. The main concern becomes the Coefficient of Thermal Expansion (CTE) of the fuel grain in these cases.

6. Chamber Pressure Tolerance: The solid fuel regression rate is only weakly dependent on chamber pressure over most of the range of useful oxidizer mass fluxes. Therefore selecting an operating pressure that optimizes the system for other performance measures is possible.

7. Tolerance to Debonding and Cracks: A crack in a hybrid fuel grain does not cause a catastrophic failure as it would in a solid. Burning down cracks is nearly impossible because there is no oxidizer flow over the fuel surface within the crack. It must be noted that large cracks are still not desirable for two

reasons: first, they act as flow trips and increase burning in that area, potentially leading to burn-through, and second, they may allow chunks of fuel to break off and either exit unburned or clog the nozzle.

8. Packaging Flexibility: Hybrids enjoy some flexibility in their packaging and can be designed to fit within many geometric constraints. The oxidizer and pressurant (if required) can be stored in any shaped pressure vessel or split between multiple vessels. The fuel is packaged in the combustion chamber. While the ballistic design will drive the aspect ratio, it negates the need for a separate fuel tank.

9. Environmental Safety: Hybrids burn relatively cleanly because of the high hydrogen-to-carbon ratios of the fuels. In most cases, very little soot is produced. They do not use perchlorates or nitrates, nor do they produce hydrochloric acid, and therefore do not contaminate the environment in which they are tested. See, for example, BluShift's MARVEL rocket engine, which is powered by a carbon-neutral, bio-derived fuel [34]. Many indoor test facilities exist where often the only evidence that a test has occurred is akin to the smell of birthday candles.

10. Low Cost: The previous list of advantages is believed to lead to a low-cost system. Most important to cost is that most of the fuels are inert, leading to increased safety, ease of handling, and decreased operational hurdles. Short development times of large-scale systems have been demonstrated, for example, AMROC's DM-01 motor was designed and tested in just over one year [35]. The raw propellants themselves are not typically expensive (e.g., at the time of this writing, the cost of paraffin is under $4 per kilogram). However, this may be because commercial (not aerospace) suppliers are typically used. Testing to ensure material repeatability and associated paperwork may significantly drive the material costs.

The grain fabrication process for hybrids is quite simple. For paraffin-based grains, the wax is melted and mixed with additives and dyes, and then the mixture is spun in a cylindrical casing until it cools. The grains can be cast in layers or all at once. This typically depends on the size of the system. Classical, polymer-based hybrid fuel grains take slightly more effort since curing is required. However, their inert nature greatly reduces hazards during fabrication compared to solids. Plastic fuel grains can be 3D printed, enabling complex geometries, interlocking fuel grain segments, or just a simple way of manufacturing. Traditional machining is also straightforward with plastics.

1.4.2 Challenges and mitigations

There are many challenges associated with hybrid rockets that need to be overcome for a successful system. Their lag in development behind liquid and solid systems has historically been linked to the low regression rate associated with classical hybrid fuels. However, higher regression rate fuels and unique injection schemes have renewed interest in hybrid rockets. Additionally, applications that require lower thrust (e.g., SmallSats) have been identified to take advantage of the low regression rate propellant combinations.

The disadvantages of hybrid rocket propulsion are listed below. When possible, suggestions for mitigation are included.

1. Low Regression Rate: Classical hybrid rockets suffer from low regression rate fuels caused by diffusion-limited mixing. High thrust levels required by most applications led to the requirement for an increased burning area commonly achieved through the use of multiple ports. Multiport hybrids, as depicted in Fig. 1.5 b or Section 1.6.7, are plagued by a host of disadvantages [36]. The

fuel grains are more difficult to design and fabricate. They either require structural supports or high-strength propellant [37], [38]. They can lose structural integrity towards the end of the burn, and chunks can break off. The volumetric loading is poor compared to a single-port design and excessive unburned mass fractions (5–10%) are possible. Either the oxidizer must be injected into each port individually, or a substantial pre-combustion chamber must be included to ensure uniform flow and even burning. Research over the past several decades has presented a number of solutions, including the use of high regression rate fuels, swirl injectors (which increase resonance time in the combustion chamber), etc. A longer list of techniques that have been attempted to increase regression rate is presented in [39]. Note: low regression rate fuels may be finding their place in small thrusters (e.g., SmallSats).

2. O/F Shift: The port area in the combustion chamber increases as the solid fuel burns. This causes the oxidizer-to-fuel (O/F) ratio to shift over the course of the burn. This can lead to operation at less than optimal mixture ratios. The O/F shift is also particularly challenging for systems where deep throttling might be required. Careful design can minimize the impact on performance (peak I_{sp}) to less than 1% [15]. On several occasions, this can be exploited for small performance increases, as will be discussed in Section 3.4.3. Regardless, this is a design complexity that needs to be considered.

3. Transient Times: Hybrid rockets have multiple transient events that are necessary for nominal operation (ignition and shutdown), some that are imposed by mission requirements (throttling), and several that are undesirable (instabilities). A fairly complete discussion of transients is available in [40]. Ignition and shutdown transients are typically acceptable unless rapid pulsing is desired. Instabilities will be discussed in Chapter 8.

4. Low Technology Readiness Level (TRL): TRL is a measure of technical maturity often used by NASA and the US government. Hybrids are at a comparatively low TRL and have not been flight proven for most practical applications. However, this has steadily been changing since the discovery of high regression rate hybrid fuels in the late 1990s. Also, SpaceShipOne, a HTPB/N_2O rocket, reached the edge of space in 2004, and SpaceShipTwo launched its first full crew (six people) in 2021. Several large-sounding rockets with potential transition to launch vehicles are currently being developed around the world. On a smaller scale, many university research programs are studying the fundamental physics of the hybrid combustion process and improving ballistic modeling capabilities.

1.5 Applications

The benefits of hybrid rockets vary by propellant combination. The high thrust achievable by liquefying fuels such as paraffin make them ideal for sounding rockets or boosters. A number of sounding rockets have been launched in the past, and launchers are under development, see Section 1.6. They are also on the brink of being utilized for space tourism. On the other end of the spectrum, one of the slowest burning fuels: acrylic, is excellent for SmallSat applications since the thrust is low enough to allow for the spacecraft to be controllable. The oxidizer can be chosen so that no thermal conditioning will be required for in-space missions. Finally, the safety associated with these systems makes them ideal for teaching opportunities.

1.6 History and current programs

Hybrid rockets have existed for nearly as long as the more common solid and liquid options.

The list below is not exhaustive but highlights some of the milestones and interesting development programs since the late 1930s. The propulsion engineer is pointed to Ref. [41] for a recent status of worldwide developments.

1.6.1 Early hybrid rockets, 1930s–1940s

The first hybrid rocket, as it is classified today, using a solid fuel and liquid or gaseous oxidizer, was tested in 1937 at I. G. Farben. The propellant combination was coal and gaseous nitrous oxide. However, this propellant combination turned out to be an unfortunate choice, as the carbon did not regress noticeably due to its exceptionally high heat of sublimation [30]. The Pacific Rocket Society had better luck in the 1940s using liquid oxygen (LOX) and Douglas fir (wood) as well as some more modernly conventional fuels such as wax loaded with carbon black and rubber. During the experimental Douglas fir (XDF) testing, the dependence of chamber pressure on oxidizer mass flow rate instead of burning surface area was recorded [42]. Note: many of the old texts on rocket development allow the author's sense of humor to shine through and are worth the read.

1.6.2 Sandpiper, 1960s

The Sandpiper rocket was developed in the 1960s by United Technology Center and Beech Aircraft for the Air Force [43]. The propellant combination 90% Plexiglas and 10% magnesium fuel with a MON-25 (75% N_2O_4 and 25% NO) oxidizer. (They also tested Inhibited Red Fuming Nitric Acid (IRFNA) as an oxidizer. The rockets were subjected to thermal cycling from $-54\,°C$ to $74\,°C$. Heavy-weight ground testing was completed (30 tests), resulting in a flight certification campaign, including eight altitude tests. Finally, three flight tests were completed (the first of which was on 12 Dec 1967), demonstrating almost 5 minutes of powered flight at speeds up to Mach 2.5 and an altitude of almost 24 km

[44]. The missile was designed to be launched from an aircraft at approximately 15 km altitude. Ground testing was completed at both ambient and altitude representative pressures.

1.6.3 Teledyne XAQM-81A firebolt supersonic aerial target, 1970s

From the late 1970s to the early 1980s, a hybrid motor was developed as an aircraft-launched interceptor target. The XAQM-81A Firebolt was based on the previous AQM-37A, a liquid bipropellant and the Sandpiper hybrid rocket program [44]. It used the same Plexiglas plus magnesium/MON-25 propellant combination. The maximum thrust was 1200 lbf (5340 N), and it was capable of being throttled at a 10:1 ratio. It was demonstrated at up to Mach 4.3 and reached over 100,000 ft (30 km). See Fig. 1.6.

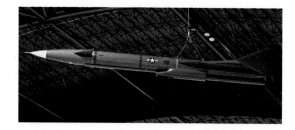

FIGURE 1.6   Interceptor Target: XAQM-81A Firebolt, photo: National Museum of the United States Air Force.

1.6.4 Booster scale hybrid rockets

AMROC

The American Rocket Company (AMROC) and the Hybrid Propulsion Demonstration Program demonstrated the largest scale hybrids to date at 250k lbf (1.1×10^6N) thrust. The AMROC motor used HTPB/liquid oxygen (LOx). The fuel grain was a wagon wheel design with a cylindrical center port surrounded by 15 triangular ports. The first development motor enjoyed three successful tests before suffering from a case failure on the fourth test. The motor was relatively stable, showing average ground performance of up to 231,900 lbf (1.0×10^6 N) of thrust and 244 s of I_{sp}, which would have translated to 272,300 lbf and 286 s under vacuum conditions [35]. AMROC sold their intellectual property rights to SpaceDev [45], later bought by Sierra Nevada Corporation.

HDPD, 1990s

The goal of the Hybrid Propulsion Demonstration Program (HPDP) was to develop hybrid rocket technology at a large scale in order to enable commercialization. The hybrid vehicle used HTPB/N_2O propellant. Four sounding rockets were launched from Wallops Flight Facility, two of which achieved altitudes over 100,000 ft (30 km). The three-year development activity had less than a $6M budget [43].

1.6.5 HYSR, 2002

Lockheed Martin built and launched the sounding rocket under a Space Act Agreement with Marshall Space Flight Center. The HTPB/LOx hybrid rocket had 264 kN (60,000 lbf) of thrust. Internal head end heaters were used for staged combustion [38]. It was only flown once, on 18 Dec 2002, but attained a 70 km altitude [46]. See Fig. 1.7.

FIGURE 1.7   HYSR ground test motor.

1.6.6 SpaceShipOne (2004) and SpaceShipTwo (2010–2021)

SpaceShipOne completed the first privately funded human spaceflight (> 100 km) in 2004.

It went on to win the Ansari X prize by then being the first privately funded spacecraft to reach an altitude of at least 100 km twice within two weeks while carrying the weight equivalent to three people on board and not replacing more than 10% of the dry mass (non-propellant mass) between the flights. This rocket is launched from the White Knight carrier aircraft at an altitude of about 14 km. Scaled Composites built the HTPB/N_2O hybrid rocket motor. It had a burn time of about 1.5 minutes.

In 2007, a N_2O tank exploded during a cold flow test at Scaled Composites. The area near the tank was not clear, and three employees were killed, highlighting the need to respect and understand these propellants and to clearly follow safety procedures during all testing. Scaled released N_2O safety guidelines after this accident [380] and oxidizer safety will be discussed in Chapter 11.

FIGURE 1.8   Second spaceflight of Virgin Galactic SpaceShipTwo.

The Spaceship Company, now owned by Virgin Galactic, has gone on to build SpaceShipTwo in order to provide commercial spaceflights for tourism. See Fig. 1.8. The hybrid motor continues with the same propellant combination as the first, though other options have been tested. The US Federal Aviation Administration's Office of Commercial Space Transportation cleared Virgin Galactic to launch commercial spaceflight participants in June 2021. The first full crew of six, including the founder, Sir Richard Branson, flew

FIGURE 1.9   Multirow and multiport fuel grains for the Falcon development effort [38].

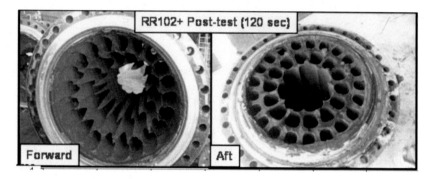

FIGURE 1.10   Multirow and multiport fuel grain after 120 s burn.

to space (86 km) in the USS Unity spacecraft on 11 July 2021.

1.6.7 Falcon

This Lockheed Martin effort under a DARPA contract endeavored to reduce the length-to-diameter (L/D) ratio of hybrid rockets to be more competitive with other solid and liquid rockets in the market. It is well documented in Ref. [38]. To achieve this Lockheed Martin (LM) tested an increased number of rows/ports compared to a traditional wagon wheel design. A 3 row, 43 port design was tested, see Fig. 1.9. The total burn time was designed to be 220 s.

LM used liquid oxygen with a proprietary, high-strength HTPB called LMF900. They found it stronger and tolerant to a wider temperature range than paraffin. Stress predictions led the team to design the motor to preferentially burn from the center outwards to achieve high fuel utilization, with a goal of 3% residual [38]. Fig. 1.10 shows the grain after a little more than half of the total burn time. The motor was shown to be quite stable.

1.6.8 Peregrine, 2000s to 2010s

The Peregrine hybrid rocket was a collaboration between NASA Ames Research Center, Stanford University and Space Propulsion Group, see Fig. 1.11. The 49 cm diameter and 10.6 m rocket used paraffin-based fuel (SP1A) and N_2O oxidizer. It had a delivered I_{sp} of 232 s (95% C^* efficiency) and was designed to launch a 5-kg payload to 100 km. The flight weight case was made of composite material. Twenty-eight ground tests were conducted, including a redesign after a Rapid Unscheduled Disassembly (RUD) due to an overpressure in test 17 [260]. While never launched, the ground test program identified and solved many instability issues [47]. For example, an upper limit on oxidizer mass flux for flame holding of 650 kg/m^{2}s was identified during testing of this motor [260].

FIGURE 1.12   SPG test of a 30 kN paraffin/LOx motor. Photo courtesy of Brian Cantwell.

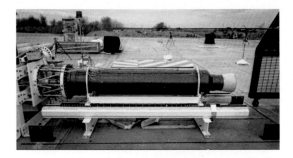

FIGURE 1.11   Peregrine Ground Test Setup.

1.6.9 Space Propulsion Group

Space Propulsion Group (SPG) drove the advancement of paraffin-based hybrid rockets in the early 2000s–2010s. During its operation, SPG successfully conducted more than 40 LOx/paraffin hybrid motor tests, resulting in a stable, efficient (up to 96%), 0.28 m diameter, 30 kN thrust motor. See Fig. 1.12. In total, SPG carried out well over 500 tests of paraffin-based fuels with several oxidizers (e.g., LOx, N_2O, MON) at several different scales [15], [32].

1.6.10 Mars Ascent Vehicle, 2010s

A potential Mars Ascent Vehicle drove the development of a 10 kN class hybrid rocket from about 2015–2019. The NASA lead team (Jet Propulsion Laboratory, Marshall Space Flight Center and Ames Research Center) employed a number of small companies and universities to complete this research. A new formulation of wax-based fuel and was tested with MON-3 and MON-25 oxidizer. This development program drove advances in this new propellant combi-

nation, low-temperature operation, hypergolic ignition using solid additives, restart capability, and Liquid Injection Thrust Vector Control (LITVC) for hybrid rockets.

The new wax-based fuel focused on a desire to operate at low temperatures on the surface of Mars. The high coefficient of thermal expansion (CTE) of the fuel led the fabrication team at MSFC to load the fuel grain into the liner under compression. It was installed at temperatures lower than it would see operationally [48]. Testing was completed at full scale by multiple companies: Whittinghill Aerospace, Space Propulsion Group and Parabilis Space. Whittinghill Aerospace completed a near full-duration burn at Mars operating temperatures (−20 °C). Liquid injection thrust vector control utilizing the oxidizer was also demonstrated. Fig. 1.13 shows the Liquid Injection Thrust Vector Control (LITVC) being exercised.

Multiple ignitions were demonstrated at full scale using TEA/TEB and GOx. This combination was not desired for flight because of the need to carry additional propellants. Two universities, Purdue and Penn State, conducted hypergolic testing to evaluate different ignition options. A number of options for hypergolic ignition with solid additives in the fuel grain were discovered during this period. These had their benefits and complications and will be discussed in Chapter 5.

FIGURE 1.14 Ground test of the Nucleus Sounding Rocket motor in June 2018. Photo: NAMMO.

FIGURE 1.13 Hybrid propulsion development test for a potential Mars Ascent Vehicle. Top: normal operation, Bottom: LITVC actuated (see shadow near nozzle outlet) Photo: NASA.

eight on the first stage and four on the second stage, to launch up to 500 kg to sun-synchronous orbit.

1.6.11 Nucleus, NAMMO

The Nucleus sounding rocket is an HTPB/ 87.5% hydrogen peroxide hybrid rocket developed by NAMMO in Norway [49]. It has about 30 kN of thrust and a total impulse of 1,000,000 Ns, over a 40 s burn time. The full flow rate of hydrogen peroxide is decomposed through a catalyst bed (turning into steam and gaseous oxygen) before entering the combustion chamber. In 2018, it was successfully launched from Andøya Space Center, reaching an altitude of 107 km. It has an I_{sp} of 270 s with a nozzle expansion ratio of 8.5, which is optimized for Earth launch. NAMMO hopes to power the North Star family of rockets using the nucleus motor [50]. See Fig. 1.14.

FIGURE 1.15 The first test of HyImpulse's 75 kN hybrid motor in 2020.

1.6.12 HyImpulse Technologies, Germany

The launch startup HyImpulse Technologies, a spin off from the German Aerospace Center (DLR) in Germany is developing a paraffin/LOx hybrid motor, see Fig. 1.15. They plan to use clusters of this turbopump driven, 75 kN motor:

Previous work by this team saw a paraffin/ N_2O rocket launched to over 32 km (106,000 feet) from the European Space and Sounding Rocket Range. The Hybrid Experimental Rocket Stuttgart (HEROS) 3 rocket was 7.5 m tall and the hybrid motor produced 10 kN of thrust. In addition to the impressive propulsion technology development for this effort, a substantial amount of carbon fiber composite structure was utilized making the rocket dry mass only 75 kg [51].

1.6.13 Gilmour Space Technologies, Australia

Gilmour Space Technologies of Australia is aiming to launch small satellites into space. They use N_2O oxidizer and a proprietary solid fuel and have completed ground tests up to their design mission duration of 110 s. They also achieved up to 91 kN of thrust in an initial (10-second) verification test of their main motor. See Fig. 1.16.

FIGURE 1.16   Gilmour's Eris Motor Test.

1.6.14 Taiwan Innovation Space (TiSPACE), Taiwan

Hapith I is a two-stage launch system made of clusters of hybrid rocket motors developed by TiSPACE. The first stage uses five hybrid rocket motors with liquid injection thrust vector control, the second stage uses four hybrid rocket motors with gimballed thrust vector control. The Hapith V is a three-stage launch system based on the Hapith I with an additional upper stage using a single, gimballed hybrid motor. Fig. 1.17 shows stills from a qualification test of the second stage cluster, including the gimballed thrust vector control.

1.6.15 Delta V Space Technologies, Turkey

The Turkish government-backed space company, Delta V, has tested multiple hybrid rocket

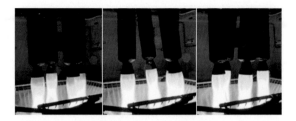

FIGURE 1.17   TiSPACE Qualification of the four-stage two hybrid rocket motors. Multiple still images from a video show the gimballed thrust vector control.

motors intending to reach the Moon. Their Sonda Roket Sistemi (SORS) sounding rocket uses a paraffin/LOx hybrid motor and has had multiple successful launches since Dec 2020, see Fig. 1.18. It has been designed to reach 100–150 km. DeltaV has also been tested with other oxidizers. The general manager of Delta V, previously founded Space Propulsion Group and contributed significantly in their hybrid rocket development.

FIGURE 1.18   Delta V's Sonda Roket Sistemi (SORS) Test Launch.

1.6.16 SpaceForest, Poland

This Polish-based company is offering low-cost, throttlable (4:1), paraffin-based/N_2O hybrid sounding rockets that can carry 50 kg up to 150 km or more [52]. Tests of their PERUN SF1000 motor have demonstrated 30 kN of thrust, see Fig. 1.19. The first demonstrator was launched from their mobile platform in 2020.

FIGURE 1.19 Test of the PERUN SF1000 motor in Poland.

1.6.17 SpaceRyde, Canada

This Canadian company is looking to launch a paraffin/LOx rocket from a stratospheric balloon. That allows the hybrid rocket motor to be optimized for vacuum performance. SpaceRyde's Grolar motor gimbals its nozzle to achieve thrust vector control and is designed to provide more the 50 kN of thrust. See Fig. 1.20.

FIGURE 1.20 SpaceRyde Hotfire Test including Gimballed Nozzle Thrust Vector Control.

1.6.18 University research and sounding rockets

There are numerous cutting-edge research programs going on at Universities around the world. Researchers at Stanford University pioneered liquefying hybrid fuels [53]. Design case studies have been completed for upper stages [54], Mars Ascent Vehicles [55], Launch Vehicles [56], etc. Also, hybrid propulsion systems

for CubeSats have been a particularly active area of research, see, for example, [57] and [58].

Universities are advancing the state of the art in propellants, e.g., strength additives improved the mechanical performance of paraffin fuels [59] and novel oxidizers [60], throttling [61], swirl [62], [63], [237], instabilities [64], mixing devices [65], [66], nozzle erosion [67], Computational Fluid Dynamic (CFD) simulations [68], and modeling of complex flows [69].

A number of sounding rockets have been launched around the world. Purdue launched a 4 kN class LDPE/Hydrogen Peroxide [70]. Students from Delft launched the Stratos II sounding rocket, which reached 21.5 km in 2015. The Hybrid Engine Development team at the University of Stuttgart launched the paraffin/N_2O, 10 kN HEROS 3 to higher than 32 km in 2016 [71]. A team of students from the University of Kwazulu-Natal (UKZN) developed and launched a paraffin/N_2O hybrid rocket, the Phoenix-1B. It was successfully launched to 18 km in 2021.

This list is nowhere near exhaustive, and new teams emerge every year. The contributions of universities will be highlighted over and over throughout this text. Academia is driving important advances in hybrid rocket propulsion through fundamental research by partnering with commercial companies and national space and/or defense agencies.

1.7 Commonly used terms

There are several items of nomenclature and standards commonly used by propulsion engineers with which the reader should be familiar. These are introduced here and will be used throughout this book.

Ablation - The erosion of a protective material due to heat. Commonly used to describe nozzle throats in this context.

Additive - A material added to the fuel to change the fuel properties (e.g., strength, regression rate, hypergolicity, performance).

Blowing - Blocking of heat transfer to the fuel grain due to the mass flux of vaporized fuel leaving the grain.

Boundary Layer - The layer of fluid near a wall (boundary) where the effects of viscosity are significant.

Characteristic velocity, C^* - A measure of performance for propulsion systems. Unlike I_{sp}, C^* does not depend on nozzle expansion ratio.

Combustion - The chemical reaction of a fuel and oxidizer.

Combustion efficiency - The measured performance divided by the theoretical performance of a propulsion system.

Diaphragm - A flow restriction, e.g., used in a combustion chamber to increase mixing or reduce instabilities.

Diffusion - The intermingling of substances by the natural movement of their particles.

Entrainment - In hybrid rockets, entrainment refers to fuel being drawn in and transported with the oxidizer flow, similar to a fuel injection system.

Fuel - A substance that combusts when mixed with an oxidizer and provided with activation energy.

Fuel Grain - The solid fuel configuration for a hybrid rocket. Typically with one or more ports for the oxidizer and combustion products to pass through.

Feed Line or Feed System - Components (e.g., tubing, valves, filters) to safely deliver liquid/gaseous propellant to the combustion chamber. This can also include a pressurant system (if required) and propellant-loading hardware. In ground systems, it would also include safety features such as pressure relief.

Ground Support Equipment (GSE) - Equipment required for testing or handling a hybrid propulsion system on the ground. Electrical Ground Safety Equipment (EGSE) refers specifically to the electronics and associated software required for propellant loading and ground testing.

Hybrid - The combination of two different elements. In this text, that will always be used to refer to a chemical propulsion system with the oxidizer and fuel stored in different phases. However, it can be applied to other combinations of propulsion systems. A common example of a different sort of hybrid propulsion system is the combination of separate chemical and electric propulsion systems on a spacecraft. The propulsion engineer should be aware of alternatives such as this so they do not cause confusion.

Hypergolic - Rocket propellants that spontaneously ignite upon contact.

Instability - Undesirable variations in thrust, typically observed through fluctuations in chamber pressure.

Injector - The hardware that introduces the liquid or gaseous propellant into the combustion chamber. It can atomize or vaporize the liquid propellant and may set the flow into the combustion chamber. The design of the injector is typically one of the most challenging aspects of new hybrid motor development.

Mass Flux, G - The mass flow of propellant (kg/s) divided by the cross-sectional area (m^2) of the flow area, typically the fuel grain port. This is a time-dependent variable. The oxidizer mass flux is only dependent on the mass flow of the oxidizer and is also commonly used in hybrid rocket propulsion.

Mixture Ratio or Oxidizer to Fuel Ratio - The mass flow rate of oxidizer divided by the mass flow rate of fuel ($\dot{m}_o/\dot{m}_f$).

Oxidizer - An oxidizing substance (i.e., a substance that accepts electrons) that combusts when mixed with fuel and provided with activation energy.

Post Combustion Chamber - Volume at the aft-end of the combustion chamber, between the end of fuel grain and the nozzle. Typically used for increasing residence time and thought to improve mixing and combustion efficiency.

Precombustion Chamber - Volume at the fore-end of the combustion chamber, between the injector and fuel grain. Typically used for propellant mixing.

Propellant - A fuel or oxidizer (or both).

Regression rate - The speed at which the solid propellant burns (m/s) within in a combustion chamber. This term can also be applied to insulation, etc.

Specific Impulse, I_{sp} - A measure of performance for propulsion systems. Unlike C^*, I_{sp} takes into account the nozzle expansion ratio.

Technology Readiness Level - TRL is a metric used by NASA and other (US) government agencies to distinguish the heritage of hardware for spaceflight. TRL 1 indicates that first principles physics has been reported, TRL 3 is proof of concept, TRL 6 is system-level verification in flight-like environments (typically required before the hardware is adopted for spaceflight), and TRL 9 is flight proven [72].

A note on units: in the United States, most scientific endeavors have switched to SI units. However, fabrication across the US aerospace industry still uses English units (e.g., inches and pounds per square inch), and many propulsion tests are still reported in psi and lbf. This mix of units can often lead to confusion. This book endeavors to use SI units wherever possible. However, both units will be included if the natural form is in English units.

1.8 Book layout

This book is divided into 11 chapters. The first part focuses on the theory providing the propulsion engineer with the tools to design a hybrid rocket propulsion system on paper. This includes an introduction to chemical rocket propulsion, covering the rocket equation, and examples of typical ΔV requirements are given in Chapter 2. Then an overview of hybrid rocket theory is provided in Chapter 3. Chapter 4 presents thermodynamics and chemistry as they apply to hybrid rocket propulsion. Focus is on understanding stoichiometry, chemical equilibrium and its limitations, and the use of enthalpy for rocket propulsion. Common propellants and additives are presented in Chapter 5. One of the most important parameters in hybrid rocket design is the regression rate. Therefore the determination and important parameters for the regression rate are given in Chapter 6.

Chapter 7 brings everything presented up to this point together and teaches the propulsion engineer how to complete a basic design, then start to take real performance and other practical considerations into account. Next, Chapter 8 introduces feed line design, combustion inabilities, and propellant budgeting.

At this point, the text transitions to practical aspects of design, to ensure the theoretical knowledge can be turned into real hardware. Chapter 9 discusses the components. Chapter 10 gives some examples of hybrid designs to reinforce the material presented in the text. Finally, the book concludes with a discussion of testing and safety practices in Chapter 11.

As the intent of this work is to be a handbook, references and further reading are introduced throughout if the reader desires more detailed information on the subjects being discussed.

CHAPTER

2

Introduction to rocket propulsion

2.1 Introduction

This chapter will briefly discuss the equations necessary for rocket design. These equations are applicable regardless of the type of chemical propulsion system. Rocket propulsion design is covered in great detail in many other texts, so the discussion here will only highlight the most important equations. The reader is pointed to several references if more background is desired, see, for example, Refs. [73], [1], or [6].

2.2 Key parameters

Rocket combustion is governed by complex physics, including propellant injection, atomization, mixing, chemical reactions, spatial variations in temperature and pressure, choking, nozzle expansion, boundary layers, heat transfer, etc. The following sections introduce equations that describe this complex behavior using ideal approximations. However, they have been shown to be reasonable approximations for a rocket motor of a typical scale. For example, the Space Shuttle Main Engines lose a massive amount of heat through the nozzle wall, but that value is tiny compared to the total heat flux through the nozzle. For a helium micro-thruster, the opposite may be true. The propulsion en-gineer is cautioned to check assumptions if the application strays too far from those used to develop the theory.

2.2.1 Thrust

Thrust is the force that propels a rocket. It is achieved by accelerating mass out of the nozzle, typically in the form of hot exhaust gases. Cold gas thrusters and water rockets operate by the same principle. Thrust is given by Eq. (2.1). The derivation of this fundamental equation from the conservation of mass and momentum is provided in Appendix A.

$$F = \dot{m}V_e + A_e(P_e - P_a) \qquad (2.1)$$

where $\dot{m}$ is the mass flow rate, V_e, A_e, and P_e are the exit velocity, cross-sectional area, and pressure, respectively, and P_a is the ambient pressure. From this equation, it can be seen that for a given mass flow rate, high thrust requires high exit velocity and/or high exit pressure. The nozzle converts the chemical energy of the propellant to kinetic energy, reducing P_e, but increasing V_e. Therefore propulsion engineers design for optimally expanded nozzles for ground-based launches, matching P_e and P_a, and the trade area ratio for nozzle mass in vacuum conditions.

Most (but not all) hybrid rockets show a small decrease in thrust over the length of the burn due to a decrease in propellant mass flow rate when the port is opened. It is called a regressive burn and will be discussed in more detail in later chapters.

There are several other equivalent ways to calculate thrust of a rocket. Eq. (2.2) uses the mass flow rate and two other important values: characteristic velocity, C^*, and thrust coefficient, C_F. These terms are defined in the following sections. Thrust can also be calculated from the specific impulse, which will be discussed in 2.2.6.

$$F = \dot{m} C^* C_F \qquad (2.2)$$

2.2.2 Thrust coefficient

The thrust coefficient is a dimensionless measurement of thrust. It is defined by Eq. (2.3).

$$C_F \equiv \frac{F}{P_{t,c} A_{th}} \qquad (2.3)$$

We can substitute Eq. (2.1) into Eq. (2.3) to obtain Eq. (2.4).

$$C_F = \frac{\dot{m} V_e + (P_e - P_a) A_e}{P_{t,c} A_{th}} \qquad (2.4)$$

If the ideal gas equation is used, $P_e = \rho_e R T_e$, along with the continuity equation, $\dot{m} = \rho_e V_e A_e$, and the definition of Mach number, $M_e \equiv \frac{V_e}{\sqrt{\gamma R T_e}}$, Eq. (2.5) can be obtained.

$$C_F = \left(\frac{P_e}{P_{t,c}}\right)\left(\frac{A_e}{A_{th}}\right)\left(\gamma M_e^2 + 1 - \frac{P_a}{P_e}\right) \qquad (2.5)$$

Eq. (2.5) can be re-written to remove the explicit dependence on exit Mach number, M_e, as shown in Eq. (2.6).

$$C_F = \left\{ \left(\frac{2\gamma^2}{\gamma-1}\right)\left(\frac{2}{\gamma+1}\right)^{\frac{\gamma+1}{\gamma-1}}\left[1-\left(\frac{P_e}{P_{t,c}}\right)^{\frac{\gamma-1}{\gamma}}\right]\right\}^{\frac{1}{2}}$$
$$+ \left(\frac{P_e}{P_{t,c}} - \frac{P_a}{P_{t,c}}\right)\frac{A_e}{A_{th}} \qquad (2.6)$$

Eq. (2.6) assumes one-dimensional, frozen equilibrium and isentropic flow. Note that it is often more convenient to write these equations (and others in this chapter) in terms of the static pressure in the combustion chamber, P_c, rather than the total pressure, $P_{t,c}$. This substitution can be made when the velocities in the chamber are sufficiently low that the static chamber pressure P_c is about equal to the total chamber pressure $P_{t,c}$. This is a reasonable approximation when the area of the combustion chamber is large compared to the nozzle throat. Measuring the static combustion chamber pressure is much more practical, hence the desire to make this substitution.

2.2.3 Propellant mass flow rate

Flow through a rocket nozzle is nearly always choked, meaning the mass flow rate out of the nozzle does not depend on the ambient pressure. The critical pressure ratio (2.7) can be used to determine if the flow is choked.

$$\frac{P_{t,c}}{P_a} \geqslant \left(\frac{\gamma+1}{2}\right)^{\frac{\gamma}{\gamma-1}} \qquad (2.7)$$

Since the range of γ for a rocket is quite narrow, about 1.2–1.67 [73], the critical pressure ratio is typically in the neighborhood of two. There are notable time periods when the nozzle is not choked, mainly transient periods, such as ignition and shutdown.

Assuming compressible, isentropic, constant heat capacity and choked flow, the mass flow rate can be found by Eq. (2.8).

$$\dot{m} = \frac{\gamma A_{th} P_{t,c}}{\sqrt{\gamma R T_{t,c}}}\left(\frac{2}{\gamma+1}\right)^{\frac{\gamma+1}{2(\gamma-1)}} \qquad (2.8)$$

The usable mass of propellant, m_p, is the integrated mass flow rate over the full burn time, per Eq. (2.9).

$$m_p = \int_{t0}^{tb} \dot{m}\, dt \qquad (2.9)$$

2.2.4 Characteristic velocity

The characteristic velocity (C^*) is a measure of the energy available from the combustion process. It is independent of nozzle characteristics. Characteristic velocity can be determined from both measured test data and calculated thermochemical parameters, see Eq. (2.10). It is often used to measure combustion efficiency because of this.

$$C^* \equiv \frac{P_{t,c} A_{th}}{\dot{m}} \qquad (2.10)$$

We can substitute in the equation for mass flow rate with the choked flow, Eq. (2.8) to obtain Eq. (2.11). Note that Eq. (2.10) assumes adiabatic and isentropic conditions between the combustion chamber and the nozzle throat.

$$C^* = \left[\frac{1}{\gamma} \left(\frac{\gamma + 1}{2} \right)^{\frac{\gamma+1}{\gamma-1}} \frac{R_u T_{t,c}}{M_w} \right]^{\frac{1}{2}} \qquad (2.11)$$

The ideal value of C^* is determined from thermochemical calculations. The approach for this is discussed in more detail in Chapter 4. For the ideal calculations, the designer should use the measured mass flow rate and nozzle throat area and determine the ideal chamber pressure under these conditions (thermochemically). The combustion efficiency can then be easily determined by comparing this ideal value for chamber pressure (which assumes complete mixing and adiabatic combustion) to that achieved during the test.

$$\eta_{C^*} = \frac{C^*|_{measured}}{C^*|_{ideal}} = \frac{P_{t,c}|_{measured}}{P_{t,c}|_{ideal}} \qquad (2.12)$$

Note that the relationship between the C^* efficiency and pressure does not account for real performance, e.g., nozzle erosion will also contribute to a decreased pressure. Also, the ideal chamber pressure is usually calculated using the measured $\dot{m}$ for the particular test.

2.2.5 Effective exhaust velocity

The effective exhaust velocity, C, is defined as the instantaneous change of momentum per unit of expelled mass, see Eq. (2.13). It is a measure of efficiency for a rocket and is related to the thrust coefficient and C^*, as described in Eq. (2.14). It can be seen, per Eq. (2.15), that for large nozzle area ratios, where M_e becomes high, the pressure contribution to the overall thrust becomes small, and $C \approx V_e$. The effective exhaust velocity is an indicator of the performance of propellant, but the term specific impulse is more commonly used now, as will be discussed in the next section.

$$C = \frac{F}{\dot{m}} \qquad (2.13)$$

$$C = C_F C^* \qquad (2.14)$$

$$C = V_e \left(1 + \frac{1}{\gamma M_e^2} \left(1 - \frac{P_a}{P_e} \right) \right) \qquad (2.15)$$

2.2.6 Specific impulse

Specific impulse (I_{sp}) is the primary measure of performance for rockets: it is the thrust per unit of propellant mass flow rate, normalized by Earth's gravitational acceleration in units of seconds, see Eq. (2.16). This normalization is somewhat arbitrary, but widely used, perhaps because it gives a unit of performance that is agnostic to the choice of units: imperial or metric. The specific impulse depends on the propellant combination as well as the area ratio of the nozzle. The importance of this term will also be highlighted in the rocket equation, Eq. (2.27).

$$I_{sp} = \frac{F}{\dot{m} g_0} = \frac{C}{g_0} \qquad (2.16)$$

The specific impulse is also proportional to the square root of the combustion products temperature divided by the molecular weight of the exhaust gases, M_w. This can be seen in Eq. (2.17) for the general form of I_{sp} and Eq. (2.18) for the ideally expanded case. Therefore to achieve high performance, the propulsion engineer wants to find a propellant combination that provides a high combustion temperature and low molecular weight combustion products. This illustrates the importance of the underlying thermodynamics in specific impulse. Here, γ is the ratio of specific heats, R_u is the universal gas constant, T_t, c is the total combustion chamber temperature, P_a is the external atmospheric pressure, and P_t, c is the total combustion chamber pressure, and P_e, T_e, and M_e are the pressure, temperature, and Mach number at the nozzle exit, respectively.

$$I_{sp} = \frac{1}{g_0}\sqrt{\frac{\gamma R_u T_e}{M_w}}\left(M_e + \frac{1}{\gamma M_e}\left(1 - \frac{P_a}{P_e}\right)\right)$$
(2.17)

For a perfectly expanded nozzle, which assumes the nozzle exit pressure equals ambient pressure, a purely thermodynamic relationship is given by (2.18).

$$I_{sp} = \frac{1}{g_0}\sqrt{\frac{2\gamma R_u T_{t,c}}{M_w(\gamma - 1)}\left[1 - \left(\frac{P_e}{P_{t,c}}\right)^{\frac{\gamma-1}{\gamma}}\right]}$$
(2.18)

As discussed in at the outset of Section 2.2, the physics in the combustion chamber is quite complex. Thermochemical information can be found initially by running a chemical equilibrium code such as Chemical Equilibrium with Applications (also known as CEA) [74]. During these calculations, the decision to use frozen or equilibrium conditions in the nozzle must be made. If Eq. (2.18) is used, the variables are typically frozen at either the combustion chamber or the nozzle throat, yielding a conservative result. If they are allowed to shift along the length of the nozzle and reach equilibrium, which can be done through computer models, a higher value for I_{sp} is achieved. It is important to note that

one-dimensional codes, such as CEA, make additional simplifying assumptions, such as the chemical reactions being faster than the flow out of the nozzle, which is not always realistic. They also ignore dimensional effects. These assumptions can become very important depending on the application. For example, hydrocarbon fuels have a rather slow reaction to form CO_2, so the reaction may not be completed while within the nozzle. The thermochemistry of propellants is further discussed in Chapter 4. Performance tables for typical hybrid propellant combinations are given in Chapter 5.

The specific impulse depends on the nozzle design. Larger area ratios generally give higher I_{sp}'s as long as flow remains attached in the nozzle. The nozzle expansion ratio is defined as the ratio of the nozzle exit cross sectional area to the nozzle throat cross sectional area (A_e/A_{th}). Flow separation depends on the ambient pressure conditions. Nozzles operating at sea level typically have small expansion ratios of about 3–4. In-space nozzles can have expansion ratios of 40–100. This will be discussed further in Section 2.3.

2.2.7 Total impulse

The total impulse (I) is the thrust (F) integrated over the burn time (t_b) with units of force times time [e.g., N s]. This is typically the parameter that propulsion engineers must design to achieve. Predictions for the total impulse can be made by specifying an average thrust and burn time; however, in practice, the thrust of a hybrid rocket will vary over the burn.

$$I = \int_{t_0}^{t_b} F\,dt$$
(2.19)

The total impulse is related to the specific impulse as given in Eq. (2.20). The designer should note that this equation assumes an average effective I_{sp}. Total impulse can also be written in terms of the average thrust coefficient and chamber pressure, per Eq. (2.21).

$$I = I_{sp} m_p g_0 \quad (2.20)$$
$$I = C_F P_{t,c} A_{th} t_b \quad (2.21)$$

2.2.8 Density impulse

It is often most useful to compare the density impulse when different propellant combinations are being considered. This accounts not only for the performance, but the volume of propellant needed to achieve it. The density impulse is given by Eq. (2.22), where ρ_p is the average density for the propellant combination given by Eq. (2.23).

$$I_d = \rho_p I_{sp} \quad (2.22)$$
$$\rho_p = \frac{m_p}{V_{total}} = \frac{\rho_O \rho_F (1 + O/F)}{\rho_O + (O/F)\rho_F} \quad (2.23)$$

2.2.9 Characteristic chamber length

The characteristic chamber length, L^*, is the length that a chamber of the same volume would have if it was a straight tube with a diameter equal to the nozzle throat diameter. It is a measure of the average distance that combustion products must travel within the combustion chamber. Therefore characteristic length can be thought of as a relative indication of residence time. L^* is defined by Eq. (2.24), where V_C is the combustion chamber volume. It is most commonly used when discussing liquid engines.

$$L^* = V_C / A_{th} \quad (2.24)$$

The longer the characteristic length, the longer the time it takes for propellant/combustion products to travel through the combustion chamber, and thus the higher the residence time. Increased L^* may correspond to the higher C^* efficiency, or equivalently, to a higher achieved C^* in the test. Note that this increase in C^* is asymptotic. After complete mixing and combustion are achieved, further increase in chamber

length will not result in an increase in performance. The goal of additional length is to enable complete mixing, which can also be achieved by other and potentially more effective means (e.g., mixing devices).

2.2.10 Relationship between key parameters

As is likely already apparent, most of the key rocket propulsion parameters discussed in this section are closely related. Table 2.1 provided here allows the reader to quickly convert between thrust, thrust coefficient, characteristic velocity, effective exhaust velocity, specific impulse, and total impulse.

2.3 Nozzles

A nozzle converts a rocket's chemical energy into kinetic energy. It is important that the nozzle design matches the conditions of use. A nozzle is ideally expanded when the exit pressure equals the ambient pressure. Therefore a nozzle designed for use at sea level (e.g., for the first stage of a launch vehicle) will be very different from that designed for an upper stage or in-space application, as introduced in 2.2.6. Fig. 2.1 shows flow through a nozzle expanding at different pressure ratios. Optimization of performance often leads to a launch vehicle starting flight with an over-expanded nozzle, moving through perfect expansion and to an under-expanded nozzle high in the atmosphere. Flow separation in the nozzle decreases the effective area ratio, reducing performance, and is avoided.

The exit Mach number for adiabatic, isentropic, and calorically perfect flow through a nozzle with area ratio R can be determined by Eq. (2.25). Note that this equation yields two solutions: a subsonic solution with $M_e < 1$, and a supersonic solution with $M_e > 1$. The supersonic

TABLE 2.1 Relationship between key propulsion parameters. Note that all inputs are assumed to be in SI units and that for the relationship with total impulse the table assumes average values e.g. average thrust, average specific impulse, etc.

Key Parameter:	As a Function of:	Thrust F	Thrust Coefficient C_F	Characteristic Velocity C^*	Effective Exhaust Velocity C	Specific Impulse I_{sp}	Total Impulse I
	Units	[N]	[-]	[m/s]	[m/s]	[s]	[Ns]
Thrust F	[N]	-	$C_F P_{t,c} A_{th}$	$\dfrac{I_{sp} g_0 P_{t,c} A_{th}}{C^*}$	$\dot{m}C$	$\dot{m}I_{sp}g_0$	$\dfrac{I}{t_b}$
Thrust Coefficient C_F	[-]	$\dfrac{F}{P_{t,c}A_{th}}$	-	$\dfrac{C}{C^*}$	$\dfrac{C}{C^*}$	$\dfrac{I_{sp}g_0}{C^*}$	$\dfrac{I}{P_{t,c}A_{th}t_b}$
Characteristic Velocity C^*	[m/s]	$\dfrac{F}{\dot{m}C_F}$	$\dfrac{C}{C_F}$	-	$\dfrac{C}{C_F}$	$\dfrac{I_{sp}g_0}{C_F}$	$\dfrac{I}{C_F m_p}$
Effective Exhaust Velocity C	[m/s]	$\dfrac{F}{\dot{m}}$	C^*C_F	C^*C_F	-	$I_{sp}g_0$	$\dfrac{I}{m_p}$
Specific Impulse I_{sp}	[s]	$\dfrac{F}{\dot{m}g_0}$	$\dfrac{C_F P_{t,c}A_{th}}{\dot{m}g_0}$	$\dfrac{C_F C^*}{g_0}$	$\dfrac{C}{g_0}$	-	$\dfrac{I}{m_p g_0}$
Total Impulse I	[Ns]	Ft_b	$C_F P_{t,c}A_{th}t_b$	$C^*C_F m_p$	Cm_p	$I_{sp}m_p g_0$	-

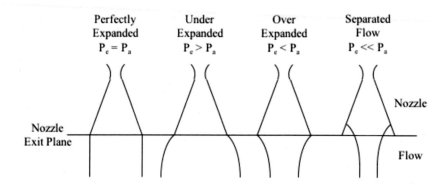

FIGURE 2.1 Flow recovery through a nozzle.

solution is the physical solution for rocket nozzle flows.

$$\frac{1}{A\!R} = \frac{A_{th}}{A_e} = \left(\frac{\gamma+1}{2}\right)^{\frac{\gamma+1}{2(\gamma-1)}} \frac{M_e}{\left(1+\frac{\gamma-1}{2}M_e^2\right)^{\frac{\gamma+1}{2(\gamma-1)}}}$$

(2.25)

The nozzle area ratio required to achieve a given exit pressure, P_e, can be calculated given the total pressure in the combustion chamber, $P_{t,c}$, and the ratio of specific heats, γ, using Eq. (2.26). This equation, like many of the equations presented in this chapter, assumes isentropic flow through the nozzle with an ideal and calorically perfect gas.

$$\mathcal{R} = \frac{A_e}{A_{th}} = \frac{\sqrt{\gamma \left(\frac{2}{\gamma+1}\right)^{\frac{\gamma+1}{\gamma-1}}}}{\left(\frac{P_e}{P_{t,c}}\right)^{\frac{1}{\gamma}} \sqrt{\frac{2\gamma}{\gamma-1}\left[1-\left(\frac{P_e}{P_{t,c}}\right)^{\frac{\gamma-1}{\gamma}}\right]}}$$

$$(2.26)$$

Moving from theoretical to practical considerations for rocket nozzles, most hybrid motors use an ablative throat. This results in throat erosion over the burn, decreasing the nozzle area ratio over time. Both mechanical and chemical erosion occurs in the nozzle. Several variables influence nozzle erosion: burn time, total pressure and temperature in the combustion chamber and concentration of oxidizing species in the exhaust gas.

Liquid rocket engines typically use regenerative cooling nozzles. See Fig. 2.2. They flow either the fuel or the oxidizer around the nozzle throat and bell to cool the metal and heat the propellant. This has two major advantages. The first is that the nozzle throat does not erode, which maintains the area ratio throughout the burn. The second is that the some of the energy that would have been otherwise lost can be recaptured and used to heat the propellant. Both of these advantages result in increased performance. However, additional plumbing and hardware are required to achieve this. Therefore the increased performance must be traded against the increased system mass and manufacturing complexity. The later may be mitigated by improving 3D printing technology. In the case of a hybrid rocket, only the oxidizer is available to remove heat from the nozzle. This configuration is typical in Russian liquid engines. However, it is more difficult to design properly since many metals burn quite well as fuels with a hot oxidizer circulating through them.

FIGURE 2.2 Regeneratively cooled nozzle as part of a liquid rocket engine. The image shows an RS-25 engine for the Space Launch System. There is no reason they cannot be used for hybrids; however, complications exist, such as being forced to use the oxidizer to cool the nozzle. Image Credit: Aerojet Rocketdyne.

2.4 Rocket equation

The rocket equation, also known as Tsiolkovsky's or the Tsiolkovsky–Moore rocket equation, Eq. (2.27), is the most fundamental relationship used in rocket design. It is derived from the conservation of momentum and describes the relationship between the mass of the rocket (m), the performance of the propellant (I_{sp}) and the change in the velocity of the rocket (ΔV). The gravity term in the rocket equation, g_0 is always defined as Earth's gravity, $9.81 \ m/s^2$, regardless of the planet on which the rocket is operating. The specific impulse times gravity is also known as the characteristic velocity, which was discussed in Section 2.2.5. The rocket equation is typically used as a first step in the design process. While the detailed masses and performance may not be known initially, it is easy to make some assumptions to understand the magnitude of the achievable performance. Later, an iterative process can be used to close the design.

$$\Delta V = I_{sp} g_0 \ln \frac{m_i}{m_f} \qquad (2.27)$$

The difference between the initial (m_i) and final (m_f) masses is the propellant expended during the burn (m_p). It is more accurate to call this the usable propellant. There is an important distinction between propellant mass and usable propellant mass. Some amount of the propellant will not be converted into kinetic energy. This includes residual and hold up on the liquid side (e.g., tanks and lines cannot be completely emptied) and residual (sliver fraction) on the solid side (e.g., uneven burning of the fuel grain leaves some fuel unburned). Additionally, there is potential for mixture ratio uncertainty, meaning the design oxidizer-to-fuel (O/F) ratio is not exactly achieved. There is also some amount of insulation/other inert materials that are expelled during the combustion process. Finally, propellants that have metal additives can form slag during the combustion process, which will be discussed in a later chapter.

Both (m_i) and (m_f) include the dry or non-propellant mass (m_d) and the payload mass (m_L). Here the non-propellant mass is taken to include the structure and other inert masses, including the unusable propellant.

2.4.1 Launch mission design

The key forces and terms used for launch mission design are briefly discussed, such that the propulsion engineer has an understanding of the underlying physics that drives rocket design. Note that the design process for launch is iterative, with the final propellant load, including a sufficient margin for day-of-launch varia-

tions. For example, to account for changes in delivered I_{sp}, burn time, atmospheric density, and wind velocities. As a first step, the propulsion engineer will complete an initial propulsion system design based on an estimated rough order of magnitude ΔV, which could be found from Section 2.5 or estimated from a Hohmann transfer calculation (see 2.5.2). This design will then be fed into a trajectory analysis tool, based on the expected thrust of the propulsion system, to give a prediction of propellant usage and trajectory achieved. These results allow the velocity losses to be determined. The propulsion engineer can then iterate to refine the design of the propulsion system and the trajectory analysis until the desired orbit is achieved.

For mission design, it is useful to first consider the free-body diagram for a rocket in flight, Fig. 2.3. In this figure, the gravitational forces are neglected from all bodies except for the planetary body from which the rocket is launching. This assumption does not hold for more sophisticated analyses. For simplicity, the motion will also be constrained to be two-dimensional, i.e., in one plane, thereby neglecting any out of plane forces acting on the rocket. An analysis using this simplification is known as a three-degree-of-freedom, or 3DOF, analysis. A more detailed six-degree-of-freedom, or 6DOF, analysis would account for out-of-plane forces and torques from thrust vector control, thrust offset, aerodynamics, etc. It is customary to first size a launch system with a 3DOF simulation and then refine it with a 6DOF analysis. Note that there are validated commercial codes that are widely utilized by industry for such analysis, see Ref. [75], as well as texts dedicated to mission design, such as Ref. [76]. The flight simulation codes all numerically integrate the rocket equations of motion. Closed-form analytical solutions of these equations are generally not possible due to the complicated nature of the forces involved. Finally, a Monte Carlo analysis is completed, dispersing the inputs of the 6DOF simulation to ensure sufficient propellant is available to achieve the desired orbit.

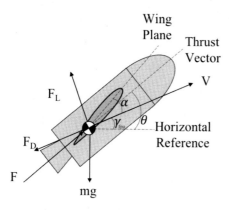

FIGURE 2.3   Rocket free-body diagram.

The key forces shown in Fig. 2.3 are the rocket's thrust, F, the aerodynamic drag, F_D, the gravitational force on the launch vehicle, mg, aerodynamic lift, F_L, and the gravitational force on the vehicle, mg. All of these terms are time-varying and nonlinear. The equation of motion in the direction aligned with flight and normal to flight is given by Eq. (2.28) and (2.29), respectively. The derivation of these equations is provided in Appendix A.

Direction of flight:

$$\frac{dV}{dt} = \frac{F}{m}cos(\theta - \gamma_{fpa}) - \frac{F_D}{m} - gsin(\gamma_{fpa}) \quad (2.28)$$

Direction normal to flight:

$$V\frac{d\gamma_{fpa}}{dt} = \frac{F}{m}sin(\theta - \gamma_{fpa}) + \frac{F_L}{m} - gcos(\gamma_{fpa}) \quad (2.29)$$

Again, note that the inputs of Eqs. (2.28) and (2.29) are time-varying and dependent on the mission profile (e.g., flight to orbit), precluding a general solution to these equations. A full 3DOF solution will solve these two equations together over time using predicted thrust profiles and models for aerodynamic forces. For the propulsion engineer, the final outcome of such analysis is generally an understanding of the total ΔV required to be delivered by the propulsion system.

The total ΔV required for launch will include the ΔV needed to overcome drag forces and gravity losses. Assuming that you need a final velocity for orbit, V_{orbit}, and starting with some initial velocity (such as the rotational velocity of the Earth), $V_{initial}$, we can integrate Eq. (2.28) to find:

$$V_{orbit} = \int_{t_0}^{t_b} \frac{Fcos(\theta - \gamma_{fpa}) - F_D}{m}dt$$
$$- \int_{t_0}^{t_b} gsin(\gamma_{fpa})dt + V_{initial} \quad (2.30)$$

This can then be rearranged to show the major components of ΔV:

$$\int_{t_0}^{t_b} \frac{F}{m}dt = \underbrace{V_{orbit} - V_{initial}} + \underbrace{\int_{t_0}^{t_b}\left(\frac{F}{m}(1 - cos(\theta - \gamma_{fpa}))\right)dt} + \underbrace{\int_{t_0}^{t_b}\left(\frac{F_D}{m}\right)dt} + \underbrace{\int_{t_0}^{t_b}\left(gsin(\gamma_{fpa})\right)dt}$$
$$(2.31)$$

$$\Delta V_{total} = \quad \Delta V_{orbit} \qquad + \Delta V_{steering} \qquad + \Delta V_{drag} \qquad + \Delta V_{gravity}$$
$$(2.32)$$

2.4.2 Gravity losses

Gravity losses only occur while the rocket is powered and are the greatest while the flight path angle is high. They are minimized by hav-

ing high thrust in a direction orthogonal to the gravity vector (i.e., low flight path angle and short burn time); however, it is not typically a practical direction in which to apply thrust, especially during launch. There are no gravity losses during coast. Eq. (2.33) gives the loss due to gravity, g.

$$\Delta V_{gravity} = \int_{t_0}^{t_{bo}} \left(g \sin(\gamma_{fpa})\right) dt \qquad (2.33)$$

2.4.3 Aerodynamic forces

There are two aerodynamic forces of concern for propulsion engineers. Aerodynamic drag is the force opposite the direction of flight and is responsible for the characteristic long and slender shape of most Earth-launched rockets. In these cases, drag is typically a small percentage of the total ΔV, see Table 2.2. Lift occurs in the direction normal to the flight path and is small for wingless rockets. The equation for the force due to drag, F_D, is given by (2.34), and that for lift, F_L, is given by and (2.35). C_D and C_L are the coefficients of drag and lift, respectively. These coefficients depend on the vehicle design, Mach number, and angle of attack. The cross-sectional area, A, is defined as the maximum cross-sectional area normal to the axis of the flight path, or the cross-sectional area of the wings (in a winged rocket or missile). Here, ρ refers to the density of the fluid that the rocket is moving through, e.g., the atmospheric density.

$$F_D = \frac{1}{2} C_D \rho A V^2 \qquad (2.34)$$

$$F_L = \frac{1}{2} C_L \rho A V^2 \qquad (2.35)$$

Aerodynamic forces can have other detrimental effects on a rocket. Notable examples include aerodynamic heating and control issues if a stage burns out in an appreciable atmosphere. Aerodynamic features such as fins, ramps, or boat tails can be included to enhance vehicle stability. Complicating the analyses required, aerodynamic forces are also affected by the rocket exhaust plume. For further reading on this subject, see Ref. [6].

2.4.4 Oberth effect

The Oberth effect states that for a given amount of propellant burned, the largest energy increase occurs where the rocket velocity is the largest. Practically, this means that the most beneficial place for a spacecraft to complete a maneuver is at the obit periapsis. High thrust or multiple short burns as the spacecraft reaches periapsis enable it to take advantage of this benefit [77].

2.5 Examples of ΔV's

Examples of the ΔV required for various launches and an approach to estimate it solar system missions are provided. In practice, ΔV calculations depend on the specific mission details and so these examples should be treated as indicative only.

2.5.1 Launch to orbit

As discussed in Section 2.4.1, the total ΔV required to be delivered by a propulsion system for launch consists of the orbital change in velocity, ΔV_{orbit}, as well as the velocity needed to overcome losses from gravity, $\Delta V_{gravity}$, steering, $\Delta V_{steering}$, and drag, ΔV_{drag}. ΔV_{orbit} is the difference between the velocity, V_{orbit}, required to inject into orbit at a given radial distance, r_{orbit}, per Fig. 2.4, and the initial velocity of the rocket prior to ignition, $V_{initial}$. It is straightforward to calculate V_{orbit} for insertion of the launch vehicle into a desired circular or elliptical orbit using Eq. (2.36) or the Vis-viva equation, Eq. (2.37), respectively.

$$V_{orbit}|_{circ} = \sqrt{\frac{Gm_{planet}}{r_{orbit}}} \qquad (2.36)$$

$$V_{orbit}|_{ellipse} = \sqrt{Gm_{planet}\left(\frac{2}{r_{orbit}} - \frac{1}{a_{orbit}}\right)} \qquad (2.37)$$

TABLE 2.2 ΔV Required for Launch to Low Earth Orbit for a Selection of Launch Vehicles. Data is taken from Ref. [1].

Launch Vehicle	Orbit Periapsis	Orbit Apoapsis	Orbit Inclination	V_{LEO}	$V_{Initial}$	$\frac{\Delta V_{orbit}}{\Delta V_{total}}$	$\frac{\Delta V_{gravity}}{\Delta V_{total}}$	$\frac{\Delta V_{steering}}{\Delta V_{total}}$	$\frac{\Delta V_{drag}}{\Delta V_{total}}$	ΔV_{total}
[-]	[km]	[km]	[deg]	[km/s]	[km/s]	[%]	[%]	[%]	[%]	[km/s]
Saturn V	176	176	28.5	7798	348	80.4	16.6	2.6	0.4	9267
Atlas I	149	607	27.4	7946	375	81.9	15.1	1.8	1.2	9243
Titan IV/Centaur	157	463	28.6	7896	352	81.9	15.7	0.7	1.7	9207
Ariane A-44L	170	170	7	7802	413	80.9	17.2	0.4	1.5	9138
Space Shuttle	-196	278	28.5	7794	395	81.4	13.4	3.9	1.2	9086
Delta 7925	175	319	33.9	7842	347	85.0	13.0	0.4	1.5	8814

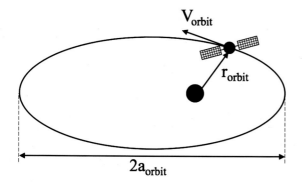

FIGURE 2.4 Elliptical orbit.

The direction and latitude from which the rocket is launched affects the total required ΔV, with less ΔV required if the launch occurs from a more quickly rotating body (since $V_{initial}$ is larger). For example, the most advantageous launch from Earth occurs due to East at the Equator taking advantage of the rotation of the planet. This provides the velocity at which the Earth is spinning, or a $\Delta V_{east} = 463.3$ m/s, towards achieving orbit. As the launch site moves up in latitude, a cosine loss is applied, i.e., at 30°, $\Delta V_{30°} = cos(30)\Delta V_{east} = 401.2$ m/s.

Example ΔV are provided in Table 2.2 for a selection of launch vehicles to reach Low Earth Orbit (LEO). It is recommended to follow the approach laid out in Section 2.4.1 to accurately determine the ΔV required for an actual launch scenario.

2.5.2 Hohmann transfers to estimate ΔV's

ΔV's required to travel about the solar system can be estimated using a Hohmann transfer, or minimum energy transfer. In the Hohmann transfer, two instantaneous propulsive maneuvers are assumed between two circular, coplanar orbits. The reference plane is critical here (e.g., heliocentric vs geocentric). This is an important distinction as Hohmann transfers only work when a single gravitational focus is assumed. Fig. 2.5 shows the transfer of a satellite from one orbit to another via the Hohmann transfer. The Hohmann transfer can be used to change orbits, e.g. raise or lower an orbit around the Earth. It should be noted that in practice, mission designers (especially for interplanetary transfers) use gravity assists and aerobraking, and other tricks to reduce the ΔV required and have to account for real issues such as drag and gravity losses that increase the ΔV required.

The periapsis and apoapsis of the Hohmann transfer ellipse correspond to the circular orbit radii of the initial and final orbits, respectively. The velocities of the circular orbit at the point at which you enter a Hohmann transfer ellipse (periapsis) and exit (apoapsis) can be found by modifying Eq. (2.36), see Eq. (2.38). In this equation, the subscript n is used to refer to the entry and exit points. In practice, it will be replaced

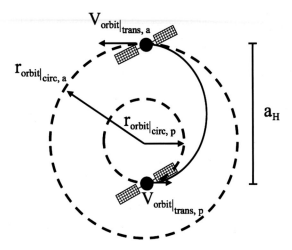

FIGURE 2.5 Hohmann transfer.

with "p" for periapsis and "a" for apoapsis.

$$V_{orbit}|_{circ,n} = \sqrt{\frac{Gm_{attract}}{r_{orbit}|_{circ}}} \qquad (2.38)$$

The velocity required for the transfer orbit can be found from the Vis-viva equation, written again here using the parameters for the Hohmann transfer, see Eq. (2.39). Here, a_H is the semimajor axis of the Hohmann transfer ellipse. This can be found by adding the radius of the periapsis and apoapsis and dividing them by two: $a_H = \frac{1}{2}(r_p + r_a)$.

$$V_{t,n} = \sqrt{Gm_{attract}\left(\frac{2}{r_{orbit}|_{circ,n}} - \frac{1}{a_H}\right)} \qquad (2.39)$$

As discussed previously, Hohmann transfers consist of two burns. The first is at the periapsis of the transfer ellipse, and the second is at the apoapsis. The ΔV at each is the difference between that required for a circular orbit at that orbital radius and the transfer velocity. This latter term, the velocity required for each side of the transfer orbit, is given by Eqs. (2.40).

$$\Delta V_p = |V_{orbit}|_{trans,p} - V_{orbit}|_{circ,p}| \qquad (2.40)$$
$$\Delta V_a = |V_{orbit}|_{circ,a} - V_{orbit}|_{trans,a}|$$

The total ΔV is can be found by adding these together, see Eq. (2.41).

$$\Delta V_{total} = \Delta V_p + \Delta V_a \qquad (2.41)$$

Example: LEO to GEO. A spacecraft is transferred from LEO (250 km) to GEO (35,780 km) using a Hohmann transfer. For this example, the following inputs are required: the mass of the Earth is $m_\oplus = 5.974 \times 10^{24}$ kg, and the radius of the Earth is $r_\oplus = 6.378 \times 10^6$ m.

The radii can be found by adding the altitudes to the radius of the Earth.

$r_{orbit}|_{LEO} = r_\oplus + 250$ km $= 6.628 \times 10^6$ m and
$r_{orbit}|_{GEO} = r_\oplus + 35,780$ km $= 4.216 \times 10^7$ m.

The velocity of a circular orbit at the defined radius about the attracting mass, in this case the Earth, can be determined by Eq. (2.38) at both LEO and GEO.

$$V_{orbit}|_{circ,LEO} = \sqrt{\frac{Gm_\oplus}{r_{orbit}|_{LEO}}} = 7756 \text{ m/s and}$$

$$V_{orbit}|_{circ,GEO} = \sqrt{\frac{Gm_\oplus}{r_{orbit}|_{GEO}}} = 3075 \text{ m/s}$$

Then the velocities required for the transfer orbit can be found from the Vis-viva equation, Eq. (2.39), where $a_H = \frac{1}{2}(r_{orbit}|_{LEO} + r_{orbit}|_{GEO})$.

$$V_{orbit}|_{trans,LEO} = \sqrt{Gm_\oplus\left(\frac{2}{r_{orbit}|_{LEO}} - \frac{1}{a_H}\right)}$$
$$= 10197 \text{ m/s}$$

$$V_{orbit}|_{trans,GEO} = \sqrt{Gm_\oplus\left(\frac{2}{r_{orbit}|_{GEO}} - \frac{1}{a_H}\right)}$$
$$= 1603 \text{ m/s}$$

Using these values, the ΔV's can be found for each part of the transfer orbit:

$$\Delta V_p = \left| V_{orbit}|_{trans,LEO} - V_{orbit}|_{circ,LEO} \right|$$
$$= 2441 \text{ m/s and}$$
$$\Delta V_a = \left| V_{orbit}|_{circ,GEO} - V_{orbit}|_{trans,GEO} \right|$$
$$= 1472 \text{ m/s}$$

Finally, the total ΔV, given by Eq. (2.41), is the following:

$$\Delta V_{total} = \Delta V_p + \Delta V_a = 3913 \text{ m/s}$$

Example: Lunar Ascent. A spacecraft is launched from the surface of the moon to a circular, lunar orbit (300 km) using a Hohmann transfer. Note, this assumes there are no obstacles in the horizontal direction, which may be difficult in practice.

The radius of the moon is $r_{orbit}|_p = r_{\mathbb{C}} = 1.738 \times 10^6$ m (periapsis), and the radius of the orbit is $r_{orbit}|_a = r_{\mathbb{C}} + 300$ km $= 2.038 \times 10^6$ m (apoapsis). Therefore the semimajor axis can be found from $a_H = \dfrac{r_{orbit}|_p + r_{orbit}|_a}{2} = 1.888 \times 10^6$ m.

Then, this problem can be broken into three velocities. The first is from the initial launch, where the Vis-viva equation (2.39) can be used. Note that the velocity of the spacecraft at launch is assumed to be zero since the rotational period of the moon is so slow (one rotation about every 27 days).

$$V_{orbit}|_{trans,p} = \sqrt{Gm_{\mathbb{C}} \left(\frac{2}{r_{orbit}|_{\mathbb{C}}} - \frac{1}{a_H} \right)}$$
$$= 1680 \text{ m/s}$$

Then the transfer orbit at apogee, using the same equation, but with values in orbit gives:

$$V_{orbit}|_{trans,a} = \sqrt{Gm_{\mathbb{C}} \left(\frac{2}{r_{orbit}|_a} - \frac{1}{a_H} \right)}$$
$$= 1489 \text{ m/s}$$

The velocity of a circular orbit at 300 km above the moon is $V_{orbit}|_{circ,a} = \sqrt{\dfrac{Gm_{\mathbb{C}}}{r_{orbit}|_a}} = 1551$ m/s.

Then, the final ΔV can be found from:

$$\Delta V_{\mathbb{C}} = V_{orbit}|_{trans,p}$$
$$+ (V_{orbit}|_{circ,a} - V_{orbit}|_{trans,a})$$
$$= 1742 \text{ m/s}$$

Example: Earth to Mars. A spacecraft is transferred from LEO (250 km) to Mars using a Hohmann transfer. This example is more difficult than the previous cases because one must switch between orbital reference frames. A patched conic approximation is shown here that breaks the trajectory into segments in the planetary and heliocentric frames. The first burn in the Earth frame sets up a hyperbolic trajectory that is matched to the elliptical transfer orbit in the heliocentric frame. The opposite is done in the Mars frame at the end of the transfer to enter orbit.

The masses and radii of Earth ($m_\oplus$, $r_\oplus$) were given in the first example. The mass of Mars is $m_\delta = 6.41 \times 10^{23}$ kg and the radius is $r_\delta = 3.3895 \times 10^6$ m. Heliocentric information will also be required for this interplanetary transfer. The orbital radii of the Earth ($R_\oplus = 1.596 \times 10^{11}$ m) and Mars ($R_\delta = 2.279 \times 10^{11}$ m) are also known. The semi major axis of the transfer orbit can be found using $a_{trans} = \dfrac{1}{2}(R_\oplus + R_\delta)$. Finally, the mass of the sun is $m_\odot = 1.989 \times 10^{30}$ kg.

The satellite is happily orbiting Earth at $V_{orbit}|_{circ,LEO}$, as calculated in a the first example. Some velocity, ΔV_1, must be added to escape Earth orbit via a hyperbolic trajectory at the moment where the tangent to the elliptical Earth - Mars transfer orbit in the heliocentric frame aligns with the asymptote to hyberbolic trajectory to Mars in the geocentric frame.

First the velocities of the Earth - Mars transfer orbit can be found:

$$V_{orbit}|_{trans,p} = \sqrt{Gm_\odot \left(\frac{2}{R_\oplus} - \frac{1}{a_{trans}} \right)} \quad (2.42)$$

$$= 31{,}279 \text{ m/s}$$

$$V_{orbit}|_{trans,a} = \sqrt{Gm_\odot \left(\frac{2}{R_\delta} - \frac{1}{a_{trans}} \right)} \quad (2.43)$$

$$= 21{,}905 \text{ m/s}$$

In order to change frames, the heliocentric velocity of the Earth is calculated:

$$V_\oplus|_{heliocentric} = \sqrt{\frac{Gm_\odot}{R_\oplus}} = 28{,}841 \text{ m/s}$$

Therefore, the required excess velocity in the geocentric frame is

$$V_{departure}|_{geocentric,\infty}$$
$$= \left| V_{orbit}|_{trans,p} - V_\oplus|_{heliocentric} \right| = 2{,}439 \text{ m/s}$$

Note: assuming an infinite radius is a simplification. One could also assume a sphere of influence instead to be more accurate.

The parameters of the elliptical orbit that match the hyperbolic trajectory in the geocentric frame can then be found.

$$a_{trans}|_{geocentric} = -\frac{Gm_\oplus}{V_{departure}^2|_{geocentric,\infty}}$$

$$= -6.705 \times 10^7 \text{ m}$$

In this case, the periapsis and velocity of the initial parking orbit are equal to those for the circular orbit (LEO) calculated in the first example.

$$r_{trans}|_{geocentric,p} = r_{orbit}|_{LEO} = 250 \text{ km} + r_\oplus$$
$$= 6.628 \times 10^6 \text{ m and}$$

$$V_{trans}|_{geocentric,circ} = V_{orbit}|_{circ,LEO}$$

$$= \sqrt{\frac{Gm_\oplus}{r_{trans}|_{geocentric,p}}}$$

$$= 7756 \text{ m/s}$$

Now everything needed to find the periapsis speed is known.

$$V_{trans}|_{geocentric,p}$$
$$= \sqrt{Gm_\oplus \left(\frac{2}{r_{trans}|_{geocentric,p}} - \frac{1}{a_{trans}|_{geocentric}} \right)}$$
$$= 11{,}237 \text{ m/s}$$

And the ΔV can be calculated

$$\Delta V_1 = \left| V_{trans}|_{geocentric,p} - V_{trans}|_{geocentric,circ} \right|$$
$$= 3480 \text{ m/s}$$

The satellite is now on its way to Mars, so the planetary reference frame needs to be converted to areocentric. First the orbital velocity of Mars in the heliocentric frame must be found:

$$V_\delta|_{heliocentric} = \sqrt{\frac{Gm_\odot}{R_\delta}} = 24{,}135 \text{ m/s}$$

Then, the excess incoming velocity at the apoapsis of the Earth - Mars transfer orbit is in the areocentric frame can be calculated.

$$V_{incoming}|_{areocentric,\infty}$$
$$= \left| V_{orbit}|_{trans,a} - V_\delta|_{heliocentric} \right| = 2{,}230 \text{ m/s}$$

The semimajor axis of the hyperbolic transfer is: $a_{trans}|_{areocentric} = -\dfrac{Gm_\delta}{V_{incoming}^2|_{areocentric,\infty}} =$ $-8.603 \times 10^6 \text{ m/s}.$

The circular velocity of the satellite in Mars orbit (areocentric) is calculated in the same way for LEO: $r_{trans}|_{areocentric,p} = 200km + r_\delta = 3.5895 \times 10^6 \text{ m}.$

Now the areocentric hyperbolic periapsis velocity can be found

$$V_{trans}|_{areocentric,p}$$

$$= \sqrt{Gm_\delta \left(\frac{2}{r_{trans}|_{areocentric,p}} - \frac{1}{a_{trans}|_{areocentric}} \right)}$$

$$= 5,368 \text{ m/s}$$

And the velocity of the satellite in circular orbit around Mars (areocentric frame) can be found using the periapsis of the hyperbolic transfer as the radius.

$$V_{trans}|_{areocentric,circ} = \sqrt{\frac{Gm_\delta}{r_{trans}|_{areocentric,p}}}$$

$$= 3,452 \text{ m/s}$$

The difference between the velocity of at the periapsis of the transfer orbit and the orbital velocity at Mars gives the magnitude of the second required burn:

$$\Delta V_2 = \left| V_{trans}|_{areocentric,p} - V_{trans}|_{areocentric,circ} \right|$$

$$= 1915 \text{ m/s}$$

Finally, the total ΔV can be found to be

$$\Delta V = \Delta V_1 + \Delta V_2 = 3,480 \text{ m/s} + 1,915 \text{ m/s}$$

$$= 5,396 \text{ m/s}$$

Note that for small bodies or for complex, multibody systems, the asymptotic assumption made here will not hold. Realistically, for interplanetary transfers, more complex trajectories and approaches to minimize ΔV (e.g. aerobraking) would be used. This approach gives the propulsion engineer a way to make a conservative estimate to inform the design process. However, it may turn out that the more complex trajectory planning may be necessary to minimize the ΔV enough to enable particular designs.

It's also important to remember that timing of Hohmann transfer maneuvers is critical. For example, in the Earth to Mars transfer, planetary alignment must be such that the Earth is at the transfer orbit periapsis for ΔV_1 and Mars is at the apoapsis for ΔV_2. This drives an approximately two month launch window every 26 months in order to reach Mars. However, the problem of determining launch windows and robust trajectories will be left to the mission designers.

2.6 Staging

Large propulsion systems, such as launch vehicles, benefit from staging. It allows the ΔV to be broken up between the stages, and the ratio of dry mass to the total vehicle mass decreases as each stage is dropped. Fig. 2.6 shows the progression of a three-stage rocket as an example, though two, or four, or more stages are possible. Multiple boosters can also be used to make up a single stage, which is often seen in the first stage of Earth launch vehicles.

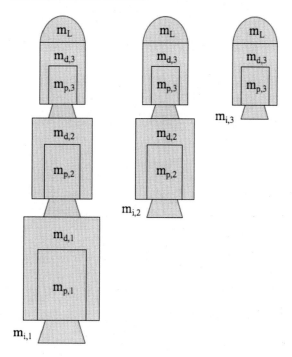

FIGURE 2.6   Staging of a three-stage rocket.

The rocket equation can be generalized for multiple stages and is presented in Eq. (2.44). In a rocket with n stages, the initial or total mass of the jth stage, $m_{i,j}$, includes the dry mass ($m_{d,j}$), which includes the structure, thermal, avionics, etc., and the propellant mass ($m_{p,j}$) for each stage, plus that of each stage above it and the payload (m_L.) The final mass $m_{f,j}$ is the initial mass minus the propellant mass for stage j.

$$\Delta V_{total} = \sum_{j=1}^{n} I_{sp,j} g_0 \ln \left(\frac{m_{i,j}}{m_{f,j}} \right) \qquad (2.44)$$

Applying (2.44) to the three-stage rocket example of Fig. 2.6 produces Eq. (2.45).

$$\Delta V = I_{sp,1} g_0 \ln \left(\frac{m_{i,1}}{m_{i,1} - m_{p,1}} \right)$$
$$+ I_{sp,2} g_0 \ln \left(\frac{m_{i,2}}{m_{i,2} - m_{p,2}} \right)$$
$$+ I_{sp,3} g_0 \ln \left(\frac{m_{i,3}}{m_{i,3} - m_{p,3}} \right) \qquad (2.45)$$

Where,

$$m_{i,1} = m_{p,1} + m_{d,1} + m_{p,2} + m_{d,2}$$
$$+ m_{p,3} + m_{d,3} + m_L$$
$$m_{i,2} = m_{p,2} + m_{d,2} + m_{p,3} + m_{d,3} + m_L \qquad (2.46)$$
$$m_{i,3} = m_{p,3} + m_{d,3} + m_L$$

It is often useful to complete initial staging design by introducing ratios that can be better estimated than exact propellant or dry masses. The payload fraction, Γ, is defined as the mass of the payload over the total mass of the rocket (Eq. (2.47)). It is a measure of the performance of the rocket, showing the percentage of the vehicle that is payload (by mass). The payload ratio, λ_j, is the mass of the payload divided by the mass of the stage lifting it (Eq. (2.48)). The propellant mass fraction, ν_j, is the percentage of the stage that is made up by the usable propellant. It is defined as the usable propellant mass over the total mass of the stage (Eq. (2.49)). Finally, the dry mass fraction (sometimes called structural

mass fraction), ϵ_j, is the dry mass divided by the total mass of the stage (Eq. (2.50)).

$$\Gamma = m_L / m_{i,1} \qquad (2.47)$$
$$\lambda_j = \frac{m_{i,(j+1)}}{m_{i,j} - m_{i,(j+1)}} \qquad (2.48)$$
$$\nu_j = \frac{m_{p,j}}{m_{p,j} + m_{d,j}} \qquad (2.49)$$
$$\epsilon_j = \frac{m_{d,j}}{m_{p,j} + m_{d,j}} = \nu_j - 1 \qquad (2.50)$$

These ratios can be estimated for propulsion systems based on historical values. This enables the propulsion engineer to get an idea of what can be achieved with the propulsion system being considered, i.e., the dry mass fraction can be used to estimate the dry mass based on the amount of propellant to solve for ΔV.

Now, using these ratios, the propellant mass fraction can be written in terms of the payload mass ratio, Eq. (2.51).

$$ln(\Gamma) = \sum_{j=1}^{n} ln \left(\frac{\lambda_n}{1 + \lambda_n} \right) \qquad (2.51)$$

Alternatively, Eq. (2.44) can be rewritten in terms of the payload fraction, structural coefficient, and propellant masses for each stage. This will allow the optimization of stages based on the ratio of propellant masses, where the stage mass ratios are as defined in Eq. (2.52). Note that the stage optimization variable, ξ_j, is introduced here for simplicity. It is defined as the ratio of the propellant mass in stage j+1 divided by that of stage 1. For a three-stage optimization, $\xi_1 = m_{p,2}/m_{p,1}$ and $\xi_2 = m_{p,3}/m_{p,1}$. Each of the stage mass ratios can then be expressed in terms of ξ_1 and ξ_2. In most cases, the optimum values of ξ_1 and ξ_2 lie at the peak of a rather flat surface, and other practical engineering considerations tend to dictate the final choices.

$$\frac{m_{i,1}}{m_{p,1}} = \left(\frac{1}{1-\Gamma} \right) \left(\frac{1}{1-\epsilon_1} + \frac{\xi_1}{1-\epsilon_2} + \frac{\xi_2}{1-\epsilon_3} \right)$$

$$\frac{m_{i,2}}{m_{p,2}} = \left(\frac{1}{1-\Gamma}\right)\left(\frac{1}{1-\epsilon_2} + \frac{\Gamma/\xi_1}{1-\epsilon_1} + \frac{\xi_2/\xi_1}{1-\epsilon_3}\right)$$

$$(2.52)$$

$$\frac{m_{i,3}}{m_{p,3}} = \left(\frac{1}{1-\Gamma}\right)\left(\frac{1}{1-\epsilon_3} + \frac{\Gamma/\xi_2}{1-\epsilon_1} + \frac{\Gamma(\xi_1/\xi_2)}{1-\epsilon_2}\right)$$

Eq. (2.45) can be rewritten as Eq. (2.53).

$$\Delta V = I_{sp,1}g_0 \ln\left(\frac{m_{i,1}/m_{p,1}}{m_{i,1}/m_{p,1} - 1}\right)$$

$$+ I_{sp,2}g_0 \ln\left(\frac{m_{i,2}/m_{p,2}}{m_{i,2}/m_{p,2} - 1}\right)$$

$$+ I_{sp,3}g_0 \ln\left(\frac{m_{i,3}/m_{p,3}}{m_{i,3}/m_{p,3} - 1}\right) \quad (2.53)$$

Then Eqs. (2.52) can be inserted into Eq. (2.53). The ΔV equation becomes a function of the known dry mass fractions, the propellant fractions (to be determined), and the given payload mass fraction for either two or three stages. With these tools, the optimum propellant mass ratios can be determined. Typically, the change in ΔV is small over a fairly large range of stage propellant mass ratios. This gives the propulsion engineer a substantial amount of flexibility in choosing the staging. There is one complication: in general, the dry mass coefficients vary with the size of the vehicle. Therefore they will actually be a function of the propellant mass, breaking this optimization. Valves come in discrete sizes and have a minimum thickness for structural materials. Practical considerations like this can be brought into the design process later. A more sophisticated model, including variable structural coefficients, based on hardware scaling laws increases the accuracy of this process and will be discussed in Chapter 7.

The propellant mass fractions for hybrid rockets are typically estimated to lie between that of liquid and solid propulsion systems. Flight systems are estimated to lie in the range of 70–80%. Solid rockets can often achieve propellant mass fractions above 90%. The propellant mass fraction increases with increasing propulsion system size, as many of the components have minimum thicknesses and number of parts (valves, etc.) Therefore small CubeSat/SmallSat hybrid systems can have propellant mass fractions closer to 50% [10].

CHAPTER

3

Hybrid rocket theory

3.1 Introduction

This chapter summarizes the theoretical basis for the combustion processes within hybrid rocket motors. The propulsion engineer is recommended to study the overview of physical phenomena that occur within the combustion chamber described in Chapter 1 before diving into this theory.

The propulsion engineer is walked through the classical derivation of regression rate developed in the 1960s. It assumes that the fuel is a flat plate and the combustion is diffusion limited within a turbulent boundary layer. About 35 years later, liquefying fuels, such as paraffin wax, were discovered to have dramatically increased regression rates. The theory was then adapted to account for entrainment of liquid drops in the flow. This derivation is also included.

A number of non-dimensional parameters are used in the upcoming derivations. The definition and equation for each parameter is given, along with typical values for hybrid rockets, to facilitate their introductions in the following sections during the derivation of the regression rate of classical and liquefying fuels.

After the theory has been derived, its application to hybrid design is introduced, e.g., simplifications for a single port motor. Chapter 7 will

then step through the details of how variables are coupled and a design could be completed.

Alternative injection techniques that affect hybrid modeling, e.g., swirling or complex flow hybrids, are noted but not discussed in detail. The propulsion engineer is pointed to alternative references for more information.

Finally, some simplifying assumptions are used to get an idea for the main transients: ignition and shutdown. A quasi-steady analysis is discussed.

3.2 Classical fuels theory

3.2.1 Dimensionless parameters

Several useful dimensionless parameters will be introduced in the following derivations. To help keep these all straight, Table 3.1 includes definitions, equations, and common values for hybrid rockets.

A couple notes on the dimensionless parameters before jumping into the derivations: first, the Damköhler number is typically assumed to be much greater than one for hybrid rockets, meaning that the chemical reactions do not limit the combustion process. However, in the low mass flux regime, Da could actually approach unity. Also, anyone familiar with Reynolds numbers for turbulent boundary lay-

TABLE 3.1 Dimensionless Numbers of Significance in Hybrid Combustion. Note: It is unlikely that the Reynolds analogy would hold in a region where reactions were taking place, i.e., above the flame.

	Definition: Ratio of	Equation	Typical Hybrid Value
Blowing	Radial to axial mass flux	$B \equiv \dfrac{(\rho V_y)_w}{\rho_e V_{x,e} \frac{C_f}{2}}$	5-15
Damköhler	Fluid diffusion to chemical reaction time scales	$Da = \dfrac{t_d}{t_k}$	$\gg 1$
Lewis	Thermal diffusion to mass diffusion	$Le = \dfrac{\kappa}{D_m} = \dfrac{Sc}{Pr}$	1, Reynolds analogy*
Prandtl	Viscous diffusion to thermal diffusion	$Pr = \dfrac{C_p \mu}{\lambda}$	1, Reynolds analogy*
Reynolds	Inertial forces to viscous forces	$Re = \dfrac{\rho u L}{\mu}$	$> 10^4$
Stanton	Heat flux into a fluid to the enthalpy at the flame	$C_H = \dfrac{\dot{Q}_w}{\rho u \Delta h}$	$\ll 1$
Weber	Inertial forces to surface tension	$We = \dfrac{\rho V^2 D}{\sigma_t}$	$\ll 1$

ers may be starting to get concerned with low value cited above (since the classical regression rate derivation is based on the assumption that the processes exist in a turbulent boundary layer. While the Reynolds numbers present in hybrid rockets are often lower than the typical laminar to turbulent transition of $10^5 - 10^6$ based on the distance from the leading edge, the destabilizing effect caused by the mass addition of the fuel and the combustion serves to reduce the transition Reynolds number. Simple evaporation has been shown to reduce the transition Reynolds number to about to 10^4 [81]. The propulsion engineer is pointed to [82] or [83] for further discussion of dimensionless parameters.

The Stanton number relates heat flux in a fluid to the thermal capacity of the fluid, in this case, the enthalpy at the flame. Two versions of the Stanton number (with and without blowing) will be used below. Without blowing, the velocity normal to the fuel grain is assumed to be zero. (Blowing will be explained later, do not worry.

The Reynolds analogy is invoked in the following derivations, which says that the Lewis and Prandtl numbers are both equal to one. This

is reasonably accurate for hybrid rocket combustion. The Prandtl number relates viscous to thermal diffusion. So, it is assumed that the viscous and thermal diffusion are roughly equal.

The Weber number will not be discussed until injectors are introduced in Chapter 7, but it is a measure of the atomization of the oxidizer entering the combustion chamber.

The subscripts c, e, and w are used to refer to conditions at the location where combustion occurs, i.e., at the flame location, the boundary layer edge and the fuel surface or "wall", respectively.

3.2.2 Regression rate

The theory for the combustion of classical hybrid rocket fuels, as summarized in this section, draws predominantly on material from Refs. [78], [79], [80], and [12]. In this approach, steady-state hybrid combustion is modeled as a turbulent diffusion flame with the reactions occurring in an infinitely thin flame zone within the boundary layer. Also, it is assumed that the presence of combustion does not significantly alter the velocity profile within the boundary

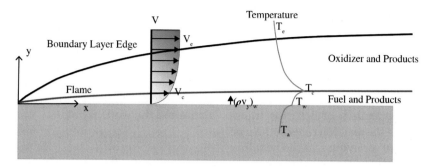

FIGURE 3.1 Schematic of diffusion-limited combustion processes in a turbulent boundary layer with blowing for a classical fuel. This figure was produced to scale using the measured boundary layer and flame location for combustion above a PMMA fuel grain at elevated combustion chamber pressure as reported in Refs. [84] and [85].

layer, e.g., the velocity profile, as shown in Fig. 3.1, is unchanged from that for turbulent flow without combustion. Heat transfer from the flame to the fuel surface dominates the combustion process. Marxman et al. [80] used a two-dimensional combustion chamber, injecting gaseous fuel from one wall and oxidizer down the axial port to validate their work. The assumption that the flame zone is infinitely thin as compared to the boundary layer size, δ, can be justified because the characteristic times associated with the diffusion dominated transport processes are much larger than those of the chemical kinetics (i.e., the Damköhler number is much greater than one).

Fig. 3.1 shows the schematic of the model presented here. A more detailed depiction of the physical processes occurring within the boundary layer was provided in Fig. 1.3 of Chapter 1. Note that while Fig. 1.3 presents a more realistic velocity profile (altered by the presence of the flame) than the simplified version shown here, Fig. 3.1 shows the assumed velocity form used for the equation derivations of this section.

Conservation of energy applied at the surface of the fuel grain produces the expression shown in Eq. (3.1). Here, h_v is the total effective heat of gasification of the solid fuel, or the total heat required to turn a unit mass of solid fuel into a gas. It includes the energy required to heat the solid fuel from ambient temperature

to the sublimation/melting temperature, the latent heat of sublimation/evaporation, h_{fg}, and, in the case of a polymer, the energy required for de-polymerization. If a melt layer is formed, the heat required to heat the melt layer from the melting temperature to the evaporation temperature and the latent heat of melting are also included in h_v. Therefore the expression for h_v is fuel dependent. V_y refers to the velocity of the gas normal to the fuel surface, see Fig. 3.1. The total heat transfer to the fuel surface (wall) per unit area, $\dot{Q}_w$, can be expressed as the addition of the convective and radiative heat transfer per unit area, as shown in Eq. (3.2).

$$\frac{\dot{Q}_w}{h_v} = \rho_F \dot{r} = (\rho V_y)_w \qquad (3.1)$$

$$\dot{Q}_w = (\dot{Q}_c)_w + (\dot{Q}_r)_w \qquad (3.2)$$

To begin, consider heat transfer that is dominated by convection, e.g., Eq. (3.2) becomes Eq. (3.3). This is a reasonable assumption for hybrid rockets that do not have metallic additives, under typical operating conditions.

$$\dot{Q}_w \simeq (\dot{Q}_c)_w \qquad (3.3)$$

The Prandtl number, Pr, is the ratio of viscous diffusion to thermal diffusion. If we can assume unity Prandtl number, $Pr = 1$, and the

Reynolds analogy (which relates turbulent momentum flux to heat flux) is valid within the turbulent boundary layer but not at the location of the flame, then we can re-write the convective heat flux at the wall, $(\dot{Q}_c)_w$ in the form of Eq. (3.4). This allows a simplification in the heat flux, making it independent of both the transport mechanism and the reaction rates in the gas. Eq. (3.4) is equivalent to saying that the heat flux is independent of both the transport mechanism and the reaction rates in the gas, and the equation is the same form as for an unreacted gas [78]. Here λ is the thermal conductivity, c_p is the specific heat at constant pressure, and h is the sum of sensible and chemical enthalpies. y is as shown in Fig. 3.1 and is the coordinate direction normal to the fuel surface. The subscript w refers to the conditions at the surface of the fuel.

$$(\dot{Q}_c)_w = \left(\frac{\lambda}{c_p} \frac{\partial h}{\partial y} \right)_w \qquad (3.4)$$

The Stanton number, C_H is defined in the same manner as Marxman and Gilbert [78] as the ratio of heat flux into a fluid to the enthalpy at the flame (Eq. (3.5)). V_x refers to the velocity parallel to the fuel surface. The subscript c throughout these equations refers to the conditions at the location where combustion occurs, i.e., at the flame location. The subscript w continues to describe the conditions at the fuel surface or "wall".

$$C_H = \frac{(\dot{Q}_c)_w}{\rho_c V_{x,c}(h_c - h_w)} \qquad (3.5)$$

Eq. (3.3), (3.4), and (3.5) can be substituted into Eq. (3.1) to yield Eq. (3.6).

$$\rho_F \dot{r} = \frac{C_H \rho_c V_{x,c}(h_c - h_w)}{h_v} \qquad (3.6)$$

If the Lewis number, the ratio of thermal diffusivity to mass diffusivity, is also assumed to be one in the boundary layer, then Eq. (3.7) can be obtained via the Reynolds analogy ($Pr = Le = $

1). It should be noted that the Reynolds analogy will not hold at the flame location. Eq. (3.7) physically implies that the convective heat transfer driven by the enthalpy difference between the flame and the wall equals the skin friction driven by the velocity difference between the flame and the wall.

$$\frac{(\dot{Q}_c)_w}{h_c - h_w} = \frac{\tau_w}{V_{x,c}} \qquad (3.7)$$

To make use of Eq. (3.7), the definition of skin-friction coefficient is given in Eq. (3.8). The skin-friction coefficient describes the skin shear stress and is used in aerodynamics to calculate drag. It decreases with increasing Reynolds number. Note that the subscript e refers to the conditions at the boundary layer edge.

$$C_f = \frac{\tau_w}{\frac{1}{2}\rho_e V_{x,e}^2} \qquad (3.8)$$

Substituting Eq. (3.8) into Eq. (3.7) and then substituting that into Eq. (3.5) gives Eq. (3.9) for the Stanton number in terms of the local skin-friction coefficient, C_f.

$$C_H = C_f \frac{\rho_e V_{x,e}^2}{2\rho_c V_{x,c}^2} \qquad (3.9)$$

The addition of mass at the wall is generally referred to as blowing and is a crucial parameter in the regression rate of classical hybrid fuels. The blowing parameter, B, is defined according to Eq. (3.10). Physically the blowing parameter can be thought of as a measure of the relative radial mass flux to axial mass flux where the skin-fiction coefficient is included for convenience. The vaporized fuel rising from the surface blocks some of the flame's heat from reaching the surface of the solid fuel.

$$B \equiv \frac{(\rho V_y)_w}{\rho_e V_{x,e} \frac{C_f}{2}} \qquad (3.10)$$

We can rewrite Eq. (3.10) by substituting Eq. (3.5), (3.8), and (3.1) (with (3.3)) into Eq. (3.10).

This produces Eq. (3.11).

$$B = \frac{V_{x,e}}{V_{x,c}} \frac{(h_c - h_w)}{h_v} \qquad (3.11)$$

Eq. (3.11) is substituted into Eq. (3.6) along with Eq. (3.9) to derive the regression rate Eq. (3.12).

$$\dot{r} = \frac{\rho_e V_{x,e} C_f B}{2\rho_F} \qquad (3.12)$$

Now it is time to introduce the Stanton number in the absence of blowing, that is the Stanton number assuming that $(\rho V_y)_w$ is zero, this is denoted by C_{H0}. This model assumes that the skin-friction coefficient under hybrid combustion does not differ greatly from ordinary boundary layer with the same level of blowing because the velocity profile near the wall is essentially the same for both. Therefore the empirical formula for the skin-friction coefficient of a turbulent boundary layer in the absence of blowing, C_{f0}, is used (Eq. (3.13)).

$$C_{f0} = 0.06 Re_x^{-0.2} \qquad (3.13)$$

where Re_x is the Reynolds number with respect to the distance x from the leading edge. The exponent equals -0.2 for a boundary layer with a thickness approximated by a one-fifth power law in the Reynolds number. This derivation takes advantage of the assumption that the Prandtl number is one or there would be a $Pr^{-2/3}$ term on the right-hand side of the equation. The constant was empirically determined based on the low Mach numbers expected in hybrids.

Eq. (3.9) and the expression for the skin-friction coefficient for a turbulent boundary layer above a flat plate without blowing, C_{f0} given in Eq. (3.13), are used to obtain Eq. (3.14).

$$\dot{r} = 0.03 V_{x,e} B Re_x^{-0.2} \frac{\rho_e}{\rho_F} \left(\frac{C_f}{C_{f0}} \right) \qquad (3.14)$$

The flat plate Reynolds number, Re_x, of Eqs. (3.13) and (3.14) is defined in the usual manner,

see Eq. (3.15).

$$Re_x = \frac{\rho_e V_{x,e} x}{\mu_e} \qquad (3.15)$$

It is convenient to recall at this time, that the mass flux in the motor can be expressed as $G = \rho_e V_{x,e}$ and substitute that into Eq. (3.14). In Ref. [79], Marxman et al. provide Eq. (3.16) as a reasonable approximation for the ratio C_f/C_{f0} for blowing parameters $5 \leqslant B \leqslant 100$, which includes the range relevant for hybrids. It is worth noting that the exponent of Eq. (3.16) was shown by Ref. [86] to be closer to -0.68 than the originally derived -0.77 for the more relevant range of $5 < B < 20$ so that is the value adopted here. Eq. (3.16) allows the regression rate to be written in the form shown in Eq. (3.17).

$$\frac{C_f}{C_{f0}} \approx 1.2 B^{-0.68} \qquad (3.16)$$

$$\dot{r} = \frac{0.036 \mu_e^{0.2} B^{.32} G^{0.8} x^{-0.2}}{\rho_F} \qquad (3.17)$$

Increasing blowing, B, reduces the slope of the turbulent boundary layer, which reduces heat transfer to the wall, $\dot{Q}_C$. The reduced heat transfer is referred to as an increase in blocking, which is seen in the equations as a decrease in C_f/C_{f0}. The regression rate still increases with increased blowing but not linearly due to the dependence on C_f/C_{f0}.

It is difficult to calculate the blowing parameter, B, directly from first principles because it depends on the propellant combination. Specifically, higher molecular weight products are less effective at blocking heat transfer than light products. Additionally, there is a difference between the flame temperature and the fuel surface temperature depending on the propellant combination. A factor of (flame temperature/fuel surface temperature)$^{0.6}$ has been suggested to account for the distortion of the velocity profile in the boundary layer due to the presence of the flame [87]. The propulsion engineer is steered towards [88] and [89] for more information on this topic. However, if the motor length to diameter is not too large, it is reasonable to treat the blowing parameter as a constant for a given propellant combination.

A constant, a, is defined for a given propellant combination, as in Eq. (3.18), and the expression for the regression rate becomes that shown in Eq. (3.19).

$$a = \frac{0.036\mu_e^{0.2}B^{0.32}}{\rho_F} \quad (3.18)$$

$$\dot{r} = aG^{0.8}x^{-0.2} \quad (3.19)$$

The commonly used form of Eq. (3.17) is as shown in Eq. (3.20) where a, n, and m are empirically determined constants for a given fuel and oxidizer combination.

$$\dot{r} = aG^n x^{-m} \quad (3.20)$$

Eq. (3.20) is significant as it indicates that the burn rate of hybrid rocket fuels is a function of only the mass flux in the combustion chamber and the axial location along the length of the fuel grain with a and n as constants that depend on the propellant combination. Unlike solid rocket motors, the burn rate of the fuel is independent of the combustion chamber pressure under typical operating conditions.

Accounting for radiative heat transfer

Radiative heat transfer is generally neglected for standard hybrid rocket operating conditions and non-metalized fuel grains. However, it should not be neglected if the motor is operating at very low oxidizer mass flux (see discussion in Section 3.2.4 and Fig. 3.2), if there are any metallic additives in the fuel grain, or if the fuel combustion is known to produce a substantial amount of soot.

Radiative heat transfer to the fuel surface, $\dot{Q}_r$, can be expressed assuming gray body radiation in the form shown in Eq. (3.21). Here, σ_B is the Stefan–Boltzmann constant, e_g and e_w are the emissivity of the gas and wall, respectively, and

a_g is the absorptivity of the gas. The gas emissivity, e_g, is a function of chamber pressure, and thus radiative heat transfer is also pressure dependent.

$$\dot{Q}_r = \sigma_B e_w \left(e_g T_c^4 - a_g T_w^4\right) \quad (3.21)$$

Very small amounts of radiative heat transfer to the fuel surface (relative to convective heat transfer to the fuel surface) can be combined with the convective regression rate expression of Eq. (3.19) using the linear relation of Eq. (3.2) and the original conservation of energy expression of Eq. (3.1).

$$\dot{r} = aG^{0.8}x^{-0.2} + \frac{\dot{Q}_r}{\rho_F h_v} \quad (3.22)$$

Substituting Eq. (3.21) into Eq. (3.22) produces Eq. (3.23). Note that Eqs. (3.22) and (3.23) are only valid if the radiative heat transfer is small enough not to significantly influence the blocking parameter C_f/C_{f0}.

$$\dot{r} = aG^{0.8}x^{-0.2} + \frac{\sigma_B e_w \left(e_g T_c^4 - a_g T_w^4\right)}{\rho_F h_v} \quad (3.23)$$

If the radiative heat transfer is large, then the regression rate equation is of the form shown in Eq. (3.24). Eq. (3.24) more correctly captures the blocking effect associated with reduced convective heat transfer to the fuel surface as the radial mass flux of vaporized fuel increases, see derivation in Ref. [1].

$$\dot{r} = \frac{\dot{Q}_c}{\rho_F h_v}\left[e^{-0.75\dot{Q}_r/\dot{Q}_c} + \frac{\dot{Q}_r}{\dot{Q}_c}\right] \quad (3.24)$$

3.2.3 Estimation of blowing parameter and flame location

The blowing parameter, B, is challenging to calculate because of its dependence on the ratio of the free-stream velocity to the velocity at the flame, $V_{x,e}/V_{x,c}$, which in turn depends on the location of the flame in the boundary layer. Marxman et al. [79] attempted to resolve this issue by re-casting the expression for $V_{x,e}/V_{x,c}$ in terms of the oxidizer-to-fuel mass ratio at the flame, O/F_c, and the chemical parameters of the system, see Eq. (3.25).

$$\frac{V_{x,c}}{V_{x,e}} = \frac{O/F_c \frac{(h_c - h_w)}{h_v}}{1 + (1 + O/F_c)\frac{(h_c - h_w)}{h_v}} \quad (3.25)$$

They assume that the velocity in the boundary layer is of the form derived in Ref. [78], given here as Eq. (3.26). Where, δ is the momentum boundary layer thickness, and y is the distance normal to the fuel surface.

$$\frac{V_x}{V_{x,e}} = \frac{\left(\frac{y}{\delta}\right)^{1/7}\left[1 + 0.5B\left(\frac{y}{\delta}\right)^{1/7}\right]}{1 + 0.5B} \quad (3.26)$$

The results of Eq. (3.25) can be used to estimate the ratio of the free-stream velocity to the velocity at the flame, $V_{x,e}/V_{x,c}$. This allows the blowing parameter, B, to be evaluated via Eq. (3.11). The blowing parameter can then be used to estimate the location of the flame within the boundary layer, according to Eq. (3.27).

$$\frac{y_c}{\delta} = \left[\frac{-1 + \sqrt{1 + 2B\frac{V_{x,c}}{V_{x,e}}(1 + 0.5B)}}{B}\right]^7 \quad (3.27)$$

3.2.4 Classical theory limitations

The theory described to this point was developed with the simplifying assumption that the fuel grain is a flat sheet with freestream flow over it. In reality, the fuel grain is three-dimensional. The fuel grain is typically contained within a cylindrical combustion chamber with hemispherical end caps where oxidizer flow and combustion all occur within fuel ports, which may be circular or wagon-wheel shaped, etc. Instead of a uniform, freestream flow beyond the boundary layer, the flow is more similar to internal pipe flow [12]. The core oxidizer flow may accelerate with axial distance along the fuel grain due to increased boundary layer growth. Depending on the length of the solid fuel grain relative to the cross section diameter of a given port, the boundary layers on either side of the port may merge.

The classical hybrid theory developed in Refs. [79] and [80] and summarized in the previous sections also assumes that the velocity in the boundary layer is relatively unaffected by either combustion or the presence of blowing from the wall. More recent research, such as that documented in Refs. [89] and [90], indicates this assumption is invalid. The hot and low-density region around the flame can have a significant effect on the velocity within the boundary layer [12].

Despite these over-simplifications, the equation for regression rate derived using the flat plate and velocity unaffected by combustion assumptions (Eq. (3.20)) has generally been shown to apply to a range of hybrid rocket motor configurations operating in a diffusion-limited regime.

A limitation of Eq. (3.20) is that it only applies when combustion is diffusion-limited and when radiation effects are insignificant. The point at which either radiation or kinetics starts to notably influence the fuel regression rate is specific to each propellant combination and motor scale [1]. If metal additives are included in fuels, then radiative heat transfer can be significant, as described above. Even for standard hydrocarbon fuels without metal additives, the assumption of diffusion-limited combustion is only valid across a central range of mass fluxes as shown schematically in Fig. 3.2. If the mass flux in the fuel port is too low, then the convective heat transfer does not dominate radiative heat transfer, and thus the radiative heat transfer term in

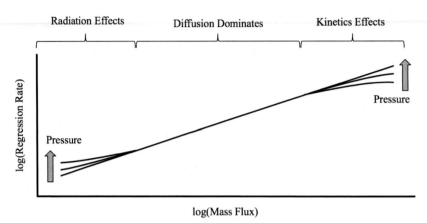

FIGURE 3.2 Illustrative figure to show the fuel regression rate vs mass flux and the regions of pressure dependence. Figure adapted from [1].

Eq. (3.24) cannot be neglected. Radiative heat transfer introduces a pressure dependency on the regression rate, further complicating modeling. At the extreme end of low-mass flux operation the propulsion engineer needs to be careful to avoid cooking the fuel grain and introducing a chuffing instability, see Chapter 8. At the other operating extreme, when the mass flux in the fuel port is too high, the reaction kinetics are no longer significantly faster than the diffusion processes within the boundary layer, and reaction kinetics must be accounted for when determining the fuel regression rate. Chemical kinetics also introduces a pressure dependency on the regression rate. At the extreme end of this high-mass flux operating regime, the propulsion engineer needs to be careful to avoid "flooding". Flooding is the limit where the combustion is no longer able to be sustained at high oxidizer mass flux either due to moving beyond the combustible O/F range or due to the flame being extinguished from fluid mixing rates exceeding the reaction rate [12].

The ratio of kinetic reaction time, t_k, to fluid diffusion time, t_d, is the parameter that dictates when the fuel regression rate starts to depend on pressure. Ref. [1] asserts that for a given propellant combination, this ratio scales with mass flux and chamber pressure per Eq. (3.28). Examination of this equation implies that at very high mass fluxes or at low chamber pressures, the fuel regression rate can become limited by kinetics and therefore start to depend on pressure. Fig. 3.3 shows that at lower chamber pressures, slower kinetics limit the fuel regression rate. At higher chamber pressures, the regression rate is consistent with diffusion-limited theory [1].

$$\frac{t_k}{t_d} \propto \frac{G^{0.8}}{P_c} \qquad (3.28)$$

3.3 Liquefying fuels

The theory for the combustion of liquefying, high regression rate fuels, shown schematically in Fig. 3.4, is summarized here based on the work by Karabeyoglu et al. published in Refs. [53], [39], [91], [92], and [93]. The solid fuel grain melts and forms a liquid layer on the surface. The shear force from the incoming oxidizer flow creates roll waves and rips droplets from the melt layer, which entrain in the oxidizer flow, dramatically increasing the regression rate, as introduced in Chapter 1.

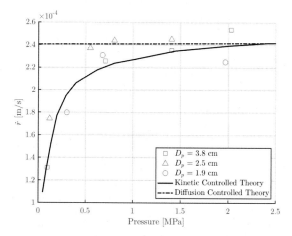

FIGURE 3.3 Regression rate at low chamber pressure for the combustion of PMMA and gaseous oxygen. All data regression rate in the figure is generated at an oxidizer mass flux of 507 kg/m²s. The fuel grain length was 0.203 m, and the burn time was 13 seconds. Figure generated using data from [1].

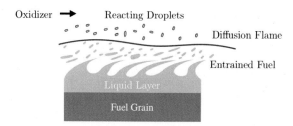

FIGURE 3.4 Schematic of the droplet entrainment mechanism for liquefying high regression rate fuels originally proposed by Karabeyoglu et al. in Ref. [39].

3.3.1 Melt layer thickness

The thickness of the melt layer above the solid portion of liquefying fuels is estimated using conservation equations for mass and energy along with some simplifying assumptions. This analysis assumes steady-state combustion and uniform thermophysical properties within the melt layer and within the solid fuel. As shown in Fig. 3.5, the steady-state assumption requires that the regression rate of the liquid layer surface matches the regression rate of the solid surface. The derivation of melt layer thickness presented

here assumes that the combustion chamber pressure is below the critical pressure of the fuel, or the fuel surface temperature is below the critical temperature, such that there is a defined liquid-vapor interface. In the super critical regime, an additional pyrolylsis layer needs to be added to the model on top of the liquid zone, see Ref. [93].

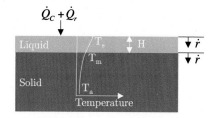

FIGURE 3.5 Schematic of model used to estimate the melt layer thickness above liquefying hybrid fuels.

Both radiative and convective heat transfer are assumed to occur in the gas phase. However, convection in the liquid layer is neglected due to the relatively high temperature gradients and low Reynolds numbers estimated within the melt layer. Penetration of radiation through the liquid layer and into the solid fuel surface is considered. The radiative field is assumed to be one-dimensional, and the liquid and solid phase fuel are assumed to have approximately flat absorption characteristics across the spectrum likely to be encountered.

Under the above assumptions, conservation of mass within a control volume encompassing the solid/liquid boundary yields Eq. (3.29). The vertical velocity in the liquid layer, $V_{y,l}$, is in the laboratory frame of reference. Since the density of the solid fuel is greater than the density of the liquid layer, Eq. (3.29) indicates that the liquid layer has bulk motion away from the solid fuel surface towards the gas interface.

$$V_{y,l} = (\rho_s/\rho_l - 1)\dot{r} \qquad (3.29)$$

Conservation of energy across the liquid-gas and liquid-solid interfaces produces the equation for the liquid layer thickness, th_l, Eq. (3.30).

Ref. [91] provides the detail of the derivation of this equation.

$$e^{-th_l/\delta_l} = \frac{h_m(a_l\delta_l - 1) + h_v(\dot{Q}_r/\dot{Q}_w)e^{-th_la_l}}{h_{ent}(a_l\delta_l - 1) + h_v(\dot{Q}_r/\dot{Q}_w)} \quad (3.30)$$

Where:

$$\dot{Q}_w = \dot{Q}_c + \dot{Q}_r \quad (3.31)$$

$$\delta_l = \frac{\kappa_l \rho_l}{\dot{r}\rho_s} \quad (3.32)$$

$$h_m = h_{fusion} + c_s(T_m - T_a) \quad (3.33)$$

$$h_{ent} = h_m + c_l(T_v - T_m) \quad (3.34)$$

$$h_v = c_l(T_v - T_m) + c_s(T_m - T_a) + h_{fusion} \\ + h_{fg}(\dot{r}_v/\dot{r}) \quad (3.35)$$

Physically, δ_l is the characteristic thermal thickness in the liquid phase, a_l is the radiative absorption coefficient of the liquid, and $a_l\delta_l$ is the ratio of thermal thickness to radiative penetration thickness. h_m, h_{ent}, and h_v are the total heat of melting, entrainment, and gasification, respectively. $\dot{Q}_r$ and $\dot{Q}_c$ are the radiative and convective heat transfer on the fuel surface. κ_l is the thermal diffusivity of the melt layer. c_l and c_s are the specific heat of the liquid and solid (per unit mass). T_a, T_m, and T_v are the fuel ambient, melting, and vaporization temperatures, respectively. h_{fg} is the latent heat of vaporization, and h_{fusion} is the latent heat of fusion.

A general explicit solution to Eq. (3.30) is not available. Instead, Karabeyoglu et al. present explicit solutions for the melt layer thickness for two limiting cases [91,93].

1. Case 1: $a_l\delta_l \gg 1$ This case is relevant for fuels that have very high radiative absorption in the melt layer. This case looks at the limiting case with all radiation absorbed at the liquid-gas interface. This case is pertinent for fuels with high infrared absorption coefficients, such as those loaded with carbon black. For these fuels, Eq. (3.30) can be simplified to the form shown in Eq. (3.36).

$$th_l = \delta_l ln\left(1 + \frac{c_l(T_v - T_m)}{h_m}\right) \quad (3.36)$$

2. Case 2: $a_l\delta_l \ll 1$ This case looks at the limit of no infrared radiation absorption in the fuel melt layer, all of the radiative heat is absorbed in the solid. Under this assumption, the implicit expression for melt layer thickness, Eq. (3.30), can be simplified to the form shown in Eq. (3.37). Optical characterization of paraffin wax, such as those from Ref. [94], indicates that for these fuels the absorption coefficient, a_l, is approximately 4 m^{-1}. Using the properties provided in Table 3.2, the thermal thickness of paraffin wax is approximately 1.9×10^{-5} m, and thus $a_l\delta_l$ is approximately 7.4×10^{-5}, which is clearly much less than 1. For this reason, neat paraffin waxes are assumed to have a melt layer thickness defined by Eq. (3.37), which Karabeyoglu et al. further simplify to Eq. (3.38) in Ref. [93] for n-alkane fuels.

$$th_l = \delta_l ln\left(1 + \frac{c_l(T_v - T_m)}{h_m - h_v(\dot{Q}_r/\dot{Q}_w)}\right) \quad (3.37)$$

$$th_l = \frac{a_t}{\dot{r}} \quad (3.38)$$

Eq. (3.37) shows that the melt layer thickness is proportional to the characteristic thermal length of the liquid (δ_l). The thickness parameter, a_t, is a function of the fuel carbon number and is introduced here to show the relationship between the melt layer thickness and regression rate. More details can be found in [93].

The thickness of the melt layer for all high regression rate fuels presented here is likely to be dictated by Eq. (3.37), rather than Eq. (3.36). Note that if a black dye with carbon black or soot is added to the fuel, such as India ink, then Case 1 would be more representative than Case 2.

The melt layer thickness of paraffin wax ($C_{32}H_{66}$) and PE wax ($C_{70}H_{142}$) are estimated using Eq. (3.38), where a_t is 8.89×10^{-7} m^2/s and 9.00×10^{-7} m^2/s for paraffin and PE, respectively [93]. The resulting estimated melt layer thicknesses are from 0.2 mm to 1.7 mm for paraffin and approximately 1.6 mm for PE. These values were calculated using a range of measured regression rates from visualization tests [85].

3.3.2 Regression rate

The rate at which fuel is entrained from the liquid layer is evaluated by Karabeyoglu et al. using experimental results from Gater and L'Ecuyer [95], linear stability theory, and theoretical studies of entrainment from the nuclear reactor field [91]. As discussed earlier in this chapter, see Fig. 3.4, nonlinear roll waves and droplet entrainment are thought to be the mechanism through which mass transfer of the liquid layer occurs. Karabeyoglu et al. propose that the entrainment rate is strongly dependent on the viscosity of the melt layer, μ_l, as well as on the thickness of the melt layer, th_l, the dynamic pressure, P_d, and the surface tension, σ_t, according to Eq. (3.39).

$$\dot{m}_{ent} \propto \frac{P_d^\alpha th_l^\beta}{\mu_l^\varphi \sigma_t^\phi}. \qquad (3.39)$$

Here, α, β, φ, and ϕ are all estimated to be in the range of 1 to 2. The entrainment mass flow rate will be used to calculate the regression rate of a liquefying fuel later in the text.

The classical hybrid rocket combustion theory summarized in Section 3.2 must be modified to be applied to liquefying high regression rate fuels. In Ref. [91], Karabeyoglu et al. propose the following three changes to the classical theory to make it applicable to liquefying fuels:

1. The effective heat of gasification that appears in the thermal blowing parameter is reduced since less energy is now required for fuel evaporation due to mass transfer as a result of mechanical entrainment of the liquid.

2. The presence of droplets between the fuel surface and the diffusion flame corresponds to a two-phase flow regime. Karabeyoglu et al. neglect the effect of the fuel droplets on the momentum and energy transfer between the flame and the fuel surface and thus write the blocking factor, Eq. (3.16), in terms of the only the gas phase mass transfer from the fuel surface, Eq. (3.40).

$$C_H/C_{H0} = \frac{2}{2 + 1.25 B_g^{0.75}} \qquad (3.40)$$

3. The surface roughness of the fuel surface and the heat transfer to the fuel from the flame front are increased as a result of the roll waves and ripples in the liquid layer.

Each of these changes leads to an increased regression rate. With these modifications, the equations governing the combustion of classical hybrid rocket motors become the series of Eqs. (3.41), (3.42), (3.43), and (3.44). These four equations can be solved to calculate the total fuel regression rate as a function of location and mass flux. Eq. (3.41) is the mathematical statement that the total regression rate, $\dot{r}$, is simply the addition of the regression rate due to entrainment, $\dot{r}_{ent}$, and the regression rate due to evaporation, $\dot{r}_v$. Eq. (3.42) is derived from conservation of energy using a control volume that spans the surface of the liquid layer. Here x refers to the axial distance along the fuel port. C_H and C_{H0} are the Stanton numbers, i.e., the ratio of heat transferred to the fluid to the thermal capacity of the fluid, with and without blowing, respectively. Eq. (3.43) is an expression for the heat transfer roughness correction parameter, Fr, originally suggested by Gater and L'Ecuyer [95]. Eq. (3.44) is derived from Eq. (3.39). The coefficient a_{ent} is not explicitly listed but is a function of the propellant properties and the average gas density in the combustion chamber. For simplicity, it is assumed that a_{ent} is a constant for a given propellant. Recall that the exponents α and β are the range of 1-2.

$$\dot{r} = \dot{r}_{ent} + \dot{r}_v \qquad (3.41)$$

$$\dot{r}_v + \dot{r}_{ent} \left[\frac{h_m}{h_{ent} + h_{fg}} + \frac{C_l(T_v - T_m)}{h_{ent} + h_{fg}} \left(\frac{\dot{r}_v}{\dot{r}} \right) \right]$$

$$= Fr \frac{0.03\mu_g^{0.2}}{\rho_F} \left(1 + \frac{\dot{Q}_r}{\dot{Q}_c} \right) B \frac{C_H}{C_{H0}} G^{0.8} x^{-0.2}$$

$$\qquad (3.42)$$

$$Fr = 1 + \frac{14.1\rho_g^{0.4}}{G^{0.8}(T_g/T_v)^{0.2}} \qquad (3.43)$$

$$\dot{r}_{ent} = a_{ent} \frac{G^{2\alpha}}{\dot{r}^\beta} \qquad (3.44)$$

These coupled equations have been shown to predict the burn rate of liquefying fuels with reasonable accuracy, see Refs. [39,93]. Again, these equations are significant in that they indicate that the burn rate of these fuels is typically independent of combustion chamber pressure, allowing that to be a free variable in the hybrid design procedure discussed in Chapter 7.

3.3.3 Properties

A selection of the material properties of liquefying high regression rate fuels, previously published in Refs. [91] and [32], are presented. These properties can be used to solve for the propellant dependent parameters in the above equations. In Ref. [32], Karabeyoglu et al. use the Asymptotic Behavior Correlation (ABC) method to predict the properties of n-alkane fuels, including paraffin waxes ($C_n H_{2(n+1)}$, with carbon numbers from approximately 28 to 32) and PolyEthylene (PE) waxes (($C_2H_4)n$, with carbon numbers from approximately 60 to 80). The predicted critical pressure and temperature of these fuels with varying carbon number are provided in Fig. 3.6. The data in this figure indicates that paraffin wax with molecular formula $C_{32}H_{66}$ has a critical pressure of 670 kPa (97 psi) and a critical temperature of 864 K. Note the fuel only have turned to vapor if it is at both the critical temperature and pressure. The predicted blowing parameter, B, of these fuels is provided in Fig. 3.7. For paraf-

fin wax, Fig. 3.7, denotes a blowing parameter of 4.75. Other pertinent properties of paraffin wax with molecular formula $C_{32}H_{66}$ are provided in Table 3.2.

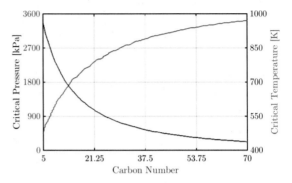

FIGURE 3.6 Critical pressure and temperature of n-alkanes versus carbon number. This figure was produced using data published in Ref. [93].

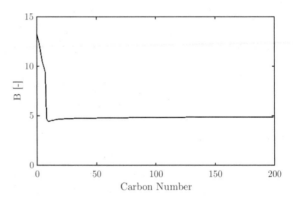

FIGURE 3.7 Blowing parameter, B, of n-alkanes vs carbon number. This figure was produced using data published in Ref. [93].

3.4 Theory applied to single port hybrid design

While multiport hybrid fuel grains have been built for numerous applications, the simplest and most efficient is a single-port hybrid fuel grain. The following section will focus on the

TABLE 3.2 Properties of paraffin wax, $C_{32}H_{66}$. Data is from Refs. [39], [93], and [96].

Property	Value	Unit
Molecular weight	450.9	[g/mol]
Density, solid phase	924.5	[kg/m^3]
Density, liquid phase	671.3	[kg/m^3]
Heat of formation	−224.2	[kcal/mol]
Melting temperature	342.3	[K]
Boiling temperature	727.4	[K]
Heat of vaporization	163.5	[J/kg]
Thermal conductivity	0.12	[W/m.K]
Specific heat, solid phase	2030	[J/kg.K]
Specific heat, liquid phase	2920	[J/kg.K]
Critical pressure	670	[kPa]
Critical temperature	864	[K]

theory derived above for a simple, single cylindrical port hybrid.

This section takes the standard regression rate equation of Eq. (3.20) and applies it to a hybrid rocket motor using a single circular port geometry. Start with Eq. (3.20), repeated here as Eq. (3.45).

$$\dot{r} = aG^n x^{-m} \qquad (3.45)$$

Recall that the propellant mass flux, G, is simply the total of local fuel and oxidizer mass flux in the hybrid fuel port, see Eq. (3.46).

$$G = \frac{\dot{m}_p}{\pi r^2} = \frac{\dot{m}_O + \dot{m}_F}{\pi r^2} \qquad (3.46)$$

Section 3.2.2 showed in the derivation of Eq. (3.45) that the dependence of regression rate on propellant mass flux, G, and the axial location along the fuel grain, x, arises from the dependence of the skin friction and heat transfer rate on Reynolds number based on distance along the port [97]. The classical hybrid combustion theory of Section 3.2.2 found the exponents $m = 0.2$ and $n = 0.8$. As will be discussed further in Chapter 6, measured values of the exponent n are a function of the selected propellants and are typically in the range of 0.3–0.8 [97], see Table 6.2. The value of n is key to understanding

the overall motor ballistics; a value of $n = 0.5$ corresponds to the special case where the fuel mass flow rate becomes independent of the fuel port cross-sectional area because the increased fuel surface area with increasing radius is perfectly balanced by the decrease in fuel regression rate from the decrease in propellant mass flux with the larger port cross-sectional area. If $n > 0.5$, then the fuel mass flow rate will tend to decrease with increasing radius (the effect on regression rate from decreasing mass flux will dominate the effect from increasing burning surface area). If $n < 0.5$, then the fuel mass flow rate will tend to increase with increasing radius (the increasing burning surface area will dominate the effect on regression rate from decreasing mass flux). The length exponent, m, is extremely difficult to measure in test. To evaluate it accurately for a given propellant combination requires a large number of tests at a range of scales [97]. It is generally believed that the theoretical value of $m = 0.2$ is an overprediction. It is common practice to assume that the spacial regression rate dependence beyond the effect of varying mass flux down the port is negligible, essentially setting the exponent $m = 0$, producing Eq. (3.47). It should be noted that some variation in regression rate over fuel grain length that is not explained by the change in total mass flux can be observed, see, for example, [17].

$$\dot{r} = aG^n \qquad (3.47)$$

The total mass flux G is often replaced with the oxidizer mass flux G_o for regression rate calculations. This is not an unreasonable assumption, and it allows the propulsion engineer to work with a much simpler equation. First, the oxidizer-to-fuel mass ratio, O/F, is generally large for hybrid rocket motor propellant combinations, as the performance tends to optimize at higher O/F, see Chapter 5. Second, the mass flow rate of the fuel, $\dot{m}_F$, actually increases along the length of the grain as more fuel is added. This leads to coupling of the fuel regression rate and the mass flux, which substantially complicates modeling. Since this approximation is made so commonly, most of the available regression rate information, specifically the empirical constants, is derived in this manner. Therefore if the O/F is not too small, it is generally reasonable to approximate Eq. (3.47) as Eq. (3.48).

$$\dot{r} = a_o G_o^n \tag{3.48}$$

$$G_o = \frac{\dot{m}_O}{\pi r^2} \tag{3.49}$$

Expansion of Eq. (3.47) using Eq. (3.46), and equating to Eq. (3.48), gives a direct relationship between a and a_o in terms of O/F, see Eq. (3.50). The fuel mass flow rate, $\dot{m}_F(x, t)$, is a function of axial length, x, along the fuel port and of time, t. Thus the empirical "constant" a_o is actually also a function of both axial length and time. Nevertheless, for hybrid rocket motors that operate at high O/F, a_o is treated as a constant and is often used for simplicity to predict hybrid performance. This simplification is used in the design equations of Section 3.4.2. For a more accurate estimate of motor ballistics, and particularly for hybrids operating at low oxidizer-to-fuel mass ratios, the initial complete regression rate equation, Eq. (3.45), can be used. This is discussed in the following section, Section 3.4.1.

$$a_o = a \left(1 + \frac{\dot{m}_F(x, t)}{\dot{m}_O}\right)^n = a \left(1 + \frac{1}{O/F(x, t)}\right)^n \tag{3.50}$$

The propulsion engineer must be careful when using regression rate data from published literature, as it is often not obvious if the researchers are presenting a or a_o. Published data does not always use the nomenclature here to distinguish between the two values. As shown by Eq. (3.50), the two values are different.

3.4.1 Complete solutions to predict performance

The full nonlinear coupled mass flow rate equations can be solved, producing a more ac-

curate estimate of motor ballistics. In this case, the first key equation is just Eq. (3.45) written explicitly in terms of propellant mass flow rate, see Eq. (3.51).

$$\dot{r} = a \left(\frac{\dot{m}_p}{\pi r^2}\right)^n x^{-m} \tag{3.51}$$

The second equation that is needed is an expression for propellant mass flow rate in terms of the surface regression rate. Start by assuming a single circular fuel port so that the incremental burning surface area, $\Delta A_b(x, t)$, across an incremental axial distance, Δx, is given by Eq. (3.52).

$$\Delta A_b(x, t) = 2\pi r \Delta x \tag{3.52}$$

Using Eq. (3.52), the incremental fuel mass injected into the port, $\Delta \dot{m}_F(x, t)$, from a port cross section of incremental axial length Δx is given by Eq. (3.53).

$$\Delta \dot{m}_F(x, t) = \Delta A_b \rho_F \dot{r}$$
$$= 2\pi r \Delta x \rho_F a \left(\frac{\dot{m}_p}{\pi r^2}\right)^n x^{-m} \tag{3.53}$$
$$\Delta \dot{m}_F(x, t) = 2a\rho_F \pi^{1-n} r^{1-2n} x^{-m} \dot{m}_p^n \Delta x$$

The total propellant mass flow rate through the fuel grain port at axial location, x, is just the sum of the fuel and oxidizer mass flow rates. Assuming that oxidizer is only injected at the fore end of the motor, then there is no change in oxidizer mass flow rate with axial length along the port, and thus $\Delta \dot{m}_p(x, t) = \Delta \dot{m}_F(x, t)$. Eq. (3.53) can therefore be rewritten as (3.54).

$$\Delta \dot{m}_p(x, t) = 2a\rho_F \pi^{1-n} r^{1-2n} x^{-m} \dot{m}_p^n \Delta x \tag{3.54}$$

Take the limit of Eq. (3.54) to obtain the an expression for the mass flow rate increase along the port, Eq. (3.55).

$$\frac{\partial \dot{m}_p}{\partial x} = 2a\rho_F \pi^{1-n} r^{1-2n} x^{-m} \dot{m}_p^n \tag{3.55}$$

Eq. (3.51) and Eq. (3.55) together are the two governing, coupled, first-order, partial differen-

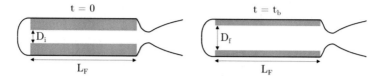

FIGURE 3.8 Schematic of a single-port hybrid rocket before and after combustion.

tial equations that need to be solved simultaneously to accurately determine the local mass flow rate and port radius.

The standard approach to solving these governing equations (Eq. (3.51) and Eq. (3.55)) is through a simple numerical integration. If a sufficiently small time step is used, then the Euler forward step method can be accurate enough for most practical purposes. An even simpler solution is available for the special case where the mass flux exponent $n = 0.5$. It is shown in Ref. [98], that for the special case when $n = 0.5$, the solution to Eqs. (3.51) and (3.55) is given by Eqs. (3.56) and (3.57). Note that r_i is the initial fuel port radius (at $t = 0$) and is a function of axial location, x.

$$r(x,t) = \left[r_i^2 + \frac{2a}{x^m \pi^{0.5}} \left(\int_0^t \sqrt{\dot{m}_O(t')}dt' + \frac{\pi^{0.5} a \rho_F x^{1-m} t}{1-m} \right) \right]^{0.5}$$ (3.56)

In Eq. (3.56), the variable, t', is used is a mathematical construction to integrate the oxidizer mass flow rate from zero to the current time step.

$$\dot{m}_p(x,t) = \left(\sqrt{\dot{m}_O} + \frac{\pi^{0.5} a \rho_F x^{1-m}}{1-m} \right)^2$$ (3.57)

For the case where the oxidizer mass flow rate is constant with time, it is also possible to solve the equations directly for any mass flux exponent, n, using a similarity solution as described in Ref. [99].

3.4.2 Simplified governing equations

Consider a cylindrical hybrid rocket fuel grain with an initial diameter, D_i, and length, L_F, as shown in Fig. 3.8. If the oxidizer mass flow rate is kept constant (e.g., via the use of a regulator and choked orifice for gaseous systems or cavitating venturi for liquid systems), then a number of simple but powerful design equations can be derived. The key simplifying assumptions for this section are:

- Constant oxidizer mass flow rate, $\dot{m}_O = $ constant.
- A cylindrical hybrid rocket fuel grain with a single cylindrical port.
- Fuel regression rate is a function of only oxidizer mass flux, per Eq. (3.48).

Fig. 3.8 shows the fuel grain before and after a burn of duration t_b seconds. Note that the initial radius, r_i, is simply half the initial fuel port diameter, D_i, and the final fuel port radius, r_f, is half the final fuel port diameter, D_f. The assumptions listed above imply that the fuel grain radius is constant across the axial length at both the beginning and end of the burn. This is not always the case, as some differential fuel grain regression rate is expected along the fuel port length, but the equations presented here generally apply well if one considers the final radius, r_f, to be equivalent to the space averaged radius, per Eq. (3.58).

$$r_f = \sqrt{\frac{m_{F,i} - m_{F,f}}{\pi L_F \rho_F} + r_i^2}$$ (3.58)

The fuel port radius at any point in time, t, during the burn can be evaluated by integrating

Eq. (3.48), see Eq. (3.59).

$$\dot{r} = a_o \frac{\dot{m}_O^n}{\pi^n r^{2n}}$$

$$\Rightarrow r^{2n} dr = a_o \frac{\dot{m}_O^n}{\pi^n} dt$$

$$\Rightarrow r(t) = \left[r_i^{2n+1} + \frac{(2n+1)a_o (\dot{m}_O)^n t}{\pi^n} \right]^{\frac{1}{2n+1}}$$

$$\text{(3.59)}$$

The surface burning area, A_b, and oxidizer mass flux, G_o, at any point in time, t, are given by Eqs. (3.60) and (3.61), respectively.

$$A_b = 2\pi r L_F \qquad \text{(3.60)}$$

$$G_o = \frac{\dot{m}_O}{\pi r^2} \qquad \text{(3.61)}$$

The instantaneous fuel regression rate, $\dot{r}$, fuel mass flow rate, $\dot{m}_F$, and oxidizer-to-fuel mass ratio, O/F can all now easily be calculated via Eqs. (3.48), (3.62), and (3.63), respectively.

$$\dot{m}_F = \rho_F A_b \dot{r} \qquad \text{(3.62)}$$

$$O/F = \frac{\dot{m}_O}{\dot{m}_F} \qquad \text{(3.63)}$$

Knowing the O/F, propellant mass flow rates, and the nozzle geometry allows the expected chamber pressure, P_c, and thrust, F, versus time also be calculated via Eqs. (3.64) and (3.65). Note that these equations assume steady-state combustion, chamber bulk fill/emptying times and fully transient phenomena are discussed in Sections 3.6.1 and 3.6.2, respectively.

$$P_c = \frac{(\dot{m}_O + \dot{m}_F)C^*_{ideal}\eta_{C^*}}{A_{th}C_d} \qquad \text{(3.64)}$$

$$F = (\dot{m}_O + \dot{m}_F)C^*_{ideal}\eta_{C^*}C_{F,ideal}\eta_n \qquad \text{(3.65)}$$

The equations presented in this section can be powerful for early-stage hybrid rocket design. This shall be made clear in Chapter 7 and in the examples of Chapter 10.

3.4.3 O/F shift

One of the advantages of hybrid propulsion is that throttling can be achieved by changing only $\dot{m}_O$. However, the disadvantage is that the oxidizer-to-fuel ratio can change with the burn time. The physics, chemistry, and thermodynamics of the system: the geometry of the port, heat transfer to the surface, generation of fuel, and oxidizer flux are interdependent and, for all values of n except $n = 0.5$, lead to a shift in O/F. The further the exponent is from $n = 0.5$, the larger the shift in O/F. This can be seen by examining the equation for instantaneous O/F derived using the simplifying assumptions in the previous section. Substitute Eqs. (3.62), (3.60), and (3.48) into (3.63) to obtain Eq. (3.66).

$$O/F = \frac{\dot{m}_O}{\dot{m}_F} = \frac{r^{2n-1}\dot{m}_O^{1-n}}{2a_o\rho_F\pi^{1-n}L_F} \qquad \text{(3.66)}$$

The drop in C^* due to O/F shift is typically too small to measure in testing, less than 0.2%. However, the efficiency can drop more substantially for propellant combinations with a large n exponent, motors with a high L_F/D, or multiport designs [100].

3.5 Regression rate enhancements

Many research efforts have focused on increasing the regression rate of classical hybrid fuels. Ref. [38] clearly lays out the desire for a reduction in the overall length-to-diameter ratio to improve marketability, and Ref. [101] suggests a 30% cost savings could be possible with a 12x increase in regression rate based on data from [38].

The propulsion engineer now knows that the regression of fuel depends on heat transfer to, and pyrolysis of, the fuel grain. The regression rate enhancing techniques presented in the next sections focus on that fact. The main options for increasing regression rate are included here. However, regression rate enhancement is a popular area of research and the propulsion engineer is pointed to Ref. [101] for a more detailed overview of this topic. It should also be noted that the authors are strong proponents of finding the right fuel/motor configuration for the right application. For example, a high regression rate fuel was selected for a potential Mars Ascent Vehicle, while the lowest possible regression rate was a better choice for a SmallSat motor. That is to say, increasing regression rate may not always be desirable for all applications.

3.5.1 Swirl injection hybrids

Swirl injection is a method used to increase regression rate and fluid residence time in hybrid rockets. To achieve the desired effect, the oxidizer can be injected through multiple ports/angles. The fuel regression rate is increased due to enhanced heat transfer. There is increased radiative heat transfer because the flame sits closer to the fuel grain. The swirling flow causes the boundary layer to be thinner [102]. Increased flow velocity (including the additional tangential component) next to the fuel wall and increased turbulence also increase convective heat transfer. The increased residence time, created by the swirling flow, may potentially improve mixing and C^* efficiency [102]. Ref. [103] saw an increase in regression rate of more than 50 percent and in combustion efficiency of 10 percent by moving to a swirl type injector.

The effect of the enhanced regression rate is typically the greatest near the head end of the fuel grain because the swirling flow weakens as fuel is added along the length of the grain without a tangential flow velocity and due to friction. The regression rate at the fore end of the motor has been demonstrated to peak at 2–3x that at the aft end [104]. Uniform regression rate enhancement has been achieved with specialized injector designs and is the focus of current research [87]. If residual swirl exists at the nozzle, it will reduce the discharge coefficient [105].

While undesirable, this reduction is not typically significant.

Not all swirling hybrids inject oxidizer from the head end. Some inject from the aft for an end burning design [106] or at multiple points along the port. The location and direction of injection would further change the regression behavior. It should be noted that the more complex swirling flow in the combustion chamber makes modeling of it much more difficult.

The configuration of the fuel grain in swirling hybrids has several advantages. Higher fuel utilization is possible since the port is typically a single, cylindrical design vs a wagon wheel. Pancake-style hybrids sandwich the liquid swirl injector between two warm fuel surfaces, which facilitates oxidizer vaporization (provided the injector provides adequate atomization) and can improve flame holding and combustion efficiency [87], see Fig. 3.11b.

Swirling flows are typically characterized by the swirl number, SN, which is a measure of the swirl intensity. The local swirl number is defined by Ref. [107] as the ratio of the axial flux of the angular momentum to the axial flux of the axial momentum. This is given in Eq. (3.67). The swirl number is a strong function of the injection geometry and generally decreases along the length of the fuel grain. However, multiple injection points along the length of the grain have been proposed (and tested) to combat this.

$$SN = \frac{\int\int \rho V_x V_\theta r^2 dr d\theta}{r_{max} \int\int \rho V_x^2 r dr d\theta} \propto \frac{V_\theta}{V_x} = \frac{M_\theta}{M_x} \quad (3.67)$$

Here, r_{max} is the cross-sectional radius at which the swirl number is being evaluated, and V is the velocity in either the axial (x) or tangential (θ) direction, respectively. The swirl number is then found to be proportional to the tangential velocity divided by the axial velocity [108] or correspondingly the ratio of the Mach numbers.

The geometric swirl number, SN_g, is a global parameter that is defined by the injector design,

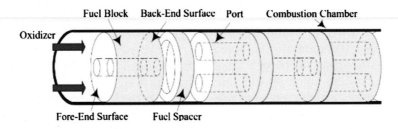

FIGURE 3.9   Unique fuel burn surfaces for the CAMUI-type hybrid rocket motor [112].

Eq. (3.68)

$$SN_g = \frac{(r_{inj} - r_{hole})r_{inj}}{N_{inj}r_{hole}^2} \qquad (3.68)$$

Here, r_{inj} is the radius of the center line of the injector exit holes, r_{hole} is the radius of the individual injector holes, and N_{inj} is the number of injector holes. Most hybrid rocket tests with swirling flow will be classified using one of these (related) definitions for swirl number, with the latter being the most common.

Those interested in modeling swirling flow are encouraged to look to Refs. [109] and [110] for hybrid rocket applications.

3.5.2 Other techniques

Injection geometry, even for standard axial injection, can impact regression modeling, see, for example, Refs. [17] and [111]. A number of alternative oxidizer injection techniques are being actively considered and generally cannot use the regression rate equations as derived in this chapter. The practical challenges and benefits of each option are left to the propulsion engineer to determine.

Impinging designs

There are other injection techniques being considered to increase the regression rate that may also require modifications to the theory presented in this chapter. In addition to swirling injection, impinging designs (e.g., Cascaded

Multi-Stage Impinging Jet (CAMUI) hybrid rocket motor) or injection of the oxidizer in other than the axial direction have been considered. In CAMUI, the oxidizer flow is reoriented multiple times throughout the burn, directly impinging on the next part of the fuel grain, see Fig. 3.9. As might be expected intuitively, there is increased localized regression at the location of flow impingement that is not captured by the previously described theory.

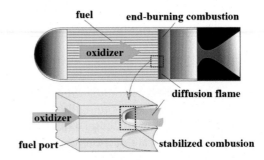

FIGURE 3.10   The many small ports of this end burning hybrid change the relationship between hybrid regression rate and pressure [115].

Pressure dependent designs

Another type of hybrid rocket that does not follow the conventionally derived regression rate theory is the multiple port, end burner proposed by Ref. [113]. In this case, the regression is actually pressure dependent, much like a solid rocket. Fig. 3.10 shows the many tiny, cylindrical ports along the fuel grain essentially acting as feed lines for the oxidizer and the combus-

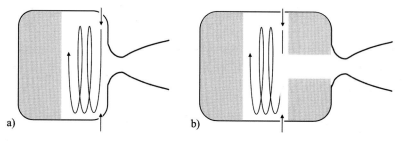

FIGURE 3.11   Vortex injection for an a) end burner and b) pancake-type hybrid rocket motor.

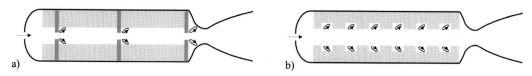

FIGURE 3.12   Methods to increase turbulence, including a) Diaphragms in a fuel grain b) Backward facing steps in the fuel grain.

tion occurring at the end of the grain. A benefit of this design is the very high volumetric loading of the fuel. An increase in the regression rate of about 9x has been demonstrated using this method [114].

Aft injection

Aft end injection has also shown an increase in regression rate [116], [117], [118], [119], [120], [121], [122]. The single disk or dual (pancake) designs have shown significant increases in the regression rate over classical hybrid motors with multiple researchers quoting $\approx 6\times$ improvement. The oxidizer is injected in a swirling flow from the aft end of the motor, where it combusts with either one or two disks of fuel, then exits through the nozzle, see Fig. 3.11. Very low L/D ratios are possible with these designs (on the order of one).

Most motors of this type have been relatively small scale. Also, the effect of any residual swirl of the exhaust gases exiting the nozzle would need to be evaluated (e.g., it could produce undesirable torques on a spacecraft).

Turbulence

A flow trip device, known commonly as a diaphragm, will create a recirculation region directly behind it. The regression rate has been shown to increase substantially downstream of the diaphragm. Fig. 3.12 shows multiple diaphragms along the length of a fuel grain. The shape, height, number, and location of diaphragms make a difference in the regression rate. For example, fewer taller diaphragms make a more substantial impact on the regression rate than many shorter ones [66]. Diaphragms can have a single or multiple holes. They can also be used to increase mixing in the post combustion chamber, e.g., the right most diaphragm in Fig. 3.12a. That particular diaphragm would not be expected to increase the regression rate but would be used to increase performance (enhance mixing). There is an increased mass associated with the addition of a (or several) diaphragm(s). However, they are often made of slower burning fuel, so some of the otherwise inert mass can therefore be captured as propellant. Another issue with these devices is that they can cause pressure oscillations [123]. Alternative ways to trip the flow, such as backward facing steps, see Fig. 3.12b, have also produced increased regression rates. This can be accomplished practically by making two fuel grains with the same outer diameter and varying inner diameters, cutting them into slices and alternating them. There are many other methods that could be (or have been) considered to achieve the same goal.

3.6 Transients

There are many transient phenomena in hybrid rocket motors that the propulsion engineer must consider, including ignition start-up, shutdown, and combustion instabilities. Typical combustion instabilities in hybrid rocket motors are discussed in Chapter 8, and some references for theoretical analysis of such transients are provided here in Section 3.6.2.

During ignition, the hybrid motor typically will initially run fuel-lean, approaching the design O/F as the motor operation moves towards steady-state combustion. The ignition transient consists of time for establishment of the combustion boundary layer and time for the establishment of the thermal profile in the solid fuel grain [40]. The quasi-steady analysis in the following section ignores these physics and instead only captures the bulk fill time for the chamber pressure and propellant mass flow rates to approach the expected steady-state behavior, assuming that the steady-state ballistics equations all apply throughout.

Shutdown of a hybrid motor is typically initiated by closing off the oxidizer flow into the combustion chamber. As a result, combustion during the shutdown transient will typically operate in a fuel-rich regime and may result in thermal soaking and sloughing of some fuel into the post combustion chamber. The approach presented in the following section ignores any changes in the fundamental ballistics equations, such as those due to the increased thermal penetration of the fuel grain, and instead just focuses on the bulk fluid emptying time from the combustion chamber.

3.6.1 Quasi-steady transients

This section focuses on quasi-steady combustion, it intended to allow the propulsion engineer to loosely predict the changes in mean motor ballistics with time. During steady-state combustion some variation in chamber pressure and thrust is typically expected, even if oxidizer mass flow rate is maintained constant. This can be seen for a single-port hybrid rocket by inspecting the equations in Section 3.4.2. The quasi-steady analysis presented here captures bulk fluid filling/emptying phenomena and gradual O/F shifts, but not combustion instabilities or rapid transients. Ignition and shutdown transients can be approximated using the method discussed here but due to the limitations of the quasi-steady model adopted should be verified via test whenever possible.

Let m_g be the total mass of gas in the combustion chamber at an instant in time. m_g is given by Eq. (3.69), where ρ_g is the average gas density in the combustion chamber, and Ψ_C is the combustion chamber volume. Differentiating Eq. (3.69) gives Eq. (3.70).

$$m_g = \rho_g \Psi_C \qquad (3.69)$$

$$\frac{dm_g}{dt} = \frac{d\left(\rho_g \Psi_C\right)}{dt} = \Psi_C \frac{d\rho_g}{dt} + \rho_g \frac{d\Psi_C}{dt} \qquad (3.70)$$

Recall that the mass flow rate of fuel is given by Eq. (3.62). The chamber volume increases as the fuel grain is converted from solid to gas, per Eq. (3.71).

$$\frac{d\Psi_C}{dt} = A_b \dot{r} \qquad (3.71)$$

To a good approximation, the chamber stagnation temperature, $T_{t,c}$, is determined by the propellant oxidizer-to-fuel ratio and tends to be relatively independent of chamber pressure, $P_{t,c}$. As discussed in Chapter 4, the gas in the combustion chamber can be treated as ideal. If the velocities in the combustion chamber are also assumed to be low, then $P_{t,c} \simeq P_c$ and $T_{t,c} \simeq T_c$. Thus the ideal gas law becomes: $P_{t,c} = \rho_g R T_{t,c}$. The change in gas density can therefore be expressed as a function of the change in chamber pressure, per Eq. (3.72).

$$\frac{d\rho_g}{dt} = \frac{1}{RT_{t,c}} \frac{dP_{t,c}}{dt} \tag{3.72}$$

Recall from compressible gas theory for choked flow that the mass flow rate out of the nozzle can be given by Eq. (3.73) (note that this was also in ideal form in Chapter 2 as Eq. (2.8)). The discharge coefficient, C_d is often approximated as 1, which is reasonable for smooth and well designed nozzle inlets. Eq. (3.73) assumes that the flow out of the nozzle is choked, which is generally the case for steady-state combustion but may not be true for a short time during ignition and shutdown transients. See Section 2.2.3, Chapter 2 for further discussions.

$$\dot{m}_{out} = \frac{\gamma C_d A_{th} P_{t,c}}{\eta_{C^*} \sqrt{\gamma R T_{t,c}}} \left(\frac{2}{\gamma+1}\right)^{\frac{\gamma+1}{2(\gamma-1)}} \tag{3.73}$$

The mass accumulation in the combustion chamber is simply the difference between the mass being added to the combustion chamber and the mass flowing out of the nozzle, see Eq. (3.74).

$$\frac{dm_g}{dt} = \dot{m}_F + \dot{m}_O - \dot{m}_{out} \tag{3.74}$$

The goal here is to find an expression for the variation in chamber pressure in time given the instantaneous conditions in the combustion chamber. Start by substituting Eq. (3.70), (3.73), and (3.62) into Eq. (3.74). This results in Eq. (3.75). The propulsion engineer is then left to substitute Eqs. (3.71), (3.72), and the ideal gas law into Eq. (3.75), resulting in Eq. (3.76).

$$V_C \frac{d\rho_g}{dt} + \rho_g \frac{dV_C}{dt}$$

$$= \rho_F A_b \dot{r} + \dot{m}_O - \frac{\gamma C_d A_{th} P_{t,c}}{\eta_{C^*} \sqrt{\gamma R T_{t,c}}} \left(\frac{2}{\gamma+1}\right)^{\frac{\gamma+1}{2(\gamma-1)}} \tag{3.75}$$

$$\frac{dP_{t,c}}{dt} = \frac{R T_{t,c}}{V_C} \left[A_b \dot{r} \left(\rho_F - \frac{P_{t,c}}{RT_{t,c}} \right) + \dot{m}_O \right.$$

$$\left. - \frac{\gamma C_d A_{th} P_{t,c}}{\eta_{C^*} \sqrt{\gamma R T_{t,c}}} \left(\frac{2}{\gamma+1}\right)^{\frac{\gamma+1}{2(\gamma-1)}} \right] \tag{3.76}$$

The instantaneous thrust, F, is a function of the propellant flow rate out of the chamber (Eq. (3.73)), and the instantaneous characteristic velocity and thrust coefficient, see Eq. (3.77).

$$F = (\dot{m}_{out}) C_{ideal}^* \eta_{C^*} C_{F,ideal} \eta_n \tag{3.77}$$

To show how these equations can be used, they are applied to a single-port hybrid motor with the assumption that regression rate can be approximated by Eq. (3.48). The instantaneous chamber volume, V_C, is now given by Eq. (3.78). Here, V_{pre} refers to the precombustion chamber volume, and V_{post} refers to the post combustion chamber volume, which are assumed to be constant (but could easily be made a function of time by assuming a constant ablation rate, for example). Eq. (3.76) then becomes Eq. (3.79).

$$V_C = \pi r^2 L_F + V_{pre} + V_{post} \tag{3.78}$$

$$\frac{dP_{t,c}}{dt} = \frac{R T_{t,c}}{V_C} \left[2\pi r L_F a_o \left(\frac{\dot{m}_O}{\pi r^2} \right)^n \left(\rho_F - \frac{P_{t,c}}{RT_{t,c}} \right) \right.$$

$$\left. + \dot{m}_O - \frac{\gamma C_d A_{th} P_{t,c}}{\eta_{C^*} \sqrt{\gamma R T_{t,c}}} \left(\frac{2}{\gamma+1}\right)^{\frac{\gamma+1}{2(\gamma-1)}} \right] \tag{3.79}$$

To utilize these equations to predict chamber pressure over time, start with knowledge of the planned oxidizer mass flow rate vs time, $\dot{m}_O$, initial chamber volume, V_C, initial chamber pressure, $P_{t,c}$, fuel grain properties (density, ρ_F, length, L_F, initial port radius, r_i, enthalpy of formation, h_f), and the burn time, t_b.

For constant oxidizer mass flow rate, the radius at any point in time can be determined by Eq. (3.59). If the oxidizer mass flow rate is not constant, then the fuel port radius can be calculated numerically using a simple forward Euler method, Eq. (3.80), and the fuel regression rate calculated from Eq. (3.48) using the oxidizer mass flux at the previous time step, t_{j-1}.

$$r_j = r_{j-1} + \dot{r}_{j-1}\Delta t_{j-1} \tag{3.80}$$

The change in chamber pressure at each time step can be calculated by Eq. (3.79), and the chamber pressure at the following time step then can be calculated via Eq. (3.81). Note that at each time step the chamber temperature, $T_{t,c}$, average gas molecular weight, M_w, and ratio of specific heats, γ, must be calculated or estimated using a chemical equilibrium combustion solver for the O/F in the chamber at that time. For simplicity, the O/F of the propellant being added to the combustion chamber at each time step can be used. In fact, since these properties usually only weakly depend on chamber pressure, it is often reasonable to use constant values for $T_{t,c}$, M_w, and γ assuming that the O/F does not vary too widely over the course of the burn. This is shown, along with an example of applying these equations to an actual hybrid hot-fire, in Section 10.4 of Chapter 10.

$$\left(P_{t,c}\right)_j = \left(P_{t,c}\right)_{j-1} + \left(\frac{dP_{t,c}}{dt}\right)_{j-1}\Delta t_{j-1} \tag{3.81}$$

As discussed previously, these equations do not capture changes to the motor ballistics from changes in the assumed fuel regression rate behavior (such as thermal start up and shutdown transients) but do capture expected variation in chamber pressure resulting from changes in oxidizer mass flow rate and characteristic bulk filling/emptying of the combustion chamber.

3.6.2 Combustion instability transients

Theoretical modeling of combustion instability transients in hybrid rocket motors is an involved topic that goes beyond the scope of this design focused book. Chapter 8 describes the various combustion instabilities that are commonly observed in hybrid rocket motors. Ref. [40] provides a theoretical formulation for analyzing combustion instabilities in hybrid rocket motors and applies this formulation successfully to evaluation of the Intrinsic Low Frequency Instability (ILFI). Ref. [47] and Ref. [124] apply a similar approach to feed-couple instabilities. In addition to these resources, the extensive liquid rocket combustion instability literature, such as Ref. [125], can often be informative resources to consult when investigating hybrid rocket combustion instabilities.

Thermodynamics and chemistry

4.1 Introduction

Thermodynamics is at the heart of rocket propulsion, dealing with the laws that govern the relationship between temperature and energy. It describes energy converted between forms, especially heat, and the direction in which it flows. It also describes if the energy is available to do work.

The relationships developed in Chapter 2 assumed knowledge of the performance parameters, e.g., specific impulse and exhaust velocity. This chapter provides an overview of the thermodynamics fundamentals that need to be applied to calculate these parameters and thereby predict hybrid motor performance. This analysis is conveniently broken into two parts by location/function: combustion in the combustion chamber, then expansion through the nozzle. Combustion, particularly the turbulent boundary layer combustion in hybrid rocket motors is theoretically rich. Much of the detail of the combustion processes in hybrid rocket motors goes well beyond the scope of this design-focused book. For interested readers, Refs. [126], [127], and [128] are recommended. Rather than delving deeply into combustion kinetics and processes, this chapter is intended to least give the reader an appreciation for the underlying theory and calculation approaches that are applied

to predict the theoretical performance of propellant combinations.

4.1.1 Combustion

Chemical equilibrium is achieved when the products and reactants are no longer changing their concentrations. This condition is used to evaluate the performance of a rocket motor. Knowledge of the combustion gas properties is necessary in order to evaluate the chemical equilibrium conditions. As an initial approximation, the combustion chamber pressure can be assumed to be constant. This is generally reasonable unless there is a large source of pressure drop (such as a diaphragm or mixing device) inside the combustion chamber. The combustion gases are assumed to behave as an ideal gas or, more accurately, a mixture of ideal gases. This is a reasonable assumption for gases at the high temperatures of typical combustion chambers. Finally, the chemical reactions are assumed to be fast relative to the fluid residence time in the combustion chamber, so chemical equilibrium can be achieved before entering the nozzle. Remember the discussion of dimensionless parameters and Table 3.1 for rationale on why these assumptions are reasonable for hybrid rockets.

4.1.2 Nozzle

After combustion, the gas enters the nozzle and, for simplicity, is often assumed to undergo isentropic and adiabatic expansion. At this point, the propulsion engineer typically assumes that the chemical composition was frozen in the combustion chamber. However, in reality, reactions will continue in the nozzle and even in the plume. Losses in the nozzle, changes to the flow from boundary layer effects, and the impact of chemical kinetics can be quite complex. In the initial design, a simple nozzle loss factor, η_n, is often applied to account for these effects. This is discussed more in Section 7.4.4.

4.2 Key terms

To dive into the physics of thermodynamics, it is helpful to define some key terms first. Fluid properties are generally categorized as intensive or extensive. Intensive properties are the same no matter the amount of the sample (per unit mole or mass). An example of an intensive quantity is specific heat capacity, c_p or c_v [J/(kg K)], which is the amount of energy required to raise the temperature of 1 kilogram of a substance by 1 degree K. Intensive properties are commonly represented by lower case letters. Other common intensive properties are specific enthalpy, h [J/kg], specific internal energy, u [J/kg], specific total energy, e [J/kg], specific volume, v [m^3/kg], and density, ρ [kg/m^3]. Note that density and specific volume are directly related per Eq. (4.1). Extensive properties depend on the mass or number of moles of a sample. They are represented by uppercase letters. Examples of extensive properties include internal energy U [J], enthalpy, H [J], volume V [m^3], and total energy, E [J]. Two widely used exceptions to the capitalization standard for properties are temperature, T, and pressure, P. Despite being capitalized, temperature and pressure are intensive. Additional exceptions to this standard

that are more specific to this book are mass, m, and number of moles, n. Lowercase letters are used for these, even though they are extensive properties to deconflict from other nomenclature in this book. For many thermodynamic calculations, the number of molecules in a mole must be known, which is given by the Avogadro constant, $N_A = 6.022141 \times 10^{23}$ mol^{-1}. Equations to relate the other properties discussed here are provided in the following sections.

$$v = \frac{1}{\rho} \qquad (4.1)$$

4.3 Ideal gas mixtures

Ideal gases are gases that can be modeled without accounting for molecular forces or the volume of the gas molecules. An equation of state is the one that relates the pressure, P, temperature, T, and volume, V (or v for specific volume), of a substance. The ideal gas equation of state, given in the form of Eq. (4.2) or Eq. (4.3), is used widely throughout this book. The ideal gas equation of state can generally be used with reasonable accuracy as long as the fluid density is not too high. As such, it can be used for most gases at low pressure and around atmospheric temperature, or it can be used at some elevated pressures when the fluid temperature is also high. Combustion processes in rockets are generally at high temperatures (they can be thousands of degrees Kelvin around the flame), and as such, the ideal gas equation of state is commonly applied to the gas within the combustion chamber.

$$P V = n R_u T = m R T \qquad (4.2)$$
$$P = \rho R T \qquad (4.3)$$

When looking at Eq. (4.2), recall that the universal constant, R_u is related to the specific gas constant, R, by Eq. (4.4), where M_w is the molecular weight of the gas. The total amount of moles is given by n and is related to the mass, m, by Eq. (4.5).

$$R = \frac{R_u}{M_w} \tag{4.4}$$

$$n = \frac{m}{M_w} \tag{4.5}$$

Caloric equations of state are equations that relate internal energy or enthalpy to temperature and volume or pressure. A general form of these equations is given by Eqs. (4.6) and (4.7).

$$u = f(T, v) \tag{4.6}$$
$$h = f(T, P) \tag{4.7}$$

The derivatives can be found to be:

$$du = \left(\frac{\partial u}{\partial T}\right)_v dT + \left(\frac{\partial u}{\partial v}\right)_T dv \tag{4.8}$$

$$dh = \left(\frac{\partial h}{\partial T}\right)_P dT + \left(\frac{\partial h}{\partial P}\right)_T dP \tag{4.9}$$

The specific heat at constant pressure and constant volume are then defined by Eqs. (4.10) and (4.11), respectively.

$$c_p \equiv \left(\frac{\partial h}{\partial T}\right)_P \tag{4.10}$$

$$c_v \equiv \left(\frac{\partial u}{\partial T}\right)_v \tag{4.11}$$

For ideal gases, the internal energy does not change with specific volume, and the enthalpy does not change with pressure. Thus the caloric equations of state for an ideal gas are given by Eqs. (4.12) and (4.13).

$$u(T) - u_{ref} = \int_{T_{ref}}^{T} c_v dT \tag{4.12}$$

$$h(T) - h_{ref} = \int_{T_{ref}}^{T} c_p dT \tag{4.13}$$

The specific heats c_p and c_v are only a function of temperature for ideal gases. A subset of

ideal gases can be considered calorically perfect across some temperature ranges. Calorically perfect gases have constant specific heat. In general, for combustion processes with large temperature changes, gases can be considered ideal but not calorically perfect. Thus the change in specific heat with temperature must be considered when using Eqs. (4.12) and (4.13). For an ideal gas, c_p and c_v are related by the gas constant, R, see Eq. (4.14).

$$c_v = c_p - R \tag{4.14}$$

Another key term to define here is the ratio of specific heats, γ, defined per Eq. (4.15). The value of γ depends on the state of the gas. The National Institute of Standards and Technology Webbook has data available for most gases of interest in hybrid propulsion [22].

$$\gamma = \frac{c_p}{c_v} \tag{4.15}$$

At this point, enough terms are defined to consider situations where different gas species are present together. This is the case inside the combustion chamber, where there are multiple different species. The relative amount of each species is generally given as a mass fraction, Y, or a mole fraction, X. The mass fraction of species i is defined per Eq. (4.16), and the mole fraction is defined per Eq. (4.17), assuming that there are N different species in the mixture.

$$Y_i = \frac{m_i}{\sum_{j=1}^{N} m_j} \tag{4.16}$$

$$X_i = \frac{n_i}{\sum_{j=1}^{N} n_j} \tag{4.17}$$

By definition the sum of all mass (or mole) fractions is 1 (Eq. (4.18)).

$$\sum_{i=1}^{N} Y_i = \sum_{i=1}^{N} X_i = 1 \tag{4.18}$$

Assuming that all species are in the gas phase, Dalton's law can be applied. Dalton's law states that the pressure of a mixture of gases is equal to the sum of the partial pressures of each constituent gas at a given temperature, Eq. (4.19). The partial pressure of each gas, P_i, is the pressure of that gas if it were isolated from the mixture at the same temperature and volume as the mixture.

Dalton's Law:

$$P = \sum_{i=1}^{N} P_i \qquad (4.19)$$

Combining Dalton's law, Eq. (4.19), with the ideal gas equation, Eq. (4.2), the rest of the state variables can be determined for the mixture (Eqs. (4.20) to (4.27)).

The total number of moles, n, is given by Eq. (4.20).

$$n = \sum_{i=1}^{N} n_i \qquad (4.20)$$

The average molecular weight of the mixture, $(M_w)_{mix}$, is essentially the total mass of the mixture divided by the total moles of the mixture, Eq. (4.21).

$$(M_w)_{mix} = \frac{m}{n} = \frac{\sum_{i=1}^{N} m_i}{\sum_{i=1}^{N} n_i} = \frac{\sum_{i=1}^{N} n_i M_{w,i}}{\sum_{i=1}^{N} n_i} \qquad (4.21)$$

$$(M_w)_{mix} = \sum_{i=1}^{N} X_i M_{w,i} \qquad (4.22)$$

$$(M_w)_{mix} = \frac{1}{\sum_{i=1}^{N} \frac{Y_i}{M_{w,i}}} \qquad (4.23)$$

It is straightforward to convert between mole and mass fraction for a given species once you know the average molecular weight, see Eqs. (4.24) and (4.25).

$$Y_i = X_i \frac{M_{w,i}}{(M_w)_{mix}} \qquad (4.24)$$

$$X_i = Y_i \frac{(M_w)_{mix}}{M_{w,i}} \qquad (4.25)$$

It is often necessary to use the specific heats and ratio of specific heats of the mixture, calculated via Eqs. (4.26) and (4.27) for c_p, specific heat and γ, respectively. Note that in these equations, the specific heat, c_p, is the specific heat per unit mass, and the specific gas constant, R, is calculated using the average molecular weight of the mixture, $(M_w)_{mix}$.

$$(c_p)_{mix} = \sum_{i=1}^{N} Y_i c_{p,i} \qquad (4.26)$$

$$(\gamma)_{mix} = \frac{(c_p)_{mix}}{(c_p)_{mix} - R} \qquad (4.27)$$

The average specific heat per unit mole of the mixture can be calculated via Eq. (4.28). The ratio of specific heats, γ, can then be calculated using this property via Eq. (4.29). Note that Eqs. (4.27) and (4.29) give equivalent results for γ.

$$(\bar{c}_p)_{mix} = \sum_{i=1}^{N} X_i \bar{c}_{p,i} \qquad (4.28)$$

$$(\gamma)_{mix} = \frac{(\bar{c}_p)_{mix}}{(\bar{c}_p)_{mix} - R_u} \qquad (4.29)$$

Other mass or mole specific mixture properties can be calculated in the same manner using either mass or mole fractions, see, for example, Eqs. (4.30), (4.31), and (4.32). Here, $\bar{h}$ [J/mol] is by definition the specific enthalpy per mole.

$$(h)_{mix} = \sum_{i=1}^{N} Y_i h_i \qquad (4.30)$$

$$(u)_{mix} = \sum_{i=1}^{N} Y_i u_i \qquad (4.31)$$

$$(\bar{h})_{mix} = \sum_{i=1}^{N} X_i \bar{h}_i \qquad (4.32)$$

4.4 Real gases

The ideal gas assumption fails at high pressures and low temperatures, though the latter is usually less applicable for hybrid rocket applications. Therefore an ideal gas assumption is commonly adopted for use in a combustion chamber. However, real gas equations should be used to calculate conditions in pressurant tanks. In this case, a different equation of state can be used, for example, the Van der Waals equation (Eq. (4.33)) or Peng–Robinson equation (Eq. (4.34)). Alternatively, a compressibility factor, Z, can be applied to the ideal gas equation, see Eq. (4.36).

The Van der Waals equation of state:

$$\left(P + \frac{a}{V_m^2}\right)(V_m - b) = R_u T \tag{4.33}$$

Where V_m is the molar volume, $M_w V/m$, a, and b are values derived experimentally for each gas.

The Peng–Robinson equation of state:

$$P_{PR} = \frac{R_u T}{v - b_{PR}} - \frac{a_{PR}\alpha_{PR}}{v^2 + 2b_{PR}v - b_{PR}^2} \tag{4.34}$$

where,

$$a_{PR} = \frac{0.45724 R_u^2 T_{crit}^2}{P_{crit}}$$
$$b_{PR} = \frac{0.07780 R_u T_{crit}}{P_{crit}}$$
$$v = M_w V/m$$
$$\alpha_{PR} = \left[1 + \left(0.37464 + 1.54226w - 0.26992w^2\right) \times \left(1 - \sqrt{\frac{T}{T_{crit}}}\right)\right]^2$$
$$\tag{4.35}$$

$$Z \equiv \frac{P}{\rho R T} \tag{4.36}$$

Using He as an example, since it is a common pressurant, Z can be calculated at a given temperature and pressure using Eq. (4.37), with $\alpha = 1.362 \times 10^{-6}\,\text{Pa}^{-1}$ from Ref. [129].

$$Z = 1 + \alpha\frac{P}{T} \tag{4.37}$$

This value for α compares favorably to NIST real gas properties (within about ±1%) over the relevant range of pressures (0.1 to 70 MPa) and temperatures (−100 to +60 °C). A compressibility factor is a convenient way to correct for real gas behavior; however, it is even more accurate to use real gas properties directly from REFPROP [130], which is a Thermodynamic and Transport Properties Database that allows the propulsion engineer to use data from NIST in their modeling or to look up the values in NIST [22].

As discussed previously, the high temperatures and relatively low pressures of the combustion chamber make ideal gas assumptions valid for most hybrid rocket motor combustion calculations. As such, the ideal gas equation of state is used throughout the remainder of this chapter. It should be noted that doing so can cause approximately 0.4% error in the calculation of C^* [131].

4.5 Enthalpy

Enthalpy is an important parameter in rocket combustion. In fact, enthalpy is key to all of the calculations discussed in the following sections. It is a thermodynamic quantity that is essentially equivalent to the total heat content of the system. Enthalpy is defined as the sum of the internal energy with the product of pressure and volume (Eq. (4.38)). Specific enthalpy, h, used most often here is the enthalpy per unit mass, as given in Eq. (4.39). Specific enthalpy per unit mole, $\bar{h}$, is also commonly used, as mentioned in

Section 4.3.

$$H \equiv U + P\mathbb{V} \qquad (4.38)$$

$$h \equiv u + Pv \qquad (4.39)$$

4.5.1 Latent heat of vaporization

The latent heat of vaporization, which is also referred to as the enthalpy of vaporization, is a key term which must be included in thermodynamic calculations for most hybrid rocket combustion processes. The latent heat of vaporization, h_{fg}, is defined as the heat required in a constant pressure process to vaporize a unit mass of liquid at a given temperature, see Eq. (4.40). Note that T and P in Eq. (4.40) refer to the saturation temperature and pressure, respectively.

$$h_{fg}(T, P) \equiv h_{vapor}(T, P) - h_{liquid}(T, P) \quad (4.40)$$

The latent heat of vaporization must be accounted for in all combustion processes where the propellants or products undergo a liquid/vapor phase change. Exampes include: oxidizers injected as a liquid into the combustion chamber will vaporize to gas before combustion, liquefying hybrid rocket fuel grains form a liquid layer that must vaporize before combustion, and if the products of combustion are sufficiently cooled, then it can be possible for water to condense, as described in Section 4.6.3.

The correct way to account for the latent heat of vaporization in enthalpy and combustion calculations will be discussed further in Sections 4.5.3, 4.6.3, 4.6.4, and 4.6.5.

4.5.2 Enthalpy of formation

The enthalpy of formation, $h_{f,i}$ of a species i, is defined as the energy stored in the bonds of the molecule at a given reference state. The term heat of formation is used interchangeably with enthalpy of formation. It is typical to use a standard reference state of $T_{ref} = 25\,°C$ and $P_{ref} = 1$ atm $= 101.325$ kPa for enthalpy of formation. This is the standard state used by NIST

[22], Chemkin [132], Cantera [133], and Chemical Equilibrium with Applications [74]. Note that at the time of this publishing, CEA had an online resource available for these calculations [134]. It is also assumed by convention that the enthalpy of formation of elements in their naturally occurring state is 0. Thus for example, the enthalpy of formation of nitrogen gas, N_2, at 25 °C and 1 atm, is 0 J/mol.

The enthalpy of formation of any other molecule is the difference (at the reference state) between the enthalpy of the molecule and the enthalpies of the reference species from which it is formed. In other words, the enthalpy of formation is the net change of enthalpy required to break the chemical bonds of the standard state reference species and form the new bonds of the molecule.

The standard enthalpy per mole of species i, $\bar{h}_i$, in terms of the enthalpy of formation is given by Eq. (4.41). The second term on the right-hand side of Eq. (4.41) refers to the sensible enthalpy, it is the enthalpy associated only with temperature change from the reference state.

$$\bar{h}_i(T) = \bar{h}_{f,i}(T_{ref}) + \left[\bar{h}_i(T) - \bar{h}_i(T_{ref})\right]_{sens} \quad (4.41)$$

For an ideal gas, the calorific equation of state (given by Eq. (4.13) and converted into molar form) is applied to Eq. (4.41), to produce Eq. (4.42) below.

$$\bar{h}_i(T) = \bar{h}_{f,i}(T_{ref}) + \int_{T_{ref}}^{T} \bar{c}_{p,i}(T)dT \quad (4.42)$$

Standard enthalpies and enthalpies of formation for many molecules are available in tabulated form from various databases such as the NIST JANAF Thermochemical Tables [135]. A subset of substances of interest are included in Table 4.1. A negative enthalpy of formation means that energy is released during formation (exothermic). A positive enthalpy of formation means energy is required for formation (endothermic). Most current thermochemical databases are relatively complete in that

TABLE 4.1 Enthalpy of Formation for Various Substances at 298.15 K and 0.1 MPa [135]. Note: * NIST JANAF reference species (heat of formation is zero).

Substance	Formula	$\bar{h}_f$ [kJ/mol]	State
Hydrogen*	H_2	0.00	gas
Oxygen*	O_2	0.00	gas
Nitrogen*	N_2	0.00	gas
Aluminum*	Al	0.00	solid (crystalline)
Graphite *	C	0.00	solid (crystalline)
Carbon Monoxide	CO	−110.53	gas
Carbon Dioxide	CO_2	−393.52	gas
Hydrogen Peroxide	H_2O_2	−136.11	gas
Nitric Oxide	NO	90.29	gas
Nitrogen Dioxide	NO_2	33.10	gas
Nitrous Oxide	N_2O	82.05	gas
Hydroxyl Radical	OH	33.99	gas
Water	H_2O	−241.83	gas
Water	H_2O	−285.83	liquid

they include a large range of molecules. However, if the molecule of interest is not included in such databases, then it is also possible to roughly approximate the enthalpy of formation using a bond additivity approach (Section 4.5.3) or to determine it by measuring the heat of reaction (Section 4.7.1). Note, again, that the enthalpy of formation is a function of the reference state at which it is defined. It is important to always ensure that the same reference assumptions (temperature, pressure· and definition of zero enthalpy) are used for all species when calculating changes in enthalpy from combustion. However, it is possible to convert the enthalpy of formation at one reference temperature to the enthalpy of formation at another reference temperature and then use it, see Ref. [136] for more information on this.

4.5.3 Bond additivity

The enthalpy of formation can be estimated using the bond additivity method. This method, while convenient, relies on the assumption that the type of bond is all that contributes to the particular thermodynamic quantity and that the energy to break a particular bond type is constant for any molecule in which that bond occurs. Therefore only the number and types of bonds are necessary to estimate the enthalpy of formation for a molecule in the gas phase using this method. The value assigned to each bond can then be applied, and the total can be found. Bond additivity is easier to use than it is accurate. For example, this method cannot distinguish between isomers (which have the same bonds but different molecular structures). The number is typically assumed to be accurate to about ±2 kcal/mol, unless the compound is heavily branched or highly electronegative, in which case it is even less accurate [137]. The saving grace for this method is that it is typically used to calculate the enthalpy of formation of a fuel, which is then used to calculate the heat of reaction. The heat of reaction is dominated by the much larger (and typically very well known) enthalpies of formation of the products (e.g., water vapor or CO_2), making the error in the enthalpy of formation of the fuel tolerable.

$$\bar{h}_{f,i,gas} = \sum \left(\Delta \bar{h}_{Bonds,i} \right) \qquad (4.43)$$

$$\bar{h}_{f,i,liquid} = \sum \left(\Delta \bar{h}_{Bonds,i} \right) - \bar{h}_{fg,i} \qquad (4.44)$$

$$\bar{h}_{f,i,solid} = \sum \left(\Delta \bar{h}_{Bonds,i} \right) - \bar{h}_{fusion,i} - \bar{h}_{fg,i}$$
$$(4.45)$$

Remember that the notation $\bar{h}$ means that the enthalpy is per mole. So Eq. (4.43), gives the heat of formation of gas i, per mole, calculated from the number of each type of bond, described above. If the species of interest is in a liquid or gaseous form, the latent heat of fusion/melting and/or vaporization needs to be subtracted. Remember that the latent heat of vaporization, $\bar{h}_{fg,i}$, is the heat required to convert a gas to a liquid at constant pressure (see Section 4.5.1). The latent heat of fusion, $\bar{h}_{fusion,i}$, is the heat required to convert (melt) a substance from a solid to a liquid. Note that while called the heat of fusion here, some texts refer to it as heat of melting. Both can be used interchangeably. Table 4.2 shows the contributions to the heat capacity, entropy, and heat of formation for some simple hydrocarbon bonds. A full table is available in [137].

TABLE 4.2 Partial Bond Contributions of Gas Phase Species at 298 K and 1 bar [137]. Note: C_{db} is the vinyl group carbon atom (tetravalent, e.g., it has four covalent bonds).

Bond Type	$\bar{c}_P$ [J/mol K]	$\bar{s}$ [J/mol K]	$\Delta \bar{h}_f$ [kJ/mol]
$C - H$	7.28	53.97	−16
$C - C$	8.28	−68.62	11.4
$C - F$	13.97	70.71	−219.7
$C - Cl$	19.41	82.42	−31
$C - O$	11.3	−16.74	−50.2
$O - H$	11.3	100.42	−113
$O - O$	20.5	38.07	90
$O - Cl$	23	135.98	38.1
$C - N$	8.79	−53.56	38.9
$N - H$	9.62	74.06	−10.9
$C_{db} - H$	2.6	13.8	3.2
$C_{db} - F$	4.6	18.6	−39
$C_{db} - Cl$	5.7	21.2	−5

The heat of formation of paraffin, $C_{32}H_{66}$, can be estimated by counting the number of each type of bond. Examination of the chemical structure shown in Fig. 5.5 of Chapter 5 gives 31 C-C bonds and 66 C-H bonds in one mole of paraffin ($C_{32}H_{66}$). From Table 4.2, the heats of formation for these types of bonds are $\Delta \bar{h}_{C-C} = 11.4$ kJ/mol and $\Delta \bar{h}_{C-H} = -16.0$ kJ/mol. For paraffin at 25 °C, the latent heat of fusion is $\bar{h}_{fusion} = 76.57$ kJ/mol, and the latent heat of vaporization is $\bar{h}_{f,g} = 157.73$ kJ/mol. Therefore the heat of formation of solid paraffin can be found using Eq. (4.43).

$$\Delta \bar{h}_{f,C_{32}H_{66}} = 31 \times 11.4 \text{ kJ/mol}$$
$$+ 66 \times (-16.0 \text{ kJ/mol})$$
$$- 76.57 \text{ kJ/mol} - 157.73 \text{ kJ/mol}$$
$$\Delta \bar{h}_{f,C_{32}H_{66}} = -937 \text{ kJ/mol}$$

Note that the original calculation and data were in kcal/mol, so due to rounding error from the unit conversion and the number of significant figures used, this value is slightly lower than that given in Chapter 5.

The approach used to estimate the heat of formation of paraffin can be used for any hydrocarbon fuel of interest, though it can be difficult to find latent heat of fusion and latent heat of vaporization for some fuels. The next higher-order approximation can be made by breaking the molecule into groups with known partial contributions to the heat of formation. This typically reduces the errors by about a quarter [137]. The heat of formation and heat of gasification of solid alkanes is estimated as a function of carbon number in Ref. [93] using Asymptotic Behavior Correlations (aka the ABC method). The heat of formation can also be found empirically, starting with bomb calorimetry measurements of the heat of reaction as discussed in Section 4.6.6.

4.6 Stoichiometry and complete combustion

4.6.1 Stoichiometry

A mixture is said to be stoichiometric when all of the fuel and oxidizer react with each other.

Complete combustion is achieved by assuming the fuel and oxidizer are balanced such that combustion of the propellants forms only the major products (e.g., CO_2, H_2O, and N_2). The stoichiometric oxidizer-to-fuel ratio ($O/F|_{stoic}$) is found from this mixture. For example, Eq. (4.46) shows that 48.5 moles of O_2 are required to react a single mole of paraffin stoichiometrically. The O/F is more commonly given in terms of mass, which for this example, is about 3.4. The same process can be followed for any propellant combination, giving a good starting point to estimate an optimal mixture ratio.

$$C_{32}H_{66} + 48.5O_2 \rightarrow 32CO_2 + 33H_2O \quad (4.46)$$

Lean mixtures have more oxidizer than this stoichiometric value, and rich mixtures have more fuel. Peak I_{sp} and $C*$ values both occur under slightly fuel-rich conditions, though not necessarily at the same O/F. Peak performance is typically fuel rich because it reduces the molecular weight of the combustion products (more hydrogen than oxygen), which increases the I_{sp} and $C*$, see, for example, Eq. (2.17) of Chapter 2. Chapter 5 gives performance curves for various propellant combinations.

4.6.2 Oxidizer to fuel, equivalence and mixture ratios

A few different terms are commonly used throughout the propulsion and combustion communities to refer to the relative amounts of fuel and oxidizer being combusted. The oxidizer-to-fuel ratio, O/F, as discussed in the previous section, is the mass ratio of oxidizer to fuel in the combustion chamber. This is exactly the same as the mixture ratio, MR, the two terms are interchangeable, as is made clear in Eq. (4.47). Note that MR and O/F do not themselves give any indication as to whether the mixture is lean (more oxidizer) or rich (more fuel) than the stoi-

chiometric oxidizer-to-fuel ratio, $O/F|_{stoic}$.

$$O/F = MR = \frac{\dot{m}_O}{\dot{m}_F} \quad (4.47)$$

An alternative term, the equivalence ratio, Φ, is instead used to indicate whether a mixture is stoichiometric, lean, or rich. The equivalence ratio, Φ, is given in Eq. (4.48). It is 1 by definition for a stoichiometric mixture, greater than 1 for rich mixtures, and less than 1 for lean mixtures.

$$\Phi = \frac{\frac{\dot{m}_F}{\dot{m}_O}}{\frac{\dot{m}_F}{\dot{m}_O}|_{stoic}} = \frac{O/F|_{stoic}}{O/F} \quad (4.48)$$

4.6.3 Heat of combustion

Hess' Law states that the enthalpy of reaction is the difference between the enthalpy of the products and the reactants, as given in Eq. (4.49). The heat of combustion, ΔH_C, is the heat released by a reaction. The heat of combustion, ΔH_C, and enthalpy of reaction, ΔH_R, are numerically the same except that they have opposite signs, see Eq. (4.49).

$$\Delta H_R = H_{prod} - H_{reac} = -\Delta H_C \quad (4.49)$$

Recall from the discussions of enthalpy and ideal gas mixtures that the total enthalpy of an ideal gas mixture is the sum of the enthalpy of its constituents, Eqs. (4.50) and (4.51).

$$H_{prod} = \sum_{prod} m_i h_i = \sum_{prod} n_i \bar{h}_i \quad (4.50)$$

$$H_{reac} = \sum_{reac} m_i h_i = \sum_{reac} n_i \bar{h}_i \quad (4.51)$$

The enthalpy of each species in the mixture can be found from the NIST JANAF tables [135] or determined from Eq. (4.42).

The state of the reactants and products is important in evaluating the heat of combustion. For example, water in the products can either be in a liquid or gaseous phase. If the water condenses to liquid, then more heat is released. Combustion literature therefore refers to the heat of combustion where all of the water is condensed to liquid as the Higher Heating Value (HHV). The heat of combustion with all of the water in gas phase is referred to as the Lower Heating Value (LHV). The Lower Heating Value is typically the more appropriate heat of combustion to use in hybrid rocket combustion. The difference between these values is the latent heat of vaporization. The phase of the reactants must also be considered when calculating the heat of combustion, as will be made clear in the following example.

4.6.4 Example 1: phase and heat of combustion

A gaseous oxidizer releases more energy than its liquid phase counterpart. To demonstrate this, consider the stoichiometric reaction of paraffin with nitrous oxide, Eq. (4.52). Assume that all of the products are at 25 °C and 1 atm and in the gas phase. Also, before the reaction, paraffin is a solid at 25 °C, and nitrous oxide is cooled to −50 °C.

The enthalpy of the reaction is given by Eq. (4.53). Nitrous oxide has high vapor pressure and can exist in both gas and liquid phases in a pressure vessel containing pure N_2O. To demonstrate the effect of phase on the heat of combustion, compare the heat of reaction of gaseous N_2O (feed line from the top of the tank) versus liquid N_2O (feed line from the bottom of the tank) when combusted with solid paraffin.

$$C_{32}H_{66} + 97N_2O \rightarrow 32CO_2 + 33H_2O + 97N_2 \quad (4.52)$$

$$\Delta \bar{h}_R = 32\bar{h}_{CO_2}(298 \text{ K}) + 33\bar{h}_{H_2O}(298 \text{ K})$$
$$+ 97\bar{h}_{N_2}(298 \text{ K}) - \bar{h}_{C_{32}H_{66}}(298 \text{ K}) \quad (4.53)$$
$$- 97\bar{h}_{N_2O}(223 \text{ K})$$

From Table 4.1, the enthalpies of our products at 25 °C are: $\bar{h}_{CO_2} = -393.52$ kJ/mol, $\bar{h}_{H_2O} =$

−241.83 kJ/mol, and $\bar{h}_{N_2} = 0.0$ kJ/mol. The enthalpy of gaseous nitrous oxide is also available in this table at 298 K: $\bar{h}_{N_2O} = 82.05$ kJ/mol. From interpolation of the data in Ref. [135], it can be seen that at −50 °C, the enthalpy of gaseous nitrous oxide is slightly higher, $\bar{h}_{N_2O}(223 \text{ K}) \simeq 82.59$ kJ/mol. The enthalpy of formation of solid paraffin was estimated to be $\bar{h}_{f,C_{32}H_{66}} = -937$ kJ/mol in Section 4.5.3.

The latent heat of vaporization of N_2O at −50 °C is −16.1 kJ/mol, per Ref. [22]. Per Eq. (4.40), the enthalpy of liquid nitrous oxide at a given temperature and pressure is the enthalpy of gaseous nitrous oxide minus the latent heat of vaporization under those conditions, see Eq. (4.54).

$$\bar{h}_{N_2O}(L, 223 \text{ K}) = 82.59 - 16.1 = 66.49 \text{ kJ/mol} \quad (4.54)$$

The enthalpy of reaction for gaseous N_2O therefore becomes Eq. (4.55), and Eq. (4.56) for liquid N_2O. Recall that the heat of combustion has the opposite sign to the enthalpy of reaction, and therefore 1.56 MJ more energy is released when gaseous N_2O is reacted with a mole of paraffin versus when liquid N_2O is reacted with a mole of paraffin. It should be clear from this example that specifying the correct phase of the reactants is therefore very important when calculating the performance of a rocket.

$$\Delta \bar{h}_R \big|_{N_2O(g)} = (32 \times -393.52) + (33 \times -241.83)$$
$$+ 0 + (1 \times 937) - (97 \times 82.59)$$
$$= -2.76 \times 10^4 \text{ kJ/mol} \quad (4.55)$$

$$\Delta \bar{h}_R \big|_{N_2O(L)} = (32 \times -393.52) + (33 \times -241.83)$$
$$+ 0 + (1 \times 937) - (97 \times 66.49)$$
$$= -2.61 \times 10^4 \text{ kJ/mol} \quad (4.56)$$

4.6.5 Example 2: evaluation of fuels

The energy released during combustion depends on the phase of the reactants as shown in the previous example. However, phase can also be important for other reasons. It is often desirable to fully vaporize the oxidizer in hybrid rockets before it reaches the fuel grain to aid in motor stability. One way of doing this is to burn an energetic propellant with a some of the oxidizer in the precombustion chamber and use the heat released to fully vaporize the rest of the oxidizer.

In this example, two possible propellants that could be used for this purpose are evaluated: TriEthylAluminum (TEA) and TriEthylBorane (TEB).

Assume that the incoming oxidizer is Liquid Oxygen (LOx) at 90 K with a heat of vaporization of 6.82 kJ/mol at 90 K. Assume that the specific heat of oxygen is constant across the temperature range of interest here at 29.1 J/mol K. The TEA and TEB are also in liquid phase and have enthalpies of formation of −217.6 kJ/mol and −188.3 kJ/mol at 25 °C, respectively. The chemical formula for TEA is $(C_2H_5)_3 Al$, and the chemical formula for TEB is $(C_2H_5)_3 B$. Assume that the TEA and TEB are used to both vaporize the oxidizer and heat it to 25 °C. Assume that the enthalpy of formation at 25 °C of aluminum oxide (Al_2O_3) is -1.67×10^3 kJ/mol, and of boric oxide (B_2O_3) is -1.28×10^3 kJ/mol. Per Table 4.1, the enthalpy of formation of carbon dioxide (CO_2) at 25 °C is −393.52 kJ/mol and the enthalpy of formation of gaseous water (H_2O) at 25 °C is −241.83 kJ/mol.

In terms of only the information provided, what is the better fuel for this purpose?

Solution

To answer this question, the propulsion engineer must determine which option will use the least amount of fuel mass to vaporize the oxidizer. To evaluate this, the amount of fuel required to vaporize 1 kg of oxygen is calculated. Assume that TEA combusts with excess

LOx occurs according to the global reaction of Eq. (4.57). Here, α refers to the total number of moles of LOx that are participating in a reaction with 1 mole of TEA. Per each mole of TEA, $\alpha - \frac{21}{2}$ moles of gaseous oxygen will be produced at 25 °C.

$$(C_2H_5)_3 Al + \alpha O_2 \rightarrow \frac{1}{2} Al_2O_3 + 6CO_2 + \frac{15}{2} H_2O$$
$$+ \left(\alpha - \frac{21}{2}\right) O_2 \qquad (4.57)$$

Assume that the temperature of all products is 25 °C [298.15 K]. Also assume that all products, except the solid oxides, are in a gas phase, and all reactants are in a liquid phase. The heat of reaction per mole of TEA is given by Eq. (4.58), where L is used to indicate if the species is in the liquid phase. To evaluate this equation, the enthalpies of all of the species need to first be determined.

$$\Delta \bar{h}_R = \frac{1}{2} \bar{h}_{Al_2O_3}(298\,\text{K}) + 6\bar{h}_{CO_2}(298\,\text{K})$$
$$+ \frac{15}{2} \bar{h}_{H_2O}(298\,\text{K}) + \left(\alpha - \frac{21}{2}\right) \bar{h}_{O_2}(298\,\text{K})$$
$$- \bar{h}_{(C_2H_5)_3Al}(L, 298\,\text{K}) - \alpha \bar{h}_{O_2}(L, 90\,\text{K})$$
$$(4.58)$$

Recall the definition of latent heat of vaporization per Eq. (4.40). Using this equation, the enthalpy per mole of LOx at 90 K is the enthalpy per mole of gaseous O_2 at 90 K minus the heat of vaporization per mole, see Eq. (4.59). Here, it is assumed that oxygen is calorically perfect ($\bar{c}_p$ is constant).

$$\bar{h}_{O_2}(L, 90\,\text{K}) = \bar{h}_{f,O_2}(298\,\text{K})$$
$$+ \int_{298}^{90} \bar{c}_{p,O_2}(T)dT - \bar{h}_{fg,O_2}$$
$$= 0 - 0.0291(298 - 90) - 6.82$$
$$= -12.87\,\text{kJ/mol}$$
$$(4.59)$$

Now the heat of reaction can be calculated from Eq. (4.58), see Eq. (4.60).

$$\Delta \bar{h}_R = \frac{1}{2} \left(-1.67 \times 10^3 \right) + 6 \left(-393.52 \right)$$
$$+ \frac{15}{2} \left(-241.83 \right) + \left(\alpha - \frac{21}{2} \right)(0) \quad (4.60)$$
$$- \left(-217.6 \right) - \alpha \left(-12.87 \right)$$
$$= -4.79 \times 10^3 + 12.87\alpha \text{ kJ/mol}$$

As a reminder, in this example, the goal is to fully vaporize the oxidizer with only the energy from the TEA. This is equivalent to looking at an adiabatic process with a net heat reaction of 0 kJ/mol. Using this information, the moles of LOx that are vaporized by 1 mole of TEA can now be found, see Eq. (4.61).

$$\Delta \bar{h}_R = 0 = -4.79 \times 10^3 + 12.87\alpha$$
$$\Rightarrow \alpha = \frac{4.79 \times 10^3}{12.87} \quad (4.61)$$
$$\Rightarrow \alpha = 372.2 \text{ mol}$$

The mass of TEA required to vaporize 1 kg of LOx can now be evaluated. From Eq. (4.61), it was determined that it takes one mole of TEA to vaporize 372.2 moles of LOx. 1 kg of LOx corresponds to $\frac{1}{.032} = 31.25$ moles of LOx. Thus to vaporize 1 kg of LOx, the moles of TEA required is: $\frac{31.25}{372.2} = 0.084$ mol. Converting 0.084 mol of TEA to mass using the molecular weight of Eq. (4.62) gives $0.084 \times 0.1142 = 0.0096$ kg.

$$M_{w,(C_2H_5)_3Al} = (2 \times 12.011 + 5 \times 1.008) \times 3$$
$$+ 26.982$$
$$= 114.2 \text{ g/mol}$$
$$(4.62)$$

To compare the two fuels, the performance of TEB also needs to be evaluated using the same approach as TEA. Eq. (4.63) gives the global reaction of TEB with excess LOx. The heat of reaction for this equation is given by Eq. (4.64), which is then evaluated numerically in Eq. (4.65).

$$(C_2H_5)_3B + \beta O_2 \rightarrow \frac{1}{2} B_2O_3 + 6CO_2$$
$$+ \frac{15}{2} H_2O + \left(\beta - \frac{21}{2} \right) O_2$$
$$(4.63)$$

$$\Delta \bar{h}_R = \frac{1}{2} \bar{h}_{B_2O_3}(298 \text{ K}) + 6\bar{h}_{CO_2}(298 \text{ K})$$
$$+ \frac{15}{2} \bar{h}_{H_2O}(298 \text{ K}) + \left(\beta - \frac{21}{2} \right) \bar{h}_{O_2}(298 \text{ K})$$
$$- \bar{h}_{(C_2H_5)_3B}(L, 298 \text{ K}) - \beta \bar{h}_{O_2}(L, 90 \text{ K})$$
$$(4.64)$$

$$\Delta \bar{h}_R = \frac{1}{2} \left(-1.28 \times 10^3 \right) + 6 \left(-393.52 \right)$$
$$+ \frac{15}{2} \left(-241.83 \right) + \left(\beta - \frac{21}{2} \right)(0) \quad (4.65)$$
$$- \left(-188.3 \right) - \beta \left(-12.87 \right)$$
$$= -4.63 \times 10^3 + 12.87\beta \text{ kJ/mol}$$

Eq. (4.65) allows us to again evaluate how many moles of LOx are vaporized from one mole of TEB. Again, this problem is looking at an adiabatic process with a net heat reaction of 0 kJ/mol. Using this, the number of moles of LOx that are vaporized by 1 mole of TEB can now be found, see Eq. (4.66).

$$\Delta \bar{h}_R = 0 = -4.63 \times 10^3 + 12.87\beta$$
$$\Rightarrow \beta = \frac{-4.63 \times 10^3}{12.87} \quad (4.66)$$
$$\Rightarrow \beta = 359.8 \text{ mol}$$

The mass of TEB required to vaporize 1 kg of LOx can now be evaluated. From Eq. (4.66), it was determined that it takes one mole of TEB to vaporize 359.8 moles of LOx. 1 kg of LOx corresponds to 31.25 moles of LOx. Thus to vaporize 1 kg of LOx, the moles of TEB required is: $\frac{31.25}{359.8} = 0.087$ mol. Converting 0.087 mol of TEB to mass using the molecular weight of Eq. (4.67) gives $.087 \times 0.098 = 0.0085$ kg.

$$M_{w,(C_2H_5)_3B} = (2 \times 12.011 + 5 \times 1.008) \times 3$$
$$+ 10.811$$
$$= 98.0 \, g/mol \qquad (4.67)$$

TEB takes less mass (8.5 g) to vaporize 1 kg of LOx than TEA (9.6 g). Thus in terms of only the information provided (and not storage density, handling constraints, etc.), TEB is a more effective fuel to vaporize LOx.

4.6.6 Measuring heat of combustion

The heat of formation can also be determined experimentally using the heat of combustion.

Heat of combustion is most practically measured using a bomb calorimeter. A small amount of the fuel in question is combusted with oxygen in a constant volume. The heat produced by this reaction, Q_C, is measured via a rise in the temperature of the surrounding water. The heat of combustion, ΔH_C, can then be calculated from Eqs. (4.68) and (4.69), where the specific heat of the calorimeter, $C_{calorimeter}$, must be known (typically provided by the manufacturer).

$$Q_C = (m c_p \Delta T)_{H_2O} + C_{calorimeter} \Delta T \qquad (4.68)$$

The molar heat of combustion can then be found by dividing Q_C by the number of moles of fuel combusted, $n_{fuel} = m_{fuel}/M_w$.

$$\Delta \bar{h}_C = q_C / n_{fuel} \qquad (4.69)$$

The heat of formation can then be determined using Hess' Law, Eq. (4.49). This assumes that the relevant product information is known. Stoichiometric combustion into major products is a reasonable assumption.

To better understand this, consider the combustion of a hydrocarbon fuel, $C_x H_y$, reacting with air at 25 °C, per Eq. (4.70).

$$C_x H_y + \left(x + \frac{y}{4}\right) O_2 + (3.76)\left(x + \frac{y}{4}\right) N_2$$
$$\rightarrow x C O_2 + \frac{y}{2} H_2 O + (3.76)\left(x + \frac{y}{4}\right) N_2 \qquad (4.70)$$

$$\bar{h}_{f,C_x H_y}(298 \, K) = \sum_{prods} n_i \bar{h}_i(298 \, K) + \Delta \bar{h}_C(298 \, K) \qquad (4.71)$$

$$\bar{h}_{f,C_x H_y}(298 \, K) = \left\{ x \bar{h}_{f,CO_2}(298 \, K) \right.$$
$$\left. + \frac{y}{2} \bar{h}_{f,H_2 O}(298 \, K) \right\} \qquad (4.72)$$
$$+ \Delta \bar{h}_C(298 \, K)$$

From Table 4.1, the heat of formation for gaseous $\bar{h}_{f,CO_2(g)} = -393.52$ kJ/mol and for $\bar{h}_{f,H_2O(g)}$ is -241.83 kJ/mol at standard temperature and pressure, 298.15 K and 0.1 MPa, respectively. These values can be used in Eq. (4.72) along with the measured heat of combustion per mole (from Eqs. (4.69) and (4.68)) to estimate the heat of formation of the hydrocarbon fuel, $C_x H_y$.

4.7 Thermodynamic laws

The laws of thermodynamics are used to derive the fundamental equations that describe the conditions in an active combustion chamber. These conditions can then be used to predict the ideal performance of a given set of propellants in a rocket.

4.7.1 First law of thermodynamics

The first law of thermodynamics is the conservation of energy, as given by Eq. (4.73).

$$dE = \delta Q - \delta W \qquad (4.73)$$

The left-hand side of the equation is the change in energy (E), and the right-hand side is the difference between the heat transfer and the work done in a combustion chamber. E is the sum of the internal energy, U, kinetic energy, $\frac{1}{2}mV^2$, and potential energy, mgh, per Eq. (4.74).

$$E = m\left(u + \frac{1}{2}V^2 + gh\right) \quad (4.74)$$

The change in energy for rocket combustion can be approximated by the change in internal energy (U). This is because the changes in velocity and height (and therefore kinetic, $\frac{1}{2}mV^2$, and potential, mgh, energies) are comparatively small inside a combustion chamber. Therefore it is assumed that $dE = dU$.

An isentropic assumption is made since the main contribution to work is $Pd\Psi$ work. Therefore work is reversible, allowing a change from a partial to a total derivative, and Eq. (4.73) becomes:

$$dU = \delta Q - Pd\Psi \quad (4.75)$$

At this point, enthalpy is reintroduced. Recall that enthalpy is defined as the energy plus the pressure times the change in volume (work) of a system (see Eq. (4.38)). It is again assumed that during combustion, the only contribution to energy is the chemical or internal energy (kinetic and potential energies are ignored). The change in enthalpy will be used to determine the energy in the propellant combination in the following sections.

$$H \equiv U + P\Psi$$

Now take the derivative of Eq. (4.38), rewritten above for convenience and Eq. (4.75) can be rewritten in terms of enthalpy. The standard enthalpy of reaction is negative for an exothermic reaction ($Q > 0$) and positive for an endothermic ($Q < 0$) reaction.

$$dH = \delta Q + \Psi dP \quad (4.76)$$

Eqs. (4.75) and (4.76) can be combined by solving for δQ, giving the first law with all the assumptions for a combustion chamber.

$$\delta Q = dH - \Psi dP = dU + Pd\Psi \quad (4.77)$$

4.7.2 Adiabatic flame temperature

At this point, the propulsion engineer will notice the dependence on δQ in both of the above forms of the first law (Eqs. (4.75) and (4.76)). It is therefore interesting to consider an adiabatic case when $\delta Q = 0$. This is typically not too far from physical as it is both desirable from a performance standpoint and necessary for the survival of the structural materials to minimize heat transfer to the walls in rocket combustion [24]. Also, in a hybrid rocket, where combustion chamber walls are composed of solid fuel, much (though not all) of the heat that might typically be lost to the chamber walls or insulation is actually transferred to the solid fuel.

This assumption allows one to solve for the adiabatic flame temperature, which is an upper bound on the combustion temperature. Now, either a constant volume or constant pressure assumption is required.

If a constant volume assumption is applied, Eq. (4.75) gives $dU = 0$, which, once integrated, says that the internal energy of the reactants equals the internal energy of the products. While true for many combustion reactions, it should be noted that some combustion chambers can increase the volume due to the physical expansion of the combustion chamber when pressurized and/or the regression of the fuel.

If a constant pressure assumption is applied, Eq. (4.76) gives $dH = 0$, which, once integrated, says the enthalpy of the reactants is equal to the enthalpy of the products. This is a reasonable assumption outside transient conditions since pressure change timescales are typically much longer than chemical reaction timescales.

To summarize:

- Adiabatic combustion at constant volume: $U_{products} - U_{reactants} = 0$

- Adiabatic combustion at constant pressure: $H_{products} - H_{reactants} = 0$

Either of these equations can be used to iteratively determine the adiabatic combustion temperature.

$$U_{products}(T_{adiabatic}, P) = U_{reactants}(T_{initial}, P)$$
$$(4.78)$$

$$H_{products}(T_{adiabatic}, P) = H_{reactants}(T_{initial}, P)$$
$$(4.79)$$

Solving for the adiabatic flame temperature is straightforward if the composition of the reactants and products is known. The initial temperature and reactants are inputs. The product composition can then be found assuming complete combustion (the reactants go to major products) or equilibrium combustion (the reactants go to a different number of products caused by dissociation and recombination based on the temperature and pressure of the reaction). Equilibrium combustion is more physical at the high temperatures relevant to combustion chambers, but is easier said than done by hand. Note that this is an upper bound for adiabatic flame temperature because as chemical effects, such as dissociation, are taken into account, energy is shifted from the sensible enthalpy to chemical bonds (heat of formation).

4.7.3 Second law of thermodynamics

In any physical combustion reaction, the reactants do not all neatly react to become only major products. The high temperatures achieved in these reactions lead to complex intermediate reactions and species. Since propulsion engineers are never afraid of a little hard work, they take the next step by evoking the second law of thermodynamics to explore the equilibrium condition and how to move past complete combustion to determine what is actually happening in the combustion chamber.

The second law of thermodynamics introduces entropy, S, which provides a direction in time for chemical and physical processes. When entropy reaches a maximum, the system can no longer perform useful work, e.g., it has reached an equilibrium.

The change in entropy is greater than or equal to the heat transfer into the system divided by the temperature. If the process under consideration is reversible, both sides of Eq. (4.80) are equal. In an irreversible process, the greater-than sign must be used.

$$T dS \geqslant \delta Q \qquad (4.80)$$

Then δQ from the first law of thermodynamics, Eq. (4.77), can be substituted into Eq. (4.80) to obtain Eq. (4.81).

$$T dS - dU - P d\Psi = T dS - dH + \Psi dP \geqslant 0 \quad (4.81)$$

When a mixture is at equilibrium, thermodynamic properties (e.g., P, T, ρ, and Y) are no longer changing. Thus during equilibrium combustion in a hybrid rocket, $dP = dT = 0$. Substitute this into Eq. (4.81) to generate Eq. (4.82). Remembering the definition of enthalpy, Eq. (4.38) allows Eq. (4.82) to be rewritten as Eq. (4.83).

$$d[TS - (U + P\Psi)] \geqslant 0 \qquad (4.82)$$
$$d[H - TS] \leqslant 0 \qquad (4.83)$$

4.7.4 Gibbs free energy

It is now useful to introduce the Gibbs free energy, G_e, which is defined in Eq. (4.84). One will often see the Gibbs free energy written simply as "G" or "g." However, those variables have already been assigned to mass flux, gravity, and the Gravitational constant in this text. Therefore the subscript "e" has been added for clarity throughout.

$$G_e \equiv H - TS \qquad (4.84)$$

The second law of thermodynamics (in Eq. (4.83)) has been conveniently written to allow for the substitution of G_e.

$$dG_e \leqslant 0 \qquad (4.85)$$

The above, Eq. (4.85), states that for constant temperature and pressure, the Gibbs free energy must either be constant or decreasing. Therefore chemical equilibrium is achieved when the Gibbs free energy is minimized, and the equilibrium condition becomes $dG_e = 0$.

The following reaction is used as an example:

$$n_A A + n_B B + ... \Longleftrightarrow n_C C + n_D D + ... \quad (4.86)$$

Now for a mixture of ideal gases, the change in the Gibbs free energy is equal to the sum of the number of moles of each species times the Gibbs function (per mol) of that species, Eq. (4.87). The subscript i, refers, in this case, to the ith chemical species being considered. A mixture of ideal gases is also an ideal gas, so partial pressures, etc., apply. For example, the molar Gibbs function of the "ith" species in a mixture can be evaluated at temperature, T, partial pressure, P_i, and at the total system volume.

$$G_e = \sum n_i \bar{g}_{e,i}(T, P) \quad (4.87)$$

where,

$$\bar{g}_{e,i}(T, P) = \bar{h}_i(T) - T\bar{s}(T, P_i) \quad (4.88)$$

Recall that the enthalpy for each species can be calculated from Eq. (4.42). The entropy of each species can be found from Eq. (4.89).

$$d\bar{s}_i(T, P_i) = \bar{c}_{p,i} \frac{dT}{T} - R_u \frac{dP_i}{P_i} \quad (4.89)$$

Solving for the derivative of the Gibbs function, using the definitions of enthalpy and entropy, and noting that $dT = 0$ near equilibrium gives Eq. (4.90).

$$dG_{e,i} = dH_i - TdS_i - S_i dT \quad (4.90)$$
$$dG_{e,i} = n_i \bar{c}_{p,i} dT - TdS_i - S_i dT$$
$$dG_{e,i} = n_i \bar{c}_{p,i} dT^{\;0} - n_i T(\bar{c}_{p,i} \frac{dT}{T}^{\;0}$$
$$- R_u \frac{dP_i}{P_i}) - S_i dT^{\;0}$$

$$dG_{e,i} = n_i R_u T \frac{dP_i}{P_i}$$

Eq. (4.90) can now be integrated to produce Eq. (4.91).

$$G_{e,i} = G^{\circ}_{e,i} + n_i R_u T ln\left(\frac{P_i}{P^{\circ}}\right) \quad (4.91)$$

Here, $G^{\circ}_{e,i}$ is the Gibbs function of the pure species, i, at standard state pressure, P°. By convention, P° is usually 1 atmosphere. $G^{\circ}_{e,i}$ is therefore a function of only temperature. $G^{\circ}_{e,i}$ can also be expressed on a per mole basis, Eq. (4.92). The NIST JANAF tables [135] tabulate known values for the useful parameters, such as the Gibbs function, at standard pressure and temperature.

$$G^{\circ}_{e,i} = n_i \bar{g}^{\circ}_{e,i} \quad (4.92)$$

As stated in Eq. (4.87), the Gibbs free energy of an ideal mixture is the sum of the Gibbs function, $G_{e,i}$, of each species. Substituting Eq. (4.91) and Eq. (4.92) into Eq. (4.87) gives Eq. (4.93).

$$G_{e,mix} = \sum n_i \left[\bar{g}^{\circ}_{e,i} + R_u T ln\left(\frac{P_i}{P^{\circ}}\right) \right] \quad (4.93)$$

Recall that the change in Gibbs free energy must equal zero at equilibrium. Going back to the example reaction, Eq. (4.86), the difference between the products and the reactants can be obtained, see Eqs. (4.94) and (4.95).

$$0 = \Delta G^{\circ}_e + R_u T \left[n_C ln\left(\frac{P_C}{P^{\circ}}\right) + n_D ln\left(\frac{P_D}{P^{\circ}}\right) + ... \right.$$
$$\left. - n_A ln\left(\frac{P_A}{P^{\circ}}\right) - n_B ln\left(\frac{P_B}{P^{\circ}}\right) - ... \right]$$
$$(4.94)$$

$$\Delta G^{\circ}_e = G^{\circ}_{e,C} + G^{\circ}_{e,D} + ... - G^{\circ}_{e,A} - G^{\circ}_{e,B} - ... \quad (4.95)$$

Rearranging Eq. (4.94) gives Eq. (4.96).

$$exp\left(\frac{-\Delta G_e^\circ}{R_u T}\right) = \left(\frac{P_C^{n_C} P_D^{n_D} \cdots}{P_A^{n_A} P_B^{n_B} \cdots}\right) \times P^{\circ(n_A + n_B - n_C - n_D + \cdots)}$$
(4.96)

Now the pressure-based equilibrium constant, K_P, is introduced given in Eq. (4.97). Note that from Eq. (4.96), K_P is also given by Eq. (4.98).

$$K_P(T) \equiv \left(\frac{P_C^{n_C} P_D^{n_D} \cdots}{P_A^{n_A} P_B^{n_B} \cdots}\right) P^{\circ(n_A + n_B - n_C - n_D + \cdots)}$$
(4.97)

$$K_P(T) = exp\left(\frac{-\Delta G_e^\circ}{R_u T}\right)$$
(4.98)

As introduced above, the equilibrium constant, K_P, is a function of temperature only as ΔG_e° is a function of only temperature. Values for the equilibrium constant for various species can be found in the NIST JANAF tables [135].

It can be more helpful to use Dalton's law to convert K_P into a function of moles. This allows the equilibrium constant to be used as an equation to find the product equilibrium composition. This is shown in the example in the next section.

4.8 Example combustion problem

The propulsion engineer can now calculate the equilibrium conditions for various combustion reactions. A simply hydrocarbon and oxygen combustion reaction will be used to walk through the steps:

Take methane (gas) reacting with gaseous oxygen in a relatively rich mixture at 1500 K.

$$CH_4 + \frac{4}{3} O_2 \Longleftrightarrow a CO_2 + b H_2O + c CO + d H_2$$

The number of carbon, hydrogen, and oxygen atoms are conserved through the reaction.

Therefore the number of moles of carbon in the reactants must equal that in the products and Table 4.3 can be populated.

TABLE 4.3 Equilibrium Reaction Example: conservation of elements.

	Reactants	Products
C	1	$a + c$
H	4	$2(b + d)$
O	8/3	$2a + b + c$

However, Table 4.3 only gives three equations for four unknowns. The one more equation is needed. Formation equations are often used. In this case, the water gas shift reaction can be used. The water gas shift reaction is a common balance between CO and H_2.

$$b H_2O + c CO \Longleftrightarrow a CO_2 + d H_2$$
(4.99)

While it is nice, well critical really, to have four equations to solve for the four unknowns, information is missing for this final reaction. While the JANAF-NIST tables can be used to find $K_p(T)$ for C, H, and O, one must find K_P at the temperature of interest for the fourth equation (in this case, we are using the water gas shift reaction to relate the products containing hydrogen to those containing carbon).

$$K_p(T) = \frac{P_{CO_2} P_{H_2}}{P_{H_2O} P_{CO}} = \frac{ad}{bc}$$

The ΔG_e method can be used to find the K_P for the balanced water gas shift equation ($a = b = c = d$). The K_p derived from the balanced water gas shift reaction can then be used to determine equilibrium mole fractions of the desired oxygen/methane reaction. Using the ΔG_e method to find K_p, ΔG_e° can be found by subtracting the number of moles of each reactant times its molar Gibbs function from the number of moles of each product times its molar Gibbs function. Note that this is for the balanced case (i.e., $a = b = c = d = 1$).

$$\Delta G_e^\circ = a\bar{g}_{e,CO_2} + d\bar{g}_{e,H_2} - b\bar{g}_{e,H_2O} - c\bar{g}_{e,CO}$$
$$= 1\bar{g}_{e,CO_2} + 1\bar{g}_{e,H_2} - 1\bar{g}_{e,H_2O} - 1\bar{g}_{e,CO}$$

Recall that $g_{e,i}(T) = \bar{h}(T) - T\bar{s}^\circ(T, P^\circ)$, which can be found from the JANAF tables, and that in the notation of those tables, $\bar{h}(T) = \Delta_f H^\circ + (H^\circ - H^\circ(T_r))$ and entropy per mole, $\bar{s}^\circ$, is S°.

Therefore $\bar{g}_{e,CO_2} = (-393.522 \text{ kJ/mol} + 61.705$ kJ/mol) $-$ (1500 K)(292.199 J/ K mol) $=$ -770.116 kJ/mol. The others can be found similarly: $\bar{g}_{e,H_2} = -231.979$, $\bar{g}_{e,H_2O} = -569.605$, and $\bar{g}_{e,CO} = -444.316$. And, finally, $\Delta G_e(1500 \text{ K}) = 11.826$ kJ/mol, which can then be used to find K_P for the water gas shift reaction using Eq. (4.98). Note that ΔG_e here is per mole of the reaction specified in Eq. (4.99).

$$K_P(T) = \exp\left(-\frac{\Delta G_e^\circ}{R_u T}\right) \quad (4.100)$$

$$K_P(1500 \text{ K}) = \exp\left(-\frac{11.826 \text{ kJ/mol}}{8.314 \text{ J K}^{-1} \text{ mol}^{-1} 1500 \text{ K}}\right)$$
$$= 0.387$$

While the balanced reaction was used to find the $K_P(1500 \text{ K})$, the water gas shift reaction is now applied to the equilibrium conditions of the oxygen/methane reaction at a certain temperature.

$$K_P(1500 \text{ K}) = \frac{ad}{cb} = 0.387$$

There are now four equations and four unknowns, and the propulsion engineer can solve for the number of moles of each product $a = 0.40$, $b = 1.27$, $c = 0.60$, $d = 0.73$ at 1500 K.

Now the mole fractions of each species, X_i, can be found.

$$X_{CO_2}^* = \frac{a}{a+b+c+d} = \frac{0.4}{3} = 0.13 \quad (4.101)$$

$$X_{H_2O}^* = \frac{b}{a+b+c+d} = \frac{1.27}{3} = 0.42$$

$$X_{CO}^* = \frac{c}{a+b+c+d} = \frac{0.6}{3} = 0.2$$

$$X_{H_2}^* = \frac{d}{a+b+c+d} = \frac{0.73}{3} = 0.24$$

The superscript * is used to represent equilibrium conditions. These mole fractions should add up to one, which they nearly do (minus the rounding error created by keeping only two significant figures). Note that in the case of products in the solid phase, e.g., carbon, they are not included in the denominator.

The only "trick" here was knowing which additional equation to use to get enough equations to solve for the unknowns. Formation reactions (for the products) are often used for this.

While this is hardly ever (really never) done by hand anymore, it is critical to understand how these calculations work and the underlying assumptions that went into them. An equilibrium solver that performs these calculations would more commonly be used, see Section 4.10.

4.9 Conditions through the nozzle

Flow through the nozzle can be quite complex with potential boundary layer effects, ongoing combustion, and when there are adverse pressure gradients or a poorly designed nozzle, the presence of shocks/flow separation. In general, nozzles are designed to avoid the presence of shocks and boundary layer effects are often assumed to be minor. Two main assumptions are then considered: frozen or shifting equilibrium.

The frozen assumption says that all combustion occurs in the combustion chamber, and the flow composition remains constant as the gas accelerates through the nozzle. The point at which the combustion is assumed to be frozen is typically chosen to be either the combustion chamber or the nozzle throat. The assumption of frozen flow is generally more valid when collisions between particles are not as frequent, and temperatures are low (low density and cool gas as the gas is expanding). This is not generally true for hybrids, and so the assumption of frozen equilibrium is conservative for hybrid rocket motors and will underpredict performance in terms of I_{sp}.

Shifting equilibrium assumes that the combustion kinetics are infinitely fast, and equilibrium is therefore achieved at each point in the nozzle. As the gas is accelerated through the nozzle, it cools and is at a lower static pressure. Shifting equilibrium re-calculates the gas equilibrium species at these new conditions at each point in the nozzle. These additional reactions release energy and therefore increase I_{sp}. In general, shifting equilibrium is not actually achieved in hybrid rocket motor nozzles, and as such, applying a shifting equilibrium assumption to the flow field will slightly overpredict performance in terms of I_{sp}.

For simplicity when performing hand calculations, frozen equilibrium and isentropic flow are generally assumed through the nozzle. In this case, the designer should use γ_{mix} as defined by Eq. (4.27) or Eq. (4.29) to calculate the exit conditions at the end of the nozzle. Commercial codes, such as Chemical Equilibrium with Applications [74], and for more fidelity: Cantera [133], which takes finite rate kinetics into account or Two Dimensional Kinetics (TDK) [138]. TDK is widely used for professional nozzle design. An appreciable change in performance estimates can be found as the fidelity of the model is increased. These commercial codes are discussed more in Section 4.10.

4.9.1 Nozzle wall temperature

Determination of the expected nozzle wall temperature, T_w, can be critical when evaluating possible nozzle materials. It is important to realize that the temperature that the nozzle wall will see, $T_{w,x}$, will be substantially larger than the static temperature, T_x, at a given nozzle location, x, but also likely to be less than the combustion chamber total temperature, $T_{t,c}$. As you move from the free steam flow towards the wall, some of the kinetic energy of the flow will be converted into heat. If the nozzle wall is assumed to

be adiabatic, then the nozzle wall temperature can be calculated to be the recovery temperature, per Eq. (4.102). Here, M_x is the free stream Mach number, Pr is the Prandtl number, and γ is the ratio of specific heats (γ_{mix} for a mixture). All of the inputs to this equation, including the local Prandtl number, can be taken from a standard chemical equilibrium code output.

$$T_{w,x} = T_{e,x} \left(1 + \zeta \frac{\gamma - 1}{2} M_x^2 \right) \quad (4.102)$$

Where

$$\zeta = \sqrt[3]{Pr} \quad (4.103)$$

Further information on this topic, including a theoretical derivation for the turbulent recovery factor of Eq. (4.103), can be found in Ref. [139].

4.10 Commercial combustion codes

4.10.1 Chemical equilibrium codes

Commercially available codes exist to determine the equilibrium combustion conditions and evaluate rocket performance. Examples of equilibrium codes include Chemical Equilibrium with Applications (CEA) [74], STANJAN indexSTANJAN [140], Rocket Propulsion Analysis (RPA) [141], Cantera [133], and Mutation++. It is important to note that codes like these generally make a number of simplifying assumptions that are not always true. As an example of this, the one-dimensional equilibrium code CEA assumes that the chemical kinetics are fast; however, the reaction to convert hydrocarbons into CO_2 can be slow compared to the fluid resonance time. Two-dimensional nozzle losses are also ignored in this one-dimensional code. Cantera can handle equilibrium calculations, finite rate kinetics, and quasi-one-dimensional finite rate kinetics.

Most equilibrium codes have large data sets specifying thermodynamic properties for a range of fuels and oxidizers. However, new reactants must still often be defined; this is especially common for hybrid rocket fuels, which are not generally in the standard databases. Inputting a new reactant generally only requires knowledge of the chemical composition and the enthalpy of the substance at its initial conditions. It is vital to ensure that the enthalpy correctly reflects the initial phase and temperature of the reactant. Note that specifying the correct phase is also important when using existing species in the databases. This is an area that can be overlooked, leading to errors in the predicted theoretical performance of their rocket design. It can sometimes be a challenge to input the correct initial phase. To illustrate this, consider an oxidizer that is in a liquid phase in the tank. If the oxidizer is injected into the combustion chamber as a liquid, energy from the combustion must be used to vaporize it, and it must be specified as a liquid reactant. However, if it is vaporized across the injector, or upstream of the injector, without using much/any heat from the combustion chamber, then it is more correct to specify it as a gas phase reactant.

As discussed in Section 4.9, the assumed combustion behavior must also be specified to evaluate the rocket performance. A frozen equilibrium assumption assumes that reactions are absent in the nozzle, resulting in a conservative estimate of performance. On the other hand, a shifting equilibrium assumption typically overpredicts the performance by 1–4% [6]. The real solution lies somewhere in between the two: $I_{sp,frozen} < I_{sp,actual} < I_{sp,equilibrium}$.

Loss coefficients are generally applied to the theoretical performance from combustion codes to account for incomplete combustion and losses in the combustion chamber and nozzle, see Chapter 7, Eqs. (2.12) in Chapter 2, and Eqs. (11.5), (11.6), (11.7), and (11.8) in Chapter 11. For a more rigorous theoretical estimate of performance, a two-dimensional kinetic code can be used, as discussed in the following section.

4.10.2 Two-dimensional kinetics

Combustion codes exist, such as Two-Dimensional Kinetics (TDK) [138], that combine chemical kinetics with fluid flow equations. These codes are powerful tools that are generally used both to optimize nozzle design and also to quantify the expected losses through a given nozzle design. Nozzle contour optimization is generally conducted for defined expansion ratios or lengths and combustion chamber conditions. The codes typically use the method of characteristics to do the optimization as described in Ref. [142]. TDK will also determine the expected losses in an already-designed nozzle due to boundary layer effects and incomplete combustion. It can also be used to predict the potential presence of shocks/or flow separation within the nozzle. Care must be taken to avoid the presence of strong shocks in the nozzle, as they can dramatically reduce performance. If strong shocks are present and cannot be designed out of existence, then TDK [138] is not recommended for performance evaluation. Instead, alternative tools focusing on shock evaluation or supersonic Computation Fluid Dynamics (CFD) codes should be used to predict performance.

Wherever possible, hybrid rocket nozzle losses should include two-dimensional and kinetic losses for a specific propellant combination and nozzle design. This could be done with a code like TDK. Ref. [48] found that previously published nozzle efficiencies in liquid rocket literature (see Section 7.4.4 in Chapter 7 for discussion of these) are not conservative for hybrid rockets. This is believed to be due to the use of hydrocarbons as hybrid rocket fuels. Hydrocarbon fuels rely on a slow reaction to create CO_2, as compared to other common products, e.g., H_2O. These combustion kinetics can account for a substantial decrease in nozzle efficiency.

CHAPTER

5

Propellants

5.1 Introduction

This chapter discusses typical hybrid propellants. This includes both classical and high regression rate fuels. The most common oxidizers are also identified and discussed. Focus will remain on using solid fuels with liquid or gaseous oxidizers. While the opposite combination is possible, known as a reverse hybrid, the known propellant combinations cannot compete with the conventional configuration, specifically because liquid oxidizers are much higher performance than known solid options. (The exception is nitronium perchlorate, which is thermally sensitive and potentially explosive [30].)

Performance, measured in specific impulse (I_{sp}), is typically a major focus of propellant selection. It is dependent on the chemistry of the propellant combination. Combinations that produce low-molecular-weight products have higher performance. To accurately determine the theoretical performance of a given propellant combination, the engineer needs to know the heat of formation of the propellants. These values can generally be found in the literature, with values for common propellants provided here in Tables 5.2 and 5.3. An experimental method to determine the heat of formation for fuels is discussed in Chapter 11. When experimental testing is not possible, then the group contribution method developed by Benson can

be used to estimate the heat of formation, see Refs. [143] and [137]. A discussion of using the bond additivity approach from this method is provided in Chapter 4.

There are several important characteristics beyond performance to consider when choosing a propellant combination for a particular application, including packaging (density and O/F ratio), environmental considerations (storage and operational temperatures and mechanical loading), storability, safety (e.g., toxic, explosive), and cost. These will be discussed in the following sections.

5.2 Fuels

There are two distinct types of hybrid fuels: classical and liquefying. Classical hybrid fuels are slower burning since their combustion processes are diffusion limited. The entrainment of low-viscosity, liquefying fuels in the oxidizer flow causes their much faster combustion. The difference in regression rates coupled with the individual strengths and weaknesses of the fuels makes them each suitable for different applications. The list of fuels typically used for hybrid rockets is fairly short, focusing on rubber, plastics, and waxes. All of these fuels are hydrocarbons, with major products, including H_2O and CO_2. The chemical reaction resulting

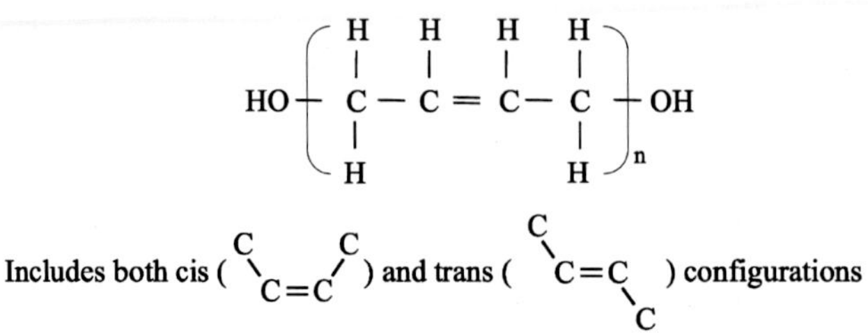

$$HO \left(\begin{array}{cccc} H & H & H & H \\ | & | & | & | \\ C & - & C & = & C & - & C \\ | & & & & | \\ H & & & & H \end{array} \right)_n OH$$

Includes both cis ($\overset{C}{\underset{C=C}{\diagdown}}\overset{C}{\diagup}$) and trans ($\overset{C}{\underset{\diagup}{\diagdown}C=C\overset{\diagdown}{\underset{C}{}}}$) configurations

FIGURE 5.1   Chemical structure of HTPB. Many different combinations are possible.

in CO_2 is slow; therefore the combustion gases may not fully react before exiting the nozzle. Water vapor is well known to promote nozzle erosion. The most common classical hybrid fuel is hydroxyl-terminated polybutadiene (HTPB). It is a polymeric, synthetic rubber also used as a binder in solid rockets. Polyethylene and polymethyl methacrylate (PMMA aka acrylic) are classical fuels, most often used in small-scale or teaching applications. PMMA is especially interesting because it is clear, so optical access and therefore insight into the combustion process can be attained. Paraffin waxes have gained a lot of popularity due to their high regression rate compared to classical hybrid fuels. Cryogenic fuels (fuels that are solid at low temperatures), such as pentane, have been tested and are crucial to the development of liquefying hybrid rocket fuels; however, they are not used in practice because of their low storage temperatures. Performance, stability, or ignition-enhancing additives such as metals, salts, and amides (e.g., aluminum, lithium, aluminum lithium hydride, sodium borohydride, and sodium amide) can easily be added to many of these fuels. Small amounts of a blackening agent, such as carbon black, are often added to hybrid rocket fuels to reduce radiative heat transfer into the fuel grain.

Solid fuel characteristics of interest for the hybrid motor designer are the fuel chemical composition, density, decomposition temperature, glass transition temperature, mechanical strength properties, coefficient of thermal expansion, heat of formation, and heat of com-

bustion. A discussion of how to experimentally determine some of these properties for a given fuel is provided in Chapter 11.

5.2.1 Classical fuels

Hydroxyl Terminated Polybutadiene (HTPB)

Hydroxyl Terminated Polybutadiene or HTPB is a type of synthetic rubber. A wide range of chemical compositions fall under this umbrella, and its performance and regression rates vary depending on the exact mixture. The general chemical structure is shown in Fig. 5.1. Therefore it is crucial to understand the mixture being used and not take performance data (or heats of formation) from literature for granted, as they vary widely due to the variety of chemical formulas that fall under the HTPB category. There are a number of reasons for this variability, such as the curative mixture ratio, relative humidity, the cure temperature and length of time, and the amount of residual gas in the cast material [144].

HTPB is commonly used as a binder in solid rocket propellant. The cross-linking (thermosetting) nature of HTPB makes it necessary to cast it in batches, typically with special tooling and under vacuum [145]. This can make it more expensive than other fuel options. However, it has been used for many years and remains one of the most common hybrid rocket propellants. Most think this is thanks to its history in solid propulsion. Thanks to this history, a decent amount of

TABLE 5.1 Material properties of ABS.

	Extruded	3D Printed
Tensile Strength, Yield	13-65 MPa (MatWeb)	28.5 MPa [151]
Elastic Modulus	1-2.65 GPa (MatWeb)	1.807 GPa [151]

data is available. HTPB is also popular in the hybrid community due to its relatively high regression rate (compared with other classical fuel alternatives). HTPB fuel grains were used in the successful SpaceShipOne test flights, which won the Ansari X Prize in 2004.

Most of the ingredients that make up HTPB are incompatible with reactive and hypergolic additives. This is because the curing process typically creates water, which reacts with the additives [146].

Polymethyl methacrylate (PMMA)

PMMA is a thermoplastic, commonly known as acrylic. PMMA is a shatter-resistant alternative to glass that is widely used in a variety of nonpropulsion applications, such as for aquariums. Therefore it is broadly available at relatively low cost. It is below its glass transition at room temperature. Pure PMMA, without any additives or contaminants, has a chemical composition of $(C_5O_2H_8)_n$. See Fig. 5.2.

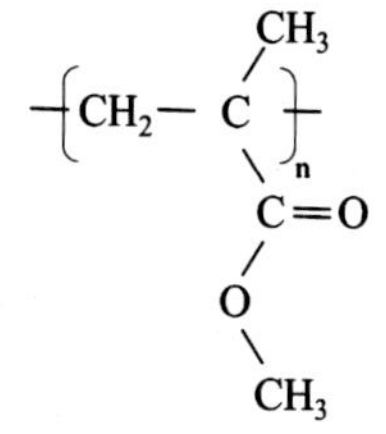

FIGURE 5.2 Chemical structure of PMMA.

PMMA has been used as a hybrid fuel since the early days of hybrid rocket propulsion [78,1]. Compared to other hybrid fuel candidates, it has high strength and density. It is a particularly slow-burning fuel [147,148], making it best suited for small applications (e.g., CubeSats and SmallSats). The presence of oxygen in its chemi-

cal composition reduces the performance of the fuel relative to other hydrocarbon options, see Fig. 5.8. However, this also causes the performance of PMMA to peak at relatively low O/F ratios, a feature that can be beneficial when using low-density oxidizers. PMMA tends to burn cleanly without producing much soot. It ablates without forming a char layer [85]. Clear PMMA has been used extensively in combustion visualization experiments, as it is strong enough to be used simultaneously as a fuel grain and a combustion chamber case [149,150,1].

Acrylonitrile butadiene styrene (ABS)

Acrylonitrile butadiene styrene (ABS) is a thermoplastic material that is widely available and is easy to machine, injection mold, and extrude. As a result, ABS is used in a variety of industries and applications, from household plumbing to automobiles and toys. More recently, ABS has become of increasing interest to the hybrid rocket community because it can easily be 3D printed via Fused Deposition Modeling (FDM). The difference in strength between extruded and printed ABS is given in Table 5.1. The ability to 3D print the fuel makes lab-scale fuel grains inexpensive to fabricate and enables complex internal geometries if desired. ABS has the added advantage that is compatible with many adhesives. It has reasonable strength properties and can be subjected to a fairly wide range of temperatures (about $-20\,^\circ$C to $+80\,^\circ$C). Its glass transition temperature is relatively high, about $105\,^\circ$C; therefore it is brittle at room temperature. As the name suggests, ABS is composed of acrylonitrile (C_3H_3N), butadiene (C_4H_6), and styrene (C_8H_8). The chemical structure is shown in Fig. 5.3. There is a wide range of chemical compositions that are lumped into the ABS designator, corresponding to different percentages of each of these polymers; however, general properties are given in Table 5.2.

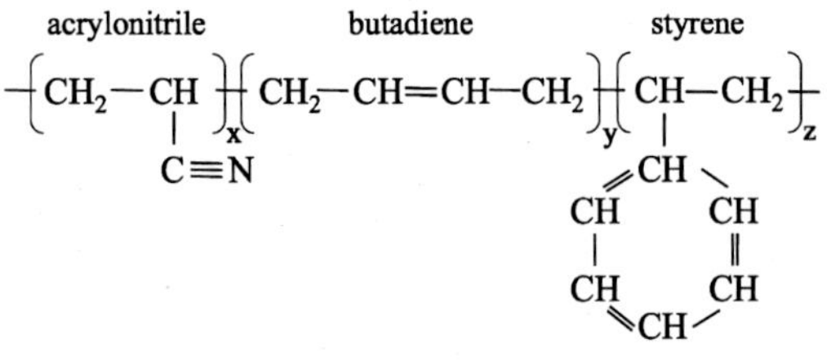

FIGURE 5.3　Chemical structure of ABS.

When exposed to heat, such as during combustion, ABS will form a melt layer before vaporizing [144]. However, this melt layer does not have the properties required to form unstable roll waves for droplet entrainment, and, therefore, ABS is not a high regression rate liquefying fuel. In fact, the regression rate of an ABS fuel (Stratasys ABS-M30: approximately 50:43:7 monomer mole ratio) with nitrous oxide was shown to be comparable to that of HTPB [144]. The performance of ABS was observed to be slightly less than HTPB. However, the performance of ABS in a hybrid rocket and the O/F ratio, which is needed to optimize this performance, vary with changing ABS compositions [144]. Thus engineers seeking to work with this fuel should be sure that they understand the chemistry of ABS they are supplied with and modify their design accordingly.

The layered pattern caused by 3D printing from ABS enables a unique ignition mechanism [152]. The structure of the additively manufactured ABS concentrates electrical charge to produce localized arcing between the printed material layers with gap distances on the order of microns [153]. This arcing pyrolyzes some of the ABS fuel and can provide sufficient energy to ignite the hybrid motor in the presence of oxidizer flow.

Polyethylene

There are several types of polyethylene (PE) that may be used for hybrid fuel, including High-Density PE (HDPE), Low-Density PE (LDPE), Ultra-High-Molecular-Weight PE (UHMW), and polyethylene wax. These fuels produce a melt layer on their surface [154]. However, the viscosity of the melt layer in HDPE, LDPE, and UHMW polyethylene fuel grains is too high to allow droplet entrainment, and thus these fuels are considered classical hybrid rocket fuels. Classical polyethylene fuels tend to form a char layer on the surface of their melt layer as they burn. Polyethylene waxes have low enough viscosity to allow droplet entrainment to occur and have been observed to be liquefying fuels [85]. These waxes can be used if an intermediate regression rate is desired. The chemical structure of polyethylene is shown in Fig. 5.4.

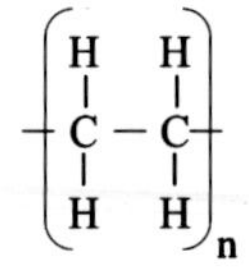

FIGURE 5.4　Chemical structure of Polyethylene.

5.2.2 Liquefying fuels

Liquefying hybrid fuels were first proposed at Stanford University in the late 1990s. Increased regression was observed in hybrid testing with cryogenic pentane as the fuel ignited the search for a fuel that exhibited the same high regression behavior and that was solid under standard atmospheric conditions. The regression rate of liquefying fuels is about 3–4 times that of conventional hybrid fuels [39].

In addition to the advantages that hybrids enjoy over solid or liquid propulsion systems, liquefying hybrid fuels (e.g., paraffin wax) have many advantages over classical fuels. These fuels form a low-viscosity and surface-tension liquid layer during the combustion process. The shear force of the oxidizer passing over the liquid fuel destabilizes the layer and entrains droplets in the flow. The high rate allows for a simple single circular port grain design and significantly improved volumetric fuel loading and increased fuel utilization (typically more than 97%).

Paraffin

Paraffin is the best known liquefying fuel. It is an alkane, with the chemical formula $C_n H_{2n+2}$. It is a crystalline hydrocarbon. The chemical structure for the commonly used hybrid fuel $C_{32}H_{66}$ is shown in Fig. 5.5.

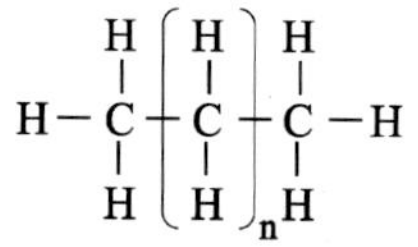

FIGURE 5.5 Chemical structure of Paraffin. ($C_{32}H_{66}$ is shown here ($n = 30$), though other carbon chain lengths are possible.)

Neat paraffin (without additives) has poor strength characteristics. A variety of strength additives have been suggested and tested, but it remains a brittle material and may require special design considerations [155]. Slump has been observed in large-scale paraffin grains under warm conditions [156]. Additionally, its coefficient of thermal expansion is quite large. Environmental conditions are an important consideration for the fuel grain design. The large CTE can make axial casting of fuel grains a challenge as the paraffin pulls away from molds as it cools. For this reason, lower carbon number paraffins are often spun cast. However, this is not always possible as the carbon number and viscosity increase. Paraffin enjoys the added flexibility of having an adjustable regression rate. Slight alterations in additive concentration can change this rate by more than a factor of two. This critical virtue can be quite beneficial in designing efficient hybrid systems with mission flexibility

since the regression rate and burn time determine the motor size.

Casting paraffin fuel grains is a relatively simple process that can be carried out at a small-scale facility. The fuel composition needed for a given mission, including structural additives, is melted and then cooled and solidified into the required grain size and shape using a centrifugal casting process designed to produce crack- and void-free grains. Neither polymerization reactions are involved, nor curing agents are required. The scrap pieces of fuel can be re-melted and reused. Paraffin is fundamentally inert, so deterioration in storage is not an issue.

A lower regression rate, a low-temperature alternative to paraffin, was developed for a potential Mars application. It is discussed here because substantial thermal measurements were completed on the initial formulation of the fuel. Heat capacity and thermal conductivity measurements for the wax-based SP7 fuel are given in [157] and, as expected, vary with temperature (over the tested range from about −80 °C to +50 °C). At 20 °C, the specific heat capacity is nearly 1.5 kJ/kgK, with the neat SP7 testing slightly higher than the aluminized fuel. At −20 °C, the specific heat capacity for both aluminized and neat SP7 is about 1.2 kJ/kgK. Thermal conductivity measurements were also made. The room temperature values for aluminized SP7 is 0.49 W/mK, and neat SP7 is 0.36 W/mK.

Strength additives for paraffin wax

Achilles heel of paraffin is its low mechanical strength. This is particularly important for large-scale systems and for motors with grain diameter ratios greater than two. Various researchers have investigated the strength properties of paraffin wax per ASTM D638-03 [158], but the results vary widely, likely due to differences in wax formulation and the presence of strengthening additives, as well as the statistical nature of brittle failure and potential differences in pull rates. References [159–161,59] list

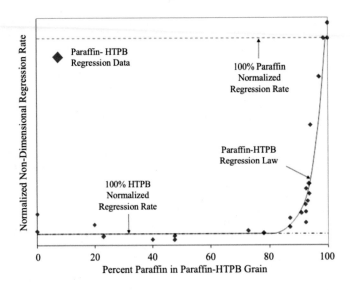

FIGURE 5.6   Regression Rate of Paraffin-HTPB Mixtures [87].

ultimate tensile strengths for pure wax between 0.8 and 3.0 MPa at ambient temperature with Ref. [59] determining a compressive strength between 2.4 and 3.3 MPa. The strength properties were observed to be strongly dependent on the strain rate and temperature, with elevated temperatures producing lower ultimate tensile strength but less strain rate dependence [59]. Additives such as Ethylene vinyl acetate (EVA) [158], [162], stearic acid [163], LDPE [158], maleic anhydride-grafted poly (styrene-ethylene/butadiene styrene) (SEBS-MA) [164], [165], and multi-walled carbon nanotubes (MWNT) [166] have been suggested to improve the mechanical properties.

Martin Chiaverini and Dan Gramer of OR-BITEC tried combining varying amounts of HTPB with paraffin to enhance its strength, postulating that the long-chain polymers would lend strength to paraffin at higher temperatures. While the HTPB did actually increase the strength of the paraffin, it significantly reduced the regression rate. Testing in a 220 N (50lbf) motor with gaseous oxygen showed that 5% HTPB by mass reduced the regression rate by about

50%. The addition of 15% HTPB by mass, caused the mixture to burn like pure HTPB (Fig. 5.6). They concluded that the HTPB, which melts at a higher temperature than the paraffin, inhibited the paraffin liquid layer and prevented it from forming droplets [87].

5.2.3 Summary of fuels

Useful design parameters for the fuels discussed in the previous sections are summarized in Table 5.2.

5.3 Oxidizers

Hybrid rocket motor oxidizer candidates are the same as those available for liquid bi-propellant engines. Luckily, the highest (known) performance oxidizers are liquids, leaving propulsion engineers with several good options. The most common oxidizers used in hybrid configurations are discussed here. Note that this is by no means a comprehensive list. Higher-performing

TABLE 5.2 Hybrid Fuel Summary. Note: * Value provided is for one mole of ABS with $x = 0.43$, $y = 0.5$, $z = 0.07$; i.e., $C_{3.85}H_{4.78}N_{0.43}$; ** Material is amorphous, temperature listed is for extrusion.

Fuel	Chemical Formula	ρ [kg/m^3]	h_f [kcal/mol]	T_{melt} [°C]	T_{glass} [°C]	Ref.
ABS	$(C_8H_8)_x(C_4H_6)_y(C_3H_3N)_z$	1057	−15*	230**	105	[144]
DCPD	$C_{10}H_{12}$	1010	52.36	33.65	170.1	[145]
HDPE	$(C_2H_4)_n$	960	−12.7	125-135		[145], [167]
HTPB	$(C_{73}OH_{103})_n$	919	13	241		[167], [168]
LDPE	$(C_2H_4)_n$	930	−12.7	106-111		[145], [167]
Paraffin	$C_{32}H_{66}$	924.5	−224.2			
PMMA	$(C_5O_2H_8)_n$	1180	−102.9	160	105	[82]

oxidizers exist, fluorines, for example; however, the disadvantages associated with these options (e.g., handling and stability) currently prohibit them from being used.

5.3.1 Oxygen

Oxygen is a very powerful oxidizer and is widely used in hybrid rocket propulsion. It is a comparatively safe oxidizer, though, as with all oxidizing agents, care needs to be taken when using it. Specific safety concerns and recommended testing protocols are discussed in Chapter 11. Oxygen is used as an oxidizer in both gaseous and liquid phases. Gaseous oxygen has slightly higher performance in terms of Isp but is hampered by its low density. It is used in small systems and many lab/university test facilities because the feed systems are relatively easy to set up. Hybrid rocket motors using gaseous oxygen also tend to be easy to stabilize and achieve high combustion efficiency.

Liquid oxygen presents a substantial improvement in storage density over gaseous oxygen; however, it is cryogenic and must be kept at low temperature or constantly vented. The boiling point of liquid oxygen is 90 K at atmospheric pressure [22]. For sounding rockets and launch vehicle applications, the increase in performance can outweigh the complication of requiring low-temperature storage and filling, and venting boil-off on the pad is a reasonable

endeavor. However, for in-space missions requiring long-term storability, the mass for insulation and power for cooling are often too high to make it a useful option. Research in zero boil-off storage for cryogenic propellants is on going, and this may change in the future. There is also continuing research into In-Situ Resource Utilization to generate oxygen from water/regolith or carbon dioxide on the moon or Mars, respectively [169–171]. Successful demonstration of this technology could make this oxidizer into a compelling propellant for in-space exploration.

Methods to increase the storage density of oxygen have been researched, including storing it in a solid form, as is done for emergency oxygen onboard aircraft [172]. However, so far, the extra tankage, energy to transform the oxygen into a useful form, and residuals associated with storing the oxygen as a solid have not made it a competitive option. There has also been substantial work performed on densifying liquid oxygen by supercooling the oxidizer below the usual operating temperature [173]. This technique is used by SpaceX on their Falcon 9 rockets and appears to be a part of their long-term exploration strategy [174].

5.3.2 Nitrous oxide

Nitrous oxide, N_2O, is a self "pressurizing oxidizer" that is also commonly used in hybrid propulsion. It is widely used in the university and research environment; however, precautions need to be taken to ensure safe utilization, see Chapter 11. The general public is most familiar with nitrous oxide through its use in medical or dental offices. The broad use of nitrous oxide makes it reasonably inexpensive and easy to access. It is, unfortunately, also a potent greenhouse gas. It has 310 times (per mole) the global warming potential of carbon dioxide over a 100-year time frame [175]. Despite this, it is generally considered a 'green' and nontoxic propellant, though it has some negative health repercussions, as discussed in Chapter 11.

Nitrous oxide is a colorless gas that is typically stored under pressure as a liquid. Nitrous oxide has high vapor pressure and so is often used in sounding rocket applications for its "self-pressurizing" ability (its high vapor pressure enables it to be used without a pressurant). However, high vapor pressure is generally undesirable for launch and in-space applications. The vapor pressure is dependent on the propellant's bulk temperature; therefore if the nonoperating temperature for a given application is high, then this will drive the tank design pressure. Modeling the blow-down behavior of nitrous oxide to predict the oxidizer mass flow rate over time is extremely complex with limited accuracy [176]. Testing at design conditions is recommended to characterize the oxidizer mass flow rate for a given tank and feed line design. Nitrous oxide has only moderate performance, see Table 5.5. It can be seen that the performance of hybrid rocket motors using nitrous oxide with a solid hydrocarbon fuel typically is optimum at high O/F ratios of around seven. Therefore the propulsion engineer should be aware that motors using this oxidizer will generally have oxidizer tanks that are significantly larger, approximately seven times the volume, than the combustion chamber [177].

An interesting benefit of nitrous oxide is that it decomposes exothermically, making it potentially suitable as a monopropellant when used with a catalyst [178–180]. However, monopropellant thrusters with nitrous oxide have not been flight proven. More research and development would need to be conducted before this oxidizer can be considered for dual use in flight.

Deflagration of nitrous oxide vapor is the primary safety concern associated with this oxidizer. The addition of diluents, such as nitrogen, helium, or oxygen, to nitrous oxide can greatly increase the ignition energy required to initiate decomposition [181] and enhance safety. The refrigerated combination of nitrous oxide with oxygen is known as Nytrox [182]. Nytrox combines the high vapor pressure of dissolved oxygen with the higher density of chilled nitrous oxide to produce a comparatively safe, nontoxic, self-pressurizing oxidizer [183]. Nytrox generally has high density and reasonable performance [182]. However, the Nytrox equilibrium will shift with changing temperature, resulting in variations in chemical composition and oxidizer tank pressure if the temperature is not tightly controlled. It is more common to supercharge the ullage with oxygen, which increases safety and enables a gas phase burn than to use a true Nytrox combination.

5.3.3 Hydrogen peroxide (HP)

Hydrogen peroxide (H_2O_2) is a commercially available chemical that is commonly used in low concentrations for household applications. It is a clear liquid miscible with water with a strong odor. H_2O_2 is typically used in an aqueous solution to improve safety and material compatibility. Typically concentrations above 87.5% are used for rocket applications. The density of HP increases with concentration [24]. The Hydrogen Peroxide Handbook is an excellent resource for this oxidizer [184].

Hydrogen peroxide is an interesting oxidizer because it can be dual use, meaning it can also be used as a monopropellant, e.g., it can be catalytically decomposed into water and oxygen with heat release. Hybrid propulsion systems utilizing H_2O_2 typically run the full oxidizer flow through a catalyst bed, see, for example, [185]. Therefore the fuel reacts with the hot by-products of the catalytic reaction: steam and gaseous oxygen. These hot gases ignite the motor and the catalytic reaction is maintained throughout combustion. Fuels, e.g., PE, are known to autoignite with the high-temperature decomposition products of HP [186]. Catalytic decomposition of HP and the reaction of the fuel with the by-products are not unique to hybrid and monopropellant applications. Some liquid systems also favor catalytic decomposition instead of direct injection [187].

There are several disadvantages to working with hydrogen peroxide. The first is that it can degrade over time (concentration decreases). Several studies at room temperature have shown degradation of up to 0.4% per year [188] and [184]. Higher temperatures lead to more rapid decomposition: 1% per year at 30 °C, 1% per week at 66 °C, 2% per day at 100 °C, and rapid decomposition at 140 °C [24]. However, storage at lower temperatures (5 °C) has resulted in negligible degradation [188]. The initial concentration, storage material and temperature affect the degradation rate. Modern hydrogen peroxide is suggested to be up to ten times more stable than that used to perform the material compatibility testing reported in the literature from the 1960s [188], [187]. Therefore it is hoped that the rate of degradation is noticeably lower than previously demonstrated. Updated testing *may* expand the list of compatible storage materials and decrease the degradation. Hydrogen peroxide also has some concerns when it comes to safety. Contaminants or additives, including water, can destabilize HP. This is discussed in Chapter 11.

5.3.4 Nitrogen tetroxide and mixed oxides of nitrogen

Nitrogen tetroxide, N_2O_4, is a storable oxidizer under Earth ambient conditions. Liquid N_2O_4 exists in equilibrium with gaseous NO_2.

Mixed Oxides of Nitrogen (MON) is a mixture of N_2O_4 and NO, where the designation after MON is the concentration of added NO, e.g., MON-25 is 75% N_2O_4 and 25% NO by mass. The addition of NO decreases the corrosive capability of the oxidizer and depresses its freezing point [23]. It is commonly used in liquid bipropellant engines because it is hypergolic with hydrazine derivatives. Recent research indicates a number of solid materials that also are hypergolic or reactive with N_2O_4 [146]. This will be discussed in Section 5.4.2.

One of the benefits of working with NTO or MON is that its extensive use in bipropellant engines means it is relatively well understood and documented for normal operating conditions. The USAF Propellant Handbook on NTO/MON [23] is an excellent reference for this oxidizer.[1] The handbook gives physical information about the oxidizer, as well as discusses compatible materials.

There are several disadvantages to working with MON. NO_2 is toxic and requires additional safety issues. When mixed with water, MON forms nitric acid. N_2O_4 has been shown to have limited compatibility with many sealing materials. Parker has developed a special material that has shown at least one year of useful life under exposure to N_2O_4. MON has been shown to have an energy deficit when used with paraffin, requiring additional energy to vaporize the liquid oxidizer and stabilize the system [48]. The relatively low vapor pressure and significantly higher enthalpy of vaporization ($h_{fg} = 414.5$ kJ/kg or approx 4x that of LOx) were cited as the reasons, and combustion with MON was harder to stabilize than previous work with LOx.

---

[1]Note that there is a typo in the equations for MON-10 in this reference.

5.3.5 Nitric acid

Nitric acid, HNO_3, was probably the most popular oxidizer for bipropellant engines from the 1950s through about 1970s when MON gained favor. While it is still used occasionally for both liquids and hybrids, it has many complications. It can be highly corrosive and toxic. It also reacts with many organic materials. However, nitric acid is hypergolic with a number of liquids and some solid additives. This will be discussed in Section 5.4.2. It has several benefits, including high density and relatively low vapor pressure. The USAF Propellant Handbook on Nitric Acid [23] is an excellent reference for this oxidizer.

Over the years, several types of nitric acid have been used in rocket propulsion, including anhydrous, White Fuming Nitric Acid (WFNA), Red Fuming Nitric Acid (RFNA), and Inhibited Red Fuming Nitric Acid (IRFNA), in order of purity. The concentrations of the main constituents determine the classification of the nitric acid. The red coloring in RFNA and IRFNA is caused by dissolved, toxic NO_2 in the mixture (about 14% by mass). Hydrofluoric acid is added to inhibit corrosion. IRFNA is the most commonly used variant today and the only one with a current MIL-spec. WFNA is unstable and was found to decompose [189].

5.3.6 Summary of oxidizers

Relevant information on the oxidizers discussed in the previous sections is compiled in Table 5.3. The values presented are under standard conditions. The National Institute of Standards and Technology Chemical Webbook [22] is an excellent reference for determining common properties such as density under nonstandard conditions.

TABLE 5.3 Common hybrid oxidizer values at standard conditions. *At 25 °C and one atmosphere except oxygen, which is at −183 °C (liquid). † At one atmosphere.

Oxidizer	Chemical Formula	Density [kg/m³]*	Heat of formation [kcal/mol]	Melting Point [°C]	Normal Boiling Point [°C]	Vapor Pressure [Pa]	Dynamic Viscosity* [cP]	Reference
Oxygen	O_2	1142	−3.08 at −183 °C	−218.9	−183	1.01×10^5 (LOx)	1.9 (LOx)	[190], [2]
Nitrous Oxide	N_2O	1226	15.5		−88	5×10^6		[6]
Hydrogen Peroxide (98-100%)	H_2O_2	1443	−44.84	−0.5	150.2	103	1.1	[190], [24], [2]
Hydrogen Peroxide (90%)	$0.9H_2O_2 + 0.1H_2O$	1388	−48.75	−11.5	145.1	180	1.24	[190], [24]
Nitric Acid (RFNA)	HNO_3	1550	−37.7	−52	64.2	1.85×10^4	1.231	[23]
Nitrogen Tetroxide/MON-3	N_2O_4	1433	−4.676	−11.2	21.2	1.2×10^5	0.396	[23], [2]
MON-25	$0.75N_2O_4 + 0.25NO$	1391			−9	1.2×10^5		[23]
Ammonium Perchlorate	NH_4ClO_4	1960	−69.42	decomposes				[190]

5.4 Additives

Solid fuel grains make using performance or stability-enhancing additives fairly straightforward. Metal particles can be added to increase performance (e.g., I_{sp} and fuel density), decrease the optimal O/F ratio, and/or enhance stability. Hypergolic additives can be mixed into the solid fuel grain to initiate combustion without a separate igniter, and reactive additives can modify the performance and stabilize the combustion (e.g., add heat to the system to more completely vaporize the oxidizer). However, hypergolic additives generally decrease performance over neat fuel, so their system benefits should be carefully accessed. Also, additives, particularly metals, can decrease the strength of the fuel. Therefore their concentration may be limited for structural integrity reasons.

Particle size matters for additives. The burning time scales with the diameter of the particle squared. Also, the heat transfer depends on the surface area of the particle. However, increased surface area can also mean an increased oxide layer. Finally, small particles increase the viscosity of the mixture, which can make producing fuel grains more difficult.

5.4.1 Aluminum

Aluminum is considered one of the best additives because it has good thermochemical properties and low cost [168]. Adding aluminum particles decreases the ideal O/F ratio while increasing the fuel density. This combination can help with packaging constraints. Adding aluminum also serves to increase the regression rate. This can lead to a further reduction in fuel grain length. Refs. [191] and [192] showed that nanosized aluminum particles increased the regression rate of HTPB and paraffin, the latter by about 60%.

The particles also serve to dampen instabilities during combustion. Aluminum will typically result in a modest increase in performance

on paper (several seconds of I_{sp}). However, it is difficult to realize this increase in practice. Aluminum oxidizes readily, and one has to ensure that the particles have not been exposed to air during production or grain manufacture. Once incorporated into the motor, it is quite difficult to combust. The melting temperature of aluminum oxide is about 2050 °C [189], and it decomposes well above that temperature. Instead of combusting, a protective liquid layer often forms around the aluminum particle, turning it into slag, which then agglomerates in the post combustion chamber and/or the nozzle. In most cases, aluminum addition is better for stability and packaging than performance. Other metals can be considered for similar reasons but are less common.

5.4.2 Hypergols

Solid hypergolic additives are an area of active research as a potential simplification for ignition (and especially of multiple ignitions) of hybrid rockets. They can be mixed into the solid fuel grain, negating the requirement for an separate ignition system. Solid hypergolic additives have been demonstrated to combust on contact with Hydrogen Peroxide, Nitric Acid, and MON oxidizers. For example, metal hydrides and amines have shown hypergolic behavior with all three oxidizers [193].

There are several important parameters to evaluate when considering a hypergolic additive for ignition. The first is ignition delay, which is defined as the time from first contact to ignition (flame). If the ignition delay is too long, the combustion chamber could be flooded with oxidizer before the reaction gets a chance to begin. The second parameter to consider is performance. Many hypergolic additives decrease specific impulse, though some can make up for this by increasing density. Hypergolic ignition is dependent on many variables, including pressure, temperature, particle size, binder, and synthesis method/supplier. Therefore the operating and storage conditions of the system must be considered. Additionally, this variation makes it difficult to apply existing data to a new design or data from a similar compound to another. Finally, the integration of the additive with the fuel binder needs to be understood. For example, the following questions (at a minimum) should be posed. Does it alter the mechanical properties? Is the additive compatible with the grain manufacturing process? Does the binder inhibit the reaction? Can the final product be handled safely?

Ignition delay is a crucial parameter in determining additive selection. Some of these additives, while hypergolic, have an unacceptably long ignition delay of 100-1000 of ms. For comparison, the common hypergolic bipropellant combination MMH/NTO has an ignition delay of about 2 ms. Delays of more than about 10 ms in liquid bipropellant engines can cause hard starts and result in a loss of hardware. Some solids have been found to have similarly short ignition delays while in loose powder form. However, the ignition delay depends on the temperature and pressure. Mixing the additive with a binder can increase the ignition delay substantially (an order of magnitude increase has been demonstrated) or completely inhibit the reaction. Paraffin is especially good at coating particles and keeping them from reacting with the oxidizer [194], though workarounds have been found, such as grinding the paraffin and pressing it with the additive [146]. Additionally, some of the most promising additives are incompatible with common binders, such as sodium amide with HTPB. It is recommended that any incompatibility testing be done with low additive concentrations, as high concentrations can often obscure the results [195].

A growing number of materials have been identified as hypergolic with common hybrid oxidizers: NTO/MON, nitric acid, and hydrogen peroxide. Table 5.4 lists many of the more promising options. Simple drop tests can be conducted to determine the ignition delay. A droplet of oxidizer is released above the additive, most commonly in powder form. These tests give a good indication of the relative reactivity with the given oxidizer. Fig. 5.7 shows an example of one of these tests [196]. However, as discussed previously, mixing the additives with the fuel (binder) can dramatically alter the ignition delay. For example, [193] found that the best case ignition delay of ethylenediamine bisborane ($C_2H_14B_2N_2$) with WFNA increased by order of magnitude, from 2.9 ± 0.3 ms to 31.7 ± 19.6 ms, when mixed with 42% HTPB. Sodium borohydride ($NaBH_4$) has been found to be hypergolic with 90% peroxide with an ignition delay of 4 ms for the pure additive and 9.5 ms for 25% $NaBH_4$ mixed in 75% PE [197]. The ignition times in real motors are expected to be substantially longer than loose powder drop tests (most commonly reported in literature) and are dependent both on the chemistry and the motor injector design.

Several classes of solid additives have been identified that are hypergolic with nitric acid, including boron-based [193], formaldehydes [198], and at least one metal hydride ($LiAlH_4$) [199].

Focused research by Purdue and Penn State identified many solid additives as hypergolic with NTO or MON. In [146], it was that found lithium amide (an inorganic amide), lithium, sodium and potassium bis(trimethylsilyl)amide (silyl amides) and ethylenediamine-bisborane (a boron compound) to be hypergolic with NTO and MON-25 across all the conditions tested: $-20\,°C$ to ambient temperature and pressures from atmospheric to 550 kPa (80 psia). The ignition delays of solid additives increase as the concentration of NO in the MON increases. An increase in ignition delay of up to 25 times was reported between NTO and MON-25 (0.3 ms to 7.4 ms for Lithium bis(trimethylsilyl)amide) [146]. Pressure also has a dramatic effect on ignition delay. Increasing pressure to 550 kPa (80 psia) for drop tests caused the decrease in ignition delay of up to 3x with NTO at 26 °C and up to 19x at $-20\,°C$ with MON-25 [146].

TABLE 5.4 Ignition delay of solid hypergolic additives. Only additives with < 500 ms delay included. The ignition delay ranged from 0.5 ms to 476.5 ms. Data is included for different pressure and temperatures: a) P, T atmospheric, b) $P = 0.55$ MPa (80psia) and $T = 26\,°C$, c) $P =$ atmospheric, $T = -10\,°C$, d) $P = 0.21$ MPa (30psia), $T = -20\,°C$, e) $P = 0.34$ MPa (50psia), $T = -20\,°C$, f) $P = 0.55$ MPa (80psia), $T = -20\,°C$, g) $P = 0.55$ MPa (80psia), $T = -25\,°C$.

Chemical	NTO/MON-3 [ms]	MON-25 [ms]	Nitric Acid [ms]	HP [ms]
Sodium amide	86.6[a] [146], < 0.5[c] [194]	4.8[g] [146]		
Lithium amide	6.4[b] [146], 100[c] [194]	126.9[d], 61.5[e], 43.6[f], 138[g] [146]		
Sodium borohydride				[197]
Sodium cyanoborohydride	12[c] [194], [194]			
Sodium bis(trimethylsilyl) amide	89[a], < 0.5[b] [146]	[146]		
Potasium bis(trimethylsilyl) amide	7.3[a], 0.3[b] [146]	3.7[d], 1.4[e], 1.1[f], 1.1[g] [146]		
Lithium bis(trimethylsilyl) amide	71.4[a], 0.3[b] [146]	15.9[d], 13.3[e], 3.1[f], 7.4[g] [146]		
Lithium Aluminum Hydride			[199]	
Azulene	7.9[b] [146]	45.2[g] [146]		
Borane trimethylamine complex	2.2[b] [146], 34[c] [194]	173.5[d], 82.7[e], 9[f], 9.3[g] [146]		
Borane tert-butylamine complex	2.2[b] [146], 363.5[c] [194]	6.2[g] [146]		
Ammonia borane	476.5[c] [194]		0.73 − 22.2[a] [195]	
Ethylenediamine bisborane	91.4[b] [146]		2.9[a] [193]	
Tagaform			150[a] [193]	
Sagaform A			5 [193]	
p-toludine/p-aminophenol			110[a] [193]	
Difurfurylidene cyclohexanone			45 − 255 [193]	
Triaminoguanidinium azotetrazolate (TAGzT)	46.75[c] [194]			
Formaldehydes (aniline, o- and m-toluidine and o-anisidine)			[198]	

FIGURE 5.7 Hypergolic reaction of PBTSA with MON [196].

There are several drawbacks to hypergolic additives. As stated earlier, many decrease performance, so the concentration needs to be minimized or the additives can be restricted to smaller volumes. They often have handling and material compatibility restrictions. Many additives are moisture sensitive, and a number are incompatible with common binders. For example, sodium amide, sodium borohydride, and potassium bis(trimethylsilyl)amide are reduc-

TABLE 5.5 Theoretical Performance Values for Common Hybrid Propellants. Results shown are for combustion chamber conditions optimized for I_{sp} and C* using CEA [74]. The results assume equilibrium combustion with fuel properties, per Table 5.2.

Ox [-]	Fuel [-]	$I_{sp,max}$					C^*_{max}				
		$I_{sp,vac}$ [s]	O/F [-]	γ [-]	T_{ad} [K]	X_{H_2O} [%]	C^* [m/s]	O/F [-]	γ [-]	T_{ad} [K]	X_{H_2O} [%]
GOx	ABS	345	2.2	1.13	3557	22	1730	1.8	1.15	3446	21
	HDPE	359	2.7	1.13	3538	31	1812	2.2	1.14	3418	29
	HTPB	357	2.3	1.13	3613	23	1801	1.9	1.15	3497	21
	Paraffin	360	2.7	1.13	3528	31	1813	2.2	1.15	3399	30
	PMMA	336	1.6	1.12	3409	30	1673	1.3	1.14	3314	31
LOx	ABS	341	2.3	1.13	3526	23	1711	1.9	1.14	3448	22
	HDPE	355	2.7	1.13	3501	32	1791	2.2	1.15	3368	30
	HTPB	353	2.4	1.13	3586	24	1780	1.9	1.15	3447	22
	Paraffin	355	2.7	1.13	3492	32	1792	2.2	1.15	3348	30
	PMMA	332	1.6	1.12	3375	31	1655	1.4	1.13	3327	32
H_2O_2	ABS	322	5.9	1.12	2891	67	1631	4.9	1.13	2894	64
	HDPE	328	7.1	1.12	2893	71	1669	5.8	1.13	2885	68
	HTPB	329	6.5	1.12	2939	68	1666	5.2	1.13	2939	64
	Paraffin	328	7.2	1.12	2887	72	1669	5.9	1.13	2884	68
	PMMA	318	4	1.12	2843	69	1608	3.5	1.13	2848	67
N_2O	ABS	308	7.5	1.14	3283	10	1580	5.6	1.16	3275	12
	HDPE	313	9	1.14	3276	15	1613	6.7	1.16	3252	16
	HTPB	313	8.2	1.14	3330	11	1614	5.9	1.16	3318	12
	Paraffin	313	9.1	1.14	3271	15	1613	6.7	1.16	3242	17
	PMMA	305	5.2	1.14	3188	15	1556	4	1.15	3175	17
N_2O_4	ABS	318	3.6	1.13	3330	19	1606	2.9	1.14	3276	20
	HDPE	327	4.4	1.13	3317	27	1663	3.5	1.14	3249	27
	HTPB	327	3.9	1.13	3391	21	1660	3	1.15	3312	21
	Paraffin	327	4.4	1.13	3311	28	1663	3.5	1.15	3234	28
	PMMA	312	2.6	1.12	3205	26	1571	2.1	1.13	3169	27

ing agents and are incompatible with chemicals required for the curing and casting of HTPB, a common hybrid fuel [146]. However, some additives, such as ethylenediamine bisborane, have been successfully combined with HTPB, even though they are suggested to be incompatible [193]. This appears possible when the additives are in high concentrations [195], but long-term compatibility testing has yet to be performed. The material properties of the new mixture need to be well understood before they can be used for a space application.

Reactive additives can also be of interest. There are materials that do not have the hyper-

golic capability sought after for ignition. However, they are usually of similar or superior performance to the selected propellants. They can add heat to the motor, which can be stabilizing, especially at the head end. Increased I_{sp} of up to 17% has been reported for more exotic additives [200]. They may also be used to increase the regression rate if desired. The change in reactivity is highly dependent on the additive. The heat generated by the additives could also play a role in decreasing ignition delay.

A side note on the use of oxygen with reactive additives: there are many pyrophoric materials, which ignite on contact with oxygen. Examples

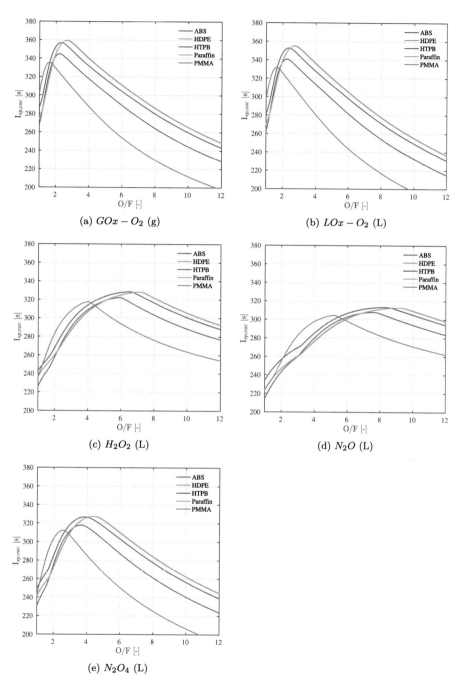

FIGURE 5.8 Ideal vacuum specific impulse versus O/F ratio of common hybrid propellant combinations. Note that the lines for paraffin and HDPE are nearly identical. The results shown assume equilibrium combustion with fuel properties, per Table 5.2. They are determined using CEA [74].

include lithium aluminum hydride ($LiAlH_4$) and liquid triethylaluminum/triethylborane (TEA/TEB). While, pyrophoric additives make ignition with oxygen straightforward, dealing with an additive that combusts on contact with oxygen (or air) makes ground handling significantly more challenging. Also, one of the major combustion products of TEA and oxygen is aluminum oxide, which has been known to clog small orifices or joints (e.g., in a gimballed nozzle) [2].

Hypergolic additives have great potential for the right applications. However, none are operating in larger than lab-scale test benches at this time. Recent small-scale testing has reported successful multiple ignitions, as well as vacuum ignitions, demonstrating that great progress is being made [146]. Stability and toxicity of the additives are the most challenging obstacles. Required concentration of the hypergolic additive is very important since many decrease the I_{sp} of the propellant combination.

5.5 Theoretical performance of common combinations

A one-dimensional chemical equilibrium calculation is a way to rapidly screen for perfor-

mance. Chemical equilibrium calculators evaluate the performance of a given propellant combination by determining the combustion products, which minimize the Gibbs free energy. This is discussed in more detail in Chapter 4. Table 5.5 and Fig. 5.8 show the results using Chemical Equilibrium with Applications (CEA) [74]. It is important to understand that one-dimensional codes assume the reactions are occurring sufficiently fast and will not account for nozzle losses, which can be significant in hybrid rocket combustion.

All the performance data is calculated using the following assumptions:

- Equilibrium
- Chamber pressure: 2.07×10^3 kPa (300 psi)
- Propellant temperature*: 293.15 K
- Nozzle area ratio: 40:1
- Fuel properties per Table 5.2

Note: * Liquid oxygen is cryogenic, so the liquid oxygen performance data is calculated with an oxidizer temperature of 90 K.

Hybrid regression rate data

6.1 Introduction

A key challenge associated with reconstructing hybrid rocket motor performance is evaluating the fuel mass flow rate over time during a burn. This chapter presents several options for regression rate reconstruction so that the fuel mass flow rate versus time can be determined for a given hot-fire test. The first approach makes use of the functional form of the equations in Chapter 3, whereas the other two approaches rely instead on test data and assumptions about C^* or η_{C^*}. The three approaches are described so the propulsion engineer can implement them if desired. A discussion of where they are most applicable is also included.

Of the three reconstruction approaches presented, only the first approach can be applied to predict the performance of a new hybrid motor design. While the regression rate equations of Chapter 3 are based on physics, the most practically used forms of these equations rely on empirical data to fill in gaps. This approach, which uses a functional form of the regression rate law derived in Chapter 3, collapses the variables dependent on the propellant combination into two empirical parameters: a_o and n. These parameters depend on the propellant combination, and typical values will be presented in Table 6.2. These values are provided here to enable the propulsion engineer to start the design process for common fuel and oxidizer combinations.

6.2 Regression rate reconstruction methods

This section is based on the work presented in Ref. [105]. The authors greatly appreciate the efforts of the co-authors of that presentation as well as the input of the AIAA Hybrid Rocket Technical Committee.

Three methods are presented as options for reconstructing the regression rate during a test. The benefits and challenges are then discussed for each. The three methods all have recommended reference(s) associated with them if the propulsion engineer feels that they need more information:

1. Regression rate law with empirical constants: Refs. [98] and [82].
2. Constant C^*: Ref. [201].
3. Constant η_{C^*}: Ref. [202].

6.2.1 Regression rate law with empirical constants

Perhaps the most common method used to predict the fuel regression rate, and therefore the fuel mass flow rate, in hybrid rockets is to apply the functional forms of Marxman's law, discussed in Chapter 3 and included here again for completeness as Eqs. (6.1) and (6.2).

$$\dot{r} = aG^n x^{-m} \qquad (6.1)$$

$$\dot{r} = a_o G_o^n \qquad (6.2)$$

This method relies upon having a previous test data set for the propellant combination under test or an analytical estimate of the regression rate coefficients. See Ref. [82] for an approach to estimate the coefficients analytically as well as a discussion of the limited accuracy of such an estimate. There is a risk that error can be introduced by the regression rate coefficients used to reconstruct the test. This is a concern even with coefficients derived from test data if the test configuration under consideration is significantly different from the test configuration(s) used to calculate the coefficients (e.g., different injection schemes, operating range of mass fluxes, scale, etc.).

The length exponent, m, of Eq. (6.1), is typically small and is challenging to measure accurately due to the need for a number of tests at a range of length scales. Thus, as discussed in Chapter 3, the value of m is often set to zero in practice. Further, for simplicity, the oxidizer mass flux, G_o, is often used to predict the fuel regression rate, via Eq. (6.2), rather than the total propellant mass flux, G, via Eq. (6.1). Using Eq. (6.2) can cause issues with accuracy if the test is operated over a large range of O/F, or when the O/F is low. This is discussed more in Section 3.4 of Chapter 3. However, since Eq. (6.2) is the form most often used in practice, it will be applied for the remainder of this section and will be used for the design process described in Chapter 7.

Applying the regression rate law to a given test to reconstruct fuel mass flow rate over time requires a different set of information for the test than the information that must be used to evaluate the regression rate coefficients for future tests. The following information is required to determine the regression rate versus time for a given test:

- Density of the fuel, ρ_F
- Oxidizer mass flow rate over time, $\dot{m}_O(t)$
- Initial fuel grain geometry (cross section and length, L_F) and expected shape evolution, for r and G_o reconstruction
- Burn time, t_b, or final port radius, r_f
- Regression rate coefficients for the propellant combination under the test: a or a_o, n, and m (if applicable)

A table of reported regression rate coefficients for a number of common fuel and oxidizers is provided in Section 6.5.

The fuel regression rate at any point in time is given by Eq. (6.2). The fuel mass flow rate can then be calculated via Eq. (6.3).

$$\dot{m}_F = A_b \rho_F \dot{r} \qquad (6.3)$$

$$\dot{m}_F = A_b \rho_F a_o G_o^n \qquad (6.4)$$

For a hybrid fuel grain with a single cylindrical port, Eq. (6.3) becomes Eq. (6.5).

$$\dot{m}_F = 2\rho_F L_F \pi^{1-n} r^{1-2n} a_o \dot{m}_O^n \qquad (6.5)$$

A single circular port fuel grain is used for the equations presented in this section, though they can readily be altered to apply to other fuel cross sections, including multiple port fuel grains.

The algorithm for estimating the fuel mass flow rate with this method is straightforward. You start by reading the oxidizer mass flow rate, $\dot{m}_O(t)$, for the test. Then the forward Euler method can be used to determine the regression rate.

Let each j time step be time Δt apart. The regression rate at each time step, $\dot{r}_j = \dot{r}(t_j)$, is given by Eq. (6.2), and the fuel mass flow rate can then be calculated via Eq. (6.5). The fuel port radius at the next time step, r_{j+1}, is then calculated via Eq. (6.6), and the time step is incremented. This is continued until the burn time is reached $t_j = t_b$, or alternatively, until the fuel port radius is calculated to be equal to or larger than the final fuel port radius at the end of the burn, $r_j = r_f$.

$$r_{j+1} = r_j + \dot{r}_j \Delta t \qquad (6.6)$$

A drawback with this method is that any inaccuracy in the regression rate coefficients used to evaluate the fuel mass flow rate will directly feed error into the results that can be compounded over time. This can be seen clearly as overall error with a mismatch in either the total mass of fuel burned, m_F (calculated fuel mass burned does not match measured mass burned), or in the burn time, t_b (the calculated burn time does not equal to the time measured in the test, though, per the discussion in Section 6.3.7, there can be ambiguity in measuring burn time). This is not a concern if the regression rate coefficients are accurate for the test configuration being considered.

The test data that needs to be collected to evaluate the regression rate coefficients, a/a_o, n, and m are:

- Density of the fuel, ρ_F
- Oxidizer mass flow rate over time, $\dot{m}_O(t)$
- Initial and final fuel mass, $m_{F,i}$ and $m_{F,f}$, or Initial and final fuel port radius, r_i and r_f
- Fuel grain geometry and expected shape evolution, for r and G_o reconstruction
- Burn time, t_b

Generally, data from multiple tests is needed to accurately evaluate the regression rate coefficients. The most common method used to evaluate the coefficients is a space-time averaged approach, as discussed in the following subsection. Alternative approaches have also been suggested, some of which utilize additional test data, such as chamber pressure trends [9], to avoid the need for multiple tests to evaluate the coefficients. It should be noted that given a set of test data, the regression rate coefficients can be selected to minimize the error in the average fuel regression rate at a given oxidizer mass flux (section below and Ref. [98]), minimize error in the predicted fuel burn time (see Ref. [9]) or minimize error in the mass of fuel burned [203]. The regression rate coefficients selected with each of these processes can be quite different, see Ref. [9], and will influence whether the coefficients can be used to extrapolate beyond the range of test data or should only be applied to tests with burn times, fluxes and scales within the test data range. It should also be noted that the local minimum, which determines the regression rate exponent, n, is typically very shallow, and that small variations in n can cause large inaccuracies in predicted test parameters, particularly when extrapolating to longer tests, different mass fluxes, and/or different test scales. In general, the space-time averaged method described below is accurate for predicting the regression rates of tests that fall within the range of the data set used to generate a_o and n.

One additional point to note about this technique is that if the exponent is assumed (e.g., $n = 0.5$ for N_2O), it is possible to obtain the a_o coefficient and therefore the regression rate law from a single test. This can be useful when making test-to-test changes (e.g., changing injectors or mixing devices). This is commonly done, as the oxidizer is thought to drive the exponent n.

Recommended use: If you have limited information and do not need to extrapolate beyond test conditions.

Space time averaged

The simplest method for determining the regression rate coefficients is an empirical space-time averaged approach. An average fuel regression rate and oxidizer mass flux is recorded for each test, and then the regression rate coefficients a_o and n that best fit this data (i.e., that minimize the error in the regression rate as a function of oxidizer mass flux) are selected.

Either the mass or the geometry of the fuel grain before and after the test must be known, along with the density of the fuel. Endpoint averaging uses the initial and final geometries combined with the burn time to give an average regression rate, see Eq. (6.7). The final fuel port radius, r_f, can either be measured (see discussion in Section 6.3.2) or can be calculated from the mass of fuel burned using Eq. (6.8).

$$\bar{\dot{r}} = \frac{r_f - r_i}{t_b} \qquad (6.7)$$

$$r_f = \sqrt{r_i^2 + \frac{(m_{F,i} - m_{F,f})}{\rho_F \pi L_F}} \qquad (6.8)$$

Where r_i and r_f are the initial and final radii of the fuel grain port, and $m_{F,i}$ and $m_{F,f}$ are the initial and final masses of the fuel.

The nonlinear evolution of oxidizer mass flux during the burn leads to a number of possible averaging approaches that can be adopted to calculate the average oxidizer mass flux, see discussion in Ref. [98]. Ref. [98] found that the most accurate averaging approach is generally to calculate the average mass flux using the average fuel port diameter, per Eq. (6.9).

$$\bar{G}_o = \frac{4 \bar{\dot{m}}_O}{\pi \left(r_i + r_f\right)^2} \qquad (6.9)$$

A major benefit of this method is its relative simplicity. To use it to collect meaningful ballistic test data, the propulsion engineer only needs:

- Mass of the fuel grain before and after each test ($m_{F,i}$ and $m_{F,f}$)
- Fuel density (ρ_F)
- Fuel grain geometry (or length, L_F, for a center perforated grain)
- Burn time (t_b).

This last item, the burn time, is typically the most difficult to obtain accurately. As will be discussed in Section 6.3.7, the definition of burn time can vary widely between researchers. Further, the injector and igniter effects are often accounted for differently between research groups. It can also be important to understand how inerts are treated, i.e., whether the mass of

burned insulation is included in the final mass. The discussion in Section 6.3 of this chapter and in Section 11.7.1 of Chapter 11 is included to provide the propulsion engineer with guidance on how to best report test data in order for it to be most useful to the hybrid rocket community.

6.2.2 Constant C^*

The second option for determining the regression rate during a burn does not use the regression rate coefficients at all, but instead assumes constant C^* during the burn, see Ref. [201]. This is a major assumption that is never completely true. However, steady-state burns with regression rate exponents, n, near 0.5, and a small variation in oxidizer mass flow rate, $\dot{m}_O$, can approach this approximation. The benefits of this method are its relatively simple, explicit approach and requirements of only a few inputs to determine the regression rate versus time.

Summarizing, the following inputs are required for this constant C^* approach:

- Oxidizer mass flow rate over time, $\dot{m}_O(t)$
- Combustion chamber pressure over time, $P_c(t)$
- Nozzle throat area (over time, if applicable), $A_{th}(t)$
- Burned fuel mass, $m_F = m_{F,i} - m_{F,f}$
- Initial fuel grain geometry and expected shape evolution, for r and G_o reconstruction.

The oxidizer mass flow rate and chamber pressure are commonly measured in propulsion testing: the former using a flow meter or calibrated venturi (see Section 11.6.5 of Chapter 11) and the latter using a pressure transducer (see Section 11.6.5 of Chapter 11). This method is the most accurate when there is not a substantial amount of nozzle erosion (e.g., for short burns, cooled nozzles, or ablating nozzles with large heat sinks).

From this point, a relatively simple routine can be written to solve for the fuel mass flow rate (see Ref. [201] for further details). It is summarized here.

First, the average, C^*, $\bar{C}^*$, is reconstructed for the test via Eq. (6.10) using posttest data. This is an approximation that relies on total propellant masses, m_O, m_F, and m_{inert}. Here, m_{inert} is the total inert mass consumed during the test.

$$\bar{C}^* = \frac{\int_0^{t_b} P_c A_{th} dt}{m_O + m_F + m_{inert}} \quad (6.10)$$

A progressive time step is used to iterate through the burn, with the time between each time step (t_j to t_{j+1}) again equal to Δt. The chamber pressure as a function of time, $P_c(t)$, and oxidizer mass flow rate, $\dot{m}_O(t)$, are inputs (data collected during a test), so these values are known at all time steps. Then, at each j time step, the propulsion engineer should:

Calculate the fuel mass flow rate, $\dot{m}_{F,j}$, at time t_j using Eq. (6.11).

$$\dot{m}_{F,j} = \frac{P_{c,j} A_{th,j}}{\bar{C}^*} - \dot{m}_{O,j} - \bar{\dot{m}}_{inert} \quad (6.11)$$

If desired, the oxidizer-to-fuel ratio, O/F_j, total propellant flux, G_j, and oxidizer mass flux, $G_{o,j}$, at time t_j can be calculated using Eqs. (6.12), (6.13) and (6.14), respectively. These values are helpful for later analysis of the data but are not required to predict fuel mass flow rate and regression rate using this method.

$$O/F_j = \frac{\dot{m}_{O,j}}{\dot{m}_{F,j}} \quad (6.12)$$

$$G_j = \frac{(\dot{m}_{O,j} + \dot{m}_{F,j})}{\pi r_j^2} \quad (6.13)$$

$$G_{o,j} = \frac{\dot{m}_{O,j}}{\pi r_j^2} \quad (6.14)$$

Calculate the fuel port regression rate, $\dot{r}_j$, at time t_j using Eq. (6.15).

$$\dot{r}_j = \frac{\dot{m}_{F,j}}{2\pi \rho_F r_j L_F} \quad (6.15)$$

Calculate the fuel port radius at the following time step, r_{j+1}, using the forward Euler method, Eq. (6.16).

$$r_{j+1} = r_j + \dot{r}_j \Delta t \quad (6.16)$$

This is then repeated until the end of the burn (all of the fuel mass is consumed).

Note that the algorithm described here assumes constant properties within a time step and uses the initial properties at that time step to calculate mass fluxes, etc. This is reasonable so long as the time steps are sufficiently small. An alternative approach can calculate the properties using the average radius across the time step $\left(\frac{r_j + r_{j+1}}{2}\right)$.

Recommended use: Steady-state burns with small variation in oxidizer mass flow rate and with propellants that have n exponent close to 0.5.

6.2.3 Constant C^* efficiency

Another option for determining the regression rate is to assume a constant C^* efficiency, η_{C^*}, see [202]. This assumption is typically correct for short periods of time and is supported by some research, see for example, [204]. However, other data suggests the efficiency does change over the burn [131]. The assumption of constant combustion efficiency can be especially problematic during transient events, such as ignition and shutdown. However, the constant η_{C^*} method is good for capturing time varying phenomena during the burn, such as throttling and O/F shift.

For this method, the propulsion engineer needs to know all of the information used by the previous method, as well as a measurement of axial thrust, $F(t)$, during the burn. The inputs required are therefore:

• Oxidizer mass flow rate over time, $\dot{m}_O(t)$
• Combustion chamber pressure over time, $P_c(t)$

- Nozzle throat area (over time, if applicable), $A_{th}(t)$
- Burned fuel mass, m_F
- Initial fuel grain geometry and expected shape evolution, for r and G_o reconstruction
- Thrust over time, $F(t)$, for I_{sp} reconstruction

All of these inputs are standard measurements that can be taken during a test, with varying degrees of difficulty. The grain evolution can be difficult to predict without prior testing or Computational Fluid Dynamics (CFD) simulations, but it is a common challenge across all three reconstruction methods discussed here.

The outputs include: the fuel mass flow rate over time, $\dot{m}_F(t)$, the regression rate over time, $\dot{r}(t)$, the oxidizer mass flux, $G_o(t)$, and finally, the specific impulse over time, $I_{sp}(t)$, based on measured thrust, $F(t)$.

To implement this method, a tolerance on convergence for fuel mass flow rate, $tol_{\dot{m}_F}$, must be selected. Ref. [202] uses $5 \times 10^{-5}/\Delta t$. Next, an initial assumption for the C^* efficiency, η_{C^*}, must be made. Then an initial condition is selected for the oxidizer-to-fuel ratio O/F_i. The average value of O/F is a good initial guess: $O/F_i = \bar{O/F} = \frac{m_O}{m_F}$. Calculations are run over the burn duration to converge on O/F. If the tolerance on total fuel mass burned is not satisfied, η_{C^*} is changed, and $O/F(t)$ is iterated on with the new η_{C^*} value.

The measured combustion chamber pressure data as a function of time, $P_c(t)$, and oxidizer mass flow rate, $\dot{m}_O(t)$, are inputs (known test data). Iterations over two variables will be completed to meet the selected convergence criteria. The first iteration, over j, is used to step forward in time. The second iteration variable used here will be referred to as k and is only used if the convergence criteria (tolerance on $\dot{m}_F$) are not met. k resets to 1 at each new timestep, $j + 1$.

The fuel mass flow rate, $\dot{m}_{F,j,k}$, is calculated using Eq. (6.17).

$$\dot{m}_{F,j,k} = \frac{\dot{m}_{O,j}}{O/F_{j,k}} \qquad (6.17)$$

Next, $C^*_{j,k}$ is calculated as a function of $O/F_{j,k}$ and $P_{c,j}$ using a chemical equilibrium solver such as CEA [74]. Then the new mass flow rate, consistent with this C^*, can be evaluated according to Eq. (6.18).

$$\dot{m}_{F,j,k+1} = \frac{P_{c,j} A_{th,j}}{\eta_{C^*} C^*_{j,k}} - \dot{m}_{O,j} \qquad (6.18)$$

Convergence on the mass flow rate can be checked by plugging the outputs of Eqs. (6.17) and (6.18) into Eq. (6.19).

$$|\dot{m}_{F,j,k} - \dot{m}_{F,j,k+1}| \le tol_{\dot{m}_F} \qquad (6.19)$$

If this condition is not satisfied, then k is iterated on using Eq. (6.20) for $O/F_{j,k+1}$. The previous two steps, from Eq. (6.17), need to be repeated using the new O/F ratio. Move on to the next step if the change in fuel mass flow rate is less than the tolerance on the fuel mass flow rate. Eq. (6.20) states that the O/F at the next step is the O/F at the previous time step plus half of the difference between the previous O/F and the one that was just estimated.

$$O/F_{j,k+1} = O/F_{j,k} + \frac{1}{2}\left(\frac{\dot{m}_{O,j}}{\dot{m}_{F,j,k+1}} - O/F_{j,k}\right) \qquad (6.20)$$

Once $\dot{m}_{F,j}$ is converged, the values of O/F_j, and C^*_j are also known from the last k iteration on fuel mass flow rate. The fuel regression rate can then be calculated from the known port geometry and the fuel mass flow rate. The oxidizer and total mass flux can also be determined. Ref. [202] has an alternative numerical approach for updating the O/F ratio, see Eq. (6.21). This equation has been successfully used in practice, but is not physical (e.g., the units are not consistent). It forces small changes in O/F, likely for forced stability of the code.

$$O/F_{j,k+1} = O/F_{j,k}$$
$$+ (\frac{\dot{m}_{O,j}}{\dot{m}_{F,j,k+1}} - O/F_{j,k})(\dot{m}_{F,j,k+1} - \dot{m}_{F,j,k})\Delta t \qquad (6.21)$$

Assuming a single circular fuel grain port gives the fuel regression rate, per Eq. (6.22). The oxidizer mass flux is calculated from Eq. (6.23), and the total mass flux is calculated from Eq. (6.24).

$$\dot{r}_j = \frac{\dot{m}_{F,j}}{2\pi \rho_F L_F r_j} \tag{6.22}$$

$$G_{o,j} = \frac{\dot{m}_{O,j}}{\pi r_j^2} \tag{6.23}$$

$$G_j = \frac{\dot{m}_{F,j} + \dot{m}_{O,j}}{\pi r_j^2} \tag{6.24}$$

The port geometry at the next time step can be calculated using the forward Euler method. For a single circular port this gives the new port radius, r_{j+1}, via Eq. (6.25).

$$r_{j+1} = r_j + \dot{r}_j \Delta t \tag{6.25}$$

Continue through this loop until the total burn time is reached, $t_j = t_b$.

At the end of the burn sum up all of the mass of fuel consumed, m_F, Eq. (6.26). If this total mass does not equal the measured mass of fuel burned (within selected convergence criteria), then η_C^* should be altered, and the process should commence again. Ref. [202] uses 10^{-4} and suggests using a small number (0.05) times the difference in fuel mass as a (nonphysical) way to iterate on η_{C*}. Repeat this loop until convergence is achieved on total burned fuel mass. For more information on this method, see Ref. [202] or [105].

$$m_F|_{calc.} = \sum \dot{m}_{F,j} \Delta t \tag{6.26}$$

The combustion efficiency has been shown to decrease over the burn time and is expected to decrease at lower chamber pressures. It has been suggested that this could be caused by decreased mixing due to changes in the port geometry over the burn combined with an increased rate of diffusion caused by decreased chamber pressure [131].

6.2.4 Comparison of methods

The pros and cons of the three methods discussed here are summarized in Table 6.1. Similarities of these methods include that they enable time reconstruction of the motor performance but require experimental measurement of the oxidizer mass flow rate, $\dot{m}_O$. All three methods, as presented here, also assumed that the regression rate is constant with fuel grain length. This assumption could be avoided with the regression rate law if the full regression rate equation, Eq. (6.1), were used instead of the simplified form of Eq. (6.2).

6.3 Important parameters

There are a number of parameters that are important to regression rate determination. As the field of hybrid rocket propulsion continues to grow, availability and potential for collaboration have increased. It is recommended that the following data be reported whenever possible to facilitate comparison with other experiments. The following sections contain suggestions for information to be included with test data and some explanation of why it is important. Error bounds should also be reported on these values, as discussed in Ref. [205] and in Section 11.7.2 of Chapter 11. Some further discussion of data reporting, including an example test data table, is in Section 11.7.1 of Chapter 11. A discussion of instrumentation for measuring test data is also covered in Chapter 11. As discussed previously, the empirical constants for the regression rate equation can vary depending on how they were derived. This section focuses on important parameters to report on to make a test data set useful as a source of regression rate information for other propulsion engineers.

TABLE 6.1   Comparison of Benefits and Challenges of Regression Rate Reconstruction Methods.

Method	Pros	Cons
Regression Rate Law	Simple Widely used Uses same formulation as an approach for predicting ballistics of new designs Can provide an accurate reconstruction of performance, including during throttling, if regression rate coefficients are well understood.	The measured fuel mass consumed for a given test may not match the reconstructed consumed fuel mass Extrapolation can be a problem Small inaccuracies in the exponent, n, lead to large variation in predicted regression rate Reported values of a/a_o and n are often scattered in literature and can lead to inaccurate results if not applied with consistent assumptions as the approach used to determine them Does not make use of the full set of test data (e.g., $P_c(t)$) The propellant combination being used must have been previously characterized in order to use the regression rate law method.
Constant C^*	Ensures that the measured and reconstructed masses match Simple explicit method, with limited need for input variables Works satisfactorily for steady-state burns with propellants with n exponent close to 0.5	Average C^* is calculated using a mathematically incorrect approximation and is not the actual average It does not fit well to burns during which the O/F varies significantly (regression rate law exponent n far from 0.5 and/or throttling) During transients, $C^*_{actual} < \bar{C}^*$, causing the reconstruction method to underestimate the fuel mass flow rate During steady-state, $C^*_{actual} > \bar{C}^*$, causing the reconstruction method to overestimate the fuel mass flow rate To be used for post-processing of experimental data not for performance prediction.
Constant η_{C*}	Enables changes of C^* as a function of O/F shift. Applicable to tests with varying combustion chamber pressure. Independent of definition of start-up and shut-down and possible throttling. Ensures that the measured and reconstructed masses match.	Assumes C^* efficiency, η_{C*}, is constant over the burn. This could be especially incorrect at start-up and shut-down depending on when the start and stop time for the reconstruction are chosen Multiple iterative loops, needing some fine tuning of the convergence criteria, and potentially down sampling of the experimental data, to converge To be used for post-processing of experimental data not for performance prediction.

6.3.1 Propellant data

In the case of propellant information, more really is more. The propulsion engineer should record/publish as much about the propellants, especially the fuel, as possible. Most propellants used for flight applications have detailed specifications setting allowable limits for the constituents in the mixture and how these require-ments should be verified. Since most of the fuels used in hybrid rocket applications are sourced from outside the aerospace industry, there is no consistent specification for their use in aerospace applications. These fuels can vary greatly from supplier to supplier. Critical propellant information includes the density (ρ_F), chemical formula/molecular weight (M_w), enthalpy of formation ($\bar{h}_f$), and whether this information was assumed (e.g., standard data sheet for the type of material) or if the specific lot was measured experimentally and if so, by what method. It is also important to list the vendor and part number if applicable. The use of any additives and the associated details (e.g., the particle size and whether the particles are likely to be oxidized) should also be reported if applicable. The oxidizer grade can make a difference, especially for hydrogen peroxide or MON, when adding H_2O or NO, respectively, makes a big difference in the performance.

6.3.2 Masses and mass flow rates

It probably goes without saying that the oxidizer mass flow rate, $\dot{m}_O$, is a necessary parameter to report. Most experiments have a way to measure this (see a discussion in Section 11.6.5 of Chapter 11) and any flight applications have at least a well calibrated/predicted value since it is necessary to predict the regression rate and thrust.

The total mass flow rate includes that of the fuel, the oxidizer and any inerts/ablatives that are burned or eroded throughout the burn. Therefore the total mass before and after the test should be included as opposed to just the change in fuel. Ablative sleeves, flow trips, mixers, etc., should also be reported.

From Section 6.2, it was clear that the physical measurements of port diameter or the amount fuel burned have a first-order impact on regression rate data. Multiple techniques exist to measure the fuel burned over a test. They vary in accuracy and cost. The methods used to measure the amount of fuel consumed during a hot-fire test include:

1. Average over grain length and burn time:
   - Mass before and after test
   - Micrometers at the fore and aft end. Several measurements are taken at each end and an average diameter is used to calculate how much fuel is left
   - Displaced water volume

2. Average over burn time:
   - Posttest destructive evaluation (e.g., fuel grain sectioned and measured at various points along the fuel length)
   - Bore crawler (e.g., [206])
   - Laser measurements (e.g., [206], [48])

3. Variable with grain length and burn time:
   - Optically, computer aided ([17], [207])
   - Ultrasonics ([208], [206])
   - X-ray measurement techniques ([209])

Any of these options can be used, and the propulsion engineers should focus on the most critical data to their problem statement. If the goal of the test campaign is to characterize the regression rate over a range of conditions and the length of the port, a more resource intense option is suggested. However, if the testing focuses on propellant utilization, an averaging method coupled with a qualitative inspection could give enough information. The optimization problem of data to resources is left to the propulsion engineer.

6.3.3 O/F ratio

The O/F (average and/or instantaneous) is particularly useful in understanding how the regression rate information could be applied in the future, particularly when assessing the data with a_o instead of a, see Ref. [98].

6.3.4 Regression rate

It is not lost on the authors that the purpose of this section is to discuss how to calculate fuel regression rate. So, it should go without saying that if the regression rate is measured directly (e.g., via discrete resistors that are exposed and burned through, or video analysis), it should be reported. Most experiments will not have this luxury, and that is completely fine. In that case, the propulsion engineer gets to enjoy the exercise of calculating it from the other data she or

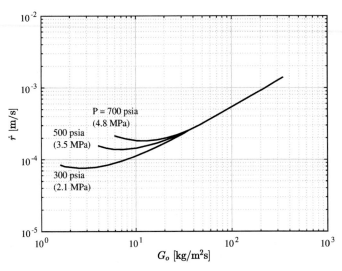

FIGURE 6.1   Regression rate variation with chamber pressure. This figure is for the combustion of oxygen with 80% PMMA and 20% Al. This figure was drawn using the data from Ref. [210].

he has collected using the approaches described in this chapter.

6.3.5 Combustion chamber pressure

If the propulsion engineer can only afford a single measurement, it is most often combustion chamber pressure, $P_c(t)$, and with good reason. This critical parameter is used to understand motor performance and efficiency. When considering the sampling frequency for this data, both the Nyquist criteria and instability reconstruction should be considered (see Section 11.6.5 of Chapter 11 for a discussion). Measurements at the aft of fuel grain are preferred, especially if there are devices in the combustion chamber that cause pressure losses. A fore-end pressure transducer could predict higher performance than is actually being achieved. For example, the Peregrine hybrid rocket motor measured a pressure difference of 0.23 MPa between the fore and aft ends [131]. However, it is understood that the aft end of a combustion chamber creates a harsher environment for the instrumentation to survive. Common methods for protect-

ing instrumentation can also attenuate data, e.g., small ports. The authors have had success with coating pressure transducer diaphragms with oxygen-safe lubricants.

Reported chamber pressure data is also useful for regression rate reconstruction. It is used explicitly in the Constant C^* (Section 6.2.2) and Constant C^* Efficiency (Section 6.2.3) methods, and some researchers also use it to determine a and n for a single test (see Ref. [9] as an example). In general, though chamber pressure can be useful for reconstruction, the fuel regression rate is not a function of chamber pressure for hybrid motors operating under normal conditions with non-metalized fuels (i.e., a_o and n are not a function of chamber pressure). However, if the combustion chamber pressure is very low, then the fuel regression rate can become limited by kinetics and therefore start to depend on pressure (see a discussion in Section 3.2.4 of Chapter 3). Also, if metal additives have been added to the fuel, then some regression rate dependence on chamber pressure is expected at low oxidizer mass flux, see Fig. 6.1.

Note that swirl or vortex injectors will have significant radial pressure gradients due to the cyclonic nature of the flow field. This can be dealt with by calculating an average pressure or selecting a reference pressure. One way to do this is by using a chemical equilibrium solver to find the pressure at which the mass flow rate at the nozzle throat matches the empirical total mass flow rate [87].

6.3.6 Physical data

Physical data can be split into several sub-categories. The first is the configuration of the combustion chamber: is there a pre- or post-combustion chamber? Is any sort of flow trip or mixing device used, and if so, where is it located? Protected surfaces, such as surfaces coated with an ablative or sheltered behind insulators, will change the available burn area and should also be reported.

Next is the nozzle type (e.g., ablative, regeneratively cooled), and if the former, the throat area over time should be reported. This is difficult data to obtain; therefore the assumption that the nozzle radius increases linearly between measurements taken before and after the test is often used. The authors have found this to be a poor assumption for treated graphite (e.g., Polymer Infiltration and Pyrolysis) but have used it successfully otherwise.

The fuel grain geometry is one of the most important parameters for regression rate reconstruction. This should include the number and shape of the port(s). The change in the port cross-sectional area over the test and the length of the fuel grain should also be reported.

6.3.7 Burn time

The calculation of burn time can be one of the biggest sources of error in regression rate calculations. This is especially true for tests in which the ignition and shutdown transients are a significant portion of the total test time, e.g.,

short-duration tests. Fig. 6.2 shows the spread in burn times that can be calculated depending on the method used. The x-axis (time [s]) is expanded in the images on the left and right, respectively, to highlight the substantial differences between the burn time calculations. The five common methods shown in that image are described below.

Method 1: chamber pressure begins to rise/fall

The burn time is from the moment the chamber pressure begins to rise to the time when it has fallen completely. The main drawback to this method is that ignition and shutdown transients can dominate the calculation.

Method 2: percentage of peak pressure

A percentage of the peak pressure is selected as the start and end point for the burn time, e.g., 90%. This method is not suitable when the chamber pressure has pressure spikes such as those due to an ignition hard start or other transient events.

Method 3: percentage of average chamber pressure over nominal burn time

A percentage of the average chamber pressure is selected as the start and end points for the burn time. This method is not suitable when pressure varies substantially during steady-state combustion, such as for a long test where the motor ballistics predict a significant change in chamber pressure.

Method 4: percentage of predicted start/end pressure

A percentage of the start or end pressure is selected to use in order to calculate the burn time. However, a model of the combustion chamber pressure based on detailed knowledge of the expected regression rate is needed.

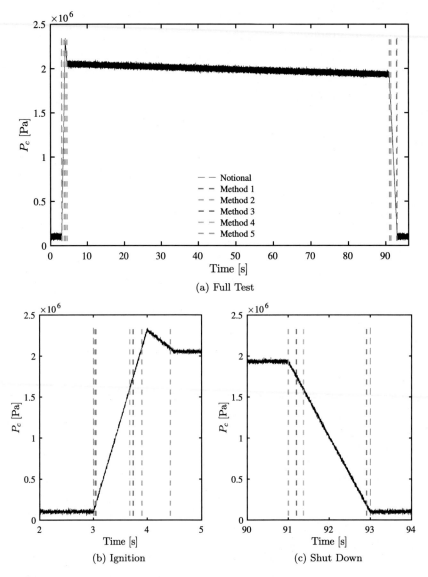

FIGURE 6.2 Comparison of various burn time calculation methods from Ref. [105]. Burn time is calculated on a notional data set to demonstrate how sensitive the calculation is to the method selected.

Method 5: bisector method

Fig. 6.3 demonstrates the method. Tangent lines are fit separately to the steady-state chamber pressure curve before shutdown (blue) and the chamber pressure curve during shutdown (green). The angle between the two curves is then bisected (purple). Finally, the intersection of the bisected angle and the chamber pressure curve (red dot) is the calculated shutdown time. This method is commonly used in solid rocket propulsion, but only to determine the shutdown time.

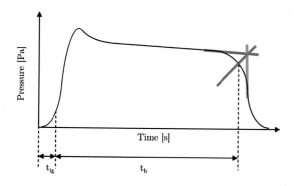

FIGURE 6.3 Bisector Method.

6.3.8 Scale and hardware pedigree

It is advisable to report the approximate scale of the hybrid rocket used to collect any regression rate data since changes to motor scale can alter the fuel regression rate, see Fig. 6.4. The rocket scale can be provided broadly in the form of reporting the thrust class of the rocket or by reporting more specific fuel grain geometry information such as the fuel length and outer diameter.

The effects of scale have been noted by many researchers (e.g., increasing port diameter leads to decreased regression rate at the same mass flux). Therefore extrapolation over large geometric scales can introduce error in the expected regression rate. It is usually desirable to do early testing with new propellant combinations at a small scale because the tests are easier/faster to conduct, and they cost less. While these tests can give a good idea of what the empirical constants might be in the regression rate equation, when the propulsion engineer is ready to move to a flight-like configuration, they often run into trouble. Ref. [211] attributed this to multicollinearity, saying that independent variables such as mass flux, port diameter, length, and pressure are highly correlated. Ref. [212] also makes recommendations for using laboratory-scale testing to simulate full-scale hybrid propulsion systems. One must be very careful about taking sub-scale data and applying it to large-scale motors and pay especially close attention to how the data reduction was completed (e.g., space-time averaged regression

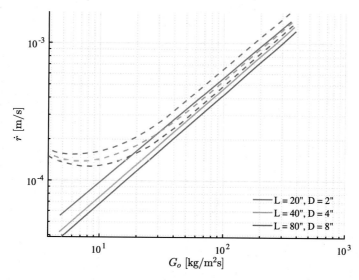

FIGURE 6.4 Regression rate variation with motor scale. This figure is for the combustion of oxygen with PMMA (solid lines) or 80% PMMA and 20% Al (dashed lines). The diameter referenced is the outer diameter. This figure was made using the data from Ref. [210].

rate), as this could drive further errors in the evolving system, making the way data is reported critical to enabling other researchers to use it.

Work has been done to improve correlations across scales. Refs. [12] and [213] postulated that the scaling effects are caused by the flame zone perturbing the velocity profile in the boundary layer after comparing their empirical results with the computational results of Ref. [214]. At larger scales, the height of flame zone is higher in the boundary layer, which decreases convective heat transfer due to a shallower temperature gradient. Distortion/acceleration of the velocity in the combustion zone would then lead to a steeper velocity gradient at smaller scales, as compared to larger ones. This would increase the regression rate at smaller scales for the same mass flux. While the effect of radiation is stronger at larger scales due to the larger gas volume to fuel surface area, it does not completely offset the decrease in convective heating [12]. A non-dimensional term, θ_f, was created to account for the distortion of the velocity profile due to the flame that was shown to improve the regression rate estimation, resulting in the semi-empirical relationship for HTPB/GOx in Eq. (6.27), where the regression rate is in m/s, and G is in kg/m^{2}s. This is explained fully in Refs. [12] and [213].

$$\frac{\rho_f \dot{r}}{G} = a\,St\,B\theta_f^b \left\{ c(\dot{Q}_r/\dot{Q}_c)^d \right.$$
$$\left. + exp\left[-c(\dot{Q}_R/\dot{Q}_c)^d \right] \right\} \qquad (6.27)$$

For HTPB/GOx, $a = 5.24 \times 10^{-4}$, $b = 0.6$, $c = 1.3$ and $d = 0.75$ [213].

Using this correlation, Chiaverini et al. were able to fit data for rubber/oxygen systems to within ±10% over 5x scales. They used data available in literature (e.g., JIRAD motor) as well as their testing at Penn State (lab-scale slab burner with x-ray capabilities for regression rate measurement). Eq. (6.27) accounts for convection and includes corrections for radiation, but does not attempt to capture radiation for metalized fuels. Therefore it is expected that by accounting for the boundary layer processes, as with the θ_f term, extrapolation of regression rate data across scales may be possible.

6.3.9 Injector geometry

The injection scheme can significantly influence the fuel regression rate and can affect the exponent n in some configurations [131]. As such, whenever it is allowable to do so, the injection geometry should be reported with regression rate data. Details of the injection geometry are often proprietary but broad categorization such as: swirl, single orifice, showerhead, radial, or circumferential injection is recommended. If swirl injection is used then the swirl number (see Eq. (3.67) and (3.68) of Chapter 3) should also be reported.

6.3.10 Secondary parameters

Additional information that should be considered for publication with regression rate data includes:

- Information on mixing devices in the combustion chamber. Ref. [131] defines these to be any geometrical feature that manipulates flow to produce turbulence in the combustion chamber (e.g., single-hole diaphragms, wagon wheel diaphragms, screens, and flow trip devices.) See, for example, Refs. [215], [216], [217], [66], and [218]. Mixing devices can result in a drop in chamber pressure by 5–10% [131].

- Information on the ignition method used in the tests. The igniter can change the combustion behavior early in the burn. As an example, a long-burning pyrotechnic device or an augmented spark igniter may introduce substantial hot gas into the combustion chamber. Section 9.9 of Chapter 9 discusses common hybrid rocket ignition schemes.

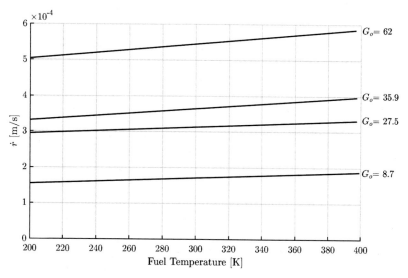

FIGURE 6.5 Regression rate variation with fuel temperature. This figure is for the combustion of gaseous oxygen with PMMA. This figure was made using the data from Ref. [219] and Ref. [1].

- Information on the initial fuel and oxidizer temperatures before combustion is also helpful to report. The initial temperature of the propellants can impact the fuel regression rate. Fig. 6.5 shows that the effect of fuel temperature on regression rate is approximately linear for the combustion of PMMA and gaseous oxygen. However, this effect is small due to the fact that the latent heat of vaporization of the solid fuel is generally substantially larger than the variation in the heat content of the fuel across typical operating temperatures [1].

- The presence of any sources of heat addition (beyond the fuel/oxidizer combustion), such as the use of heater motors or the use of hypergolic or reactive additives. Ideally, additives should be reported as part of the fuel composition but can be overlooked if they are not used throughout a burn.

6.4 Regression rate prediction method

There is only one widely used approach to predict the regression rate behavior of a hybrid rocket with a new fuel grain geometry and/or oxidizer mass flow rate: that is an application of the functional forms of Marxman's regression rate law, see Eqs. (6.1) and (6.2) (Eqs. (3.45) and (3.48) in Chapter 3). As discussed in Chapter 3, the length exponent, m, is generally considered to be 0. The values of a or a_o and n can be evaluated from a set of data using the approach described in Section 6.2.1, or else can be fit to a single test result by evaluating the fuel regression rate versus time, and oxidizer mass flux versus time using the approaches described in Sections 6.2.2 and 6.2.3 and then calculating the n exponent and empirical constant, a or a_o, that best fits the data. The regression rate reconstruction approaches of Sections 6.2.2 and 6.2.3 cannot be used directly to predict the performance of other motor configurations without first doing this fit to the functional form of Marxman's

TABLE 6.2 Published regression rate parameters for various hybrid rocket fuel and oxidizer combinations. The format used follows that of Ref. [82] though values are taken from the source references. S.I units are used here for all data (e.g. the regression rate, $\dot{r}$, is in m/s and G_o is in kg/m². The data reduction techniques are diameter averaged (DA), burn time error minimization (t_b) and other averages (OA).

Oxidizer [-]	Fuel [-]	a_o [S.I.]	n [-]	Number of Tests [-]	G_o Range [kg/m²s]	O/F [-]	P_c Range [MPa]	Data Reduction Technique [-]	Other Notes [-]	Ref. [-]
O_2	Paraffin: SP1A	1.17E-04	0.620	65	16-369	1-4	1.1-6.9	DA		[32]
O_2 (g)	HTPB: Thiokol	3.04E-05	0.681	16	38-302	-	-	-	Motor diameter 2''	[6]
O_2	HTPB + 19.7%Al	1.29E-05	0.956	2	51-230	-	1.2	OA		[220]
O_2	HTPB	9.03E-05	0.527	3	62-310	-	2	OA		[220]
O_2	HTPB + 20%GAT*	1.72E-04	0.439	5	-	-	-	-	* Guanidinium Azo Tetrazolate (GAT)	[221]
O_2	PMMA	2.11E-05	0.615	8	33-266	-	.3-2.6	-		[222]
O_2	HDPE	4.88E-05	0.469	4	77-261	3.8-5.9	.7-1.3	DA		[93]
O_2	PE Wax: Marcus 200	4.31E-05	0.712	4	48-158	2.2-3.2	.5-1.2	DA		[93]
O_2	PE Wax: Polyflo 200	2.66E-05	0.703	3	44-163	1.6-1.7	.6-1.2	DA		[93]
O_2	HTPB	5.41E-05	0.647	9	80-150	-	-	OA		[192]
O_2	HTPB	4.40E-05	0.659	6	175-320	-	-	OA		[192]
O_2	HTPB + 13%Al (nano sized)	2.43E-05	0.775	12	165-342	-	-	OA		[192]
O_2	Paraffin: FR5560 + 13%Al (nano)	1.12E-04	0.730	8	145-290	-	-	OA		[192]
O_2	Paraffin: FR5560	1.69E-04	0.600	4	63-123	1.3-1.8	-	OA		[192]
O_2	Paraffin: FR4550	8.10E-05	0.735	3	43-119	-	.7-?	DA		[93]
O_2	PMMA: Clear	8.96E-05	0.350	21	6-344	1.2-4.4	.27-1.67	t_b		[9] and [10]
O_2	PMMA: Blackened	1.44E-04	0.240	10	3-237	1.1-2.1	.65-1.53	t_b		[9] and [10]
O_2	HTPB/Escorez	2.06E-05	0.680	-	-	-	-	-		[223]
O_2	HDPE	2.34E-05	0.620	-	-	-	-	-		[223]
O_2	Paraffin	1.17E-04	0.620	-	-	-	-	-		[223]
N_2O	Paraffin	1.55E-04	0.500	-	-	-	-	-		[223]
IRFNA	80% Polybutadiene / 20% PMMA	3.18x10-5	0.560	34	7-125	-	0.4-2.8	-		[224]
N_2O	PMMA	1.31E-04	0.335	16	25-297	3.9-6.3*	2.9-4.1*	DA	* Range from subset of test data published. Actual range should be larger	[225]
N_2O	HTPB	1.88E-04	0.347	15	61-216	3.3-6.3*	1.8-4.1*	DA	* Range from subset of test data published. Actual range should be larger	[225]

continued on next page

TABLE 6.2 (continued)

Oxidizer [-]	Fuel [-]	a_o [S.I.]	n [-]	Number of Tests [-]	G_o Range [kg/m²s]	O/F [-]	P_c Range [MPa]	Data Reduction Technique [-]	Other Notes [-]	Ref. [-]
N_2O	HDPE	1.16E-04	0.331	12	30-270	6.5-8.5*	3.4-3.8*	DA	* Range from subset of test data published. Actual range should be larger	[225]
N_2O	Paraffin: SP7	7.81E-05	0.545	10	100-325	-	-	OA	D_o = 2.7". O/F Correction Applied	[226]
MON3	Paraffin: SP7	2.80E-04	0.297	8	47-164	2.3-4.9	-	OA	D_o = 2.7"	[226]
$90\%H_2O_2 + 10\%H_2O$	HDPE	7.80E-03	0.770	4	360-370	7.7-8.3	1.6	DA	250 N class motor	[227]
$98\%H_2O_2 + 2\%H_2O$	80%HTPB + 20%Al	3.94E-06	1.043	5	80-165	-	-	OA	D_o = 10 cm	[228]
$98\%H_2O_2 + 2\%H_2O$	60%HTPB + 20%$C_{14}H_{10}$ + 20%Al	4.29E-06	1.034	5	80-165	-	-	OA	D_o = 10 cm	[228]
$98\%H_2O_2 + 2\%H_2O$	60%HTPB + 28%Al + 10%Mg + 2%C	2.67E-05	0.725	5	75-150	-	-	OA	D_o = 10 cm	[228]
O_2	PVC	1.42E-04	0.51	4	56-240	1.5-1.7	~1.1	DA	Multi Section Swirl	[229]
O_2	HDPE	6.97E-05	0.561	3	24.7-44	-	0.8-1.1	OA	Double tube	[230]
O_2	HTPB	1.93E-04	0.54	34	7.3-121.8	1.2-4.9	0.08-1.9	DA	Bi vortex	[118], [12]
O_2	HTPB	4.00E-04	0.68	-	2-4	-	-	-	End-burn swirl	[231], [12]
O_2	HDPE	4.32E-05	0.53	-	30-90	-	-	-	Single step	[123], [101]
O_2	HDPE	5.85E-05	0.53	-	30-90	-	-	-	Single diaphragm	[123], [101]
O_2	PE	2.40E-05*	0.8*	3	250-700	1.2-1.7	2.3-2.5	OA	CAMUI port. * Port regression, not end faces	[232]
O_2	70%HTPB + 27.5%AP + 2.5%Fe_2O_3	1.11E-03*	0.259	3	70-281	-	1.3-4.1	DA	* Evaluated at P_c = 2.4 MPa	[233]
O_2	Paraffin	3.83E-04	0.8506	4	15.8-26.1	-	0.42-0.65	-	End-burn swirl	[120]
O_2	86% Paraffin + 14% ABS	1.75E-04	0.43	8	10-40	1.2-2.3	0.58-2.06	DA	Nested helix	[234]
N_2O	Paraffin	5.39E-04	0.36	6	100-250	2.5-5.25	0.9-3	OA	Diaphragm (1 hole) at 33% position	[235]
N_2O	Paraffin	2.93E-04	0.518	4	100-200	2-6	0.9-3	OA	Diaphragm (4 hole) at 33% position	[235]
O_2	70% Paraffin + 30% microcrystalline	8.36E-04	0.4	-	30-150	0.6-0.9	0.3	OA	Bluff body	[236]
O_2	Paraffin	4.96E-04	0.62	-	15-32	2.0	1	OA	Head end swirl	[237]

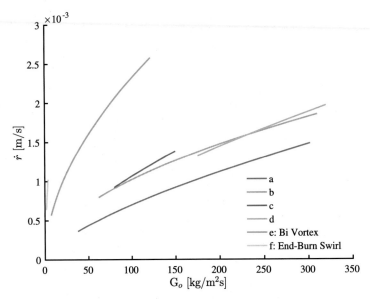

FIGURE 6.6   Regression rate versus oxidizer mass flux of HTPB fuel combusting with O_2 according to Eq. (6.2). Regression rate data is provided in Table 6.2. References: a. [6], b. [220], c. [192], d. [192], e. [118] and [12], f. [231], and [12].

regression rate law. As such, the data reported in the following section, which can be used to predict the ballistics of new hybrid rocket designs, will all be reported in terms of the empirical regression rate constants, a_o and n.

6.5 Published regression rate data

Table 6.2 includes a summary of the reported regression rate parameters in literature for common hybrid rocket fuels and oxidizers. If test data is not available a priori for the fuel and oxidizer under consideration, then the approach described in Ref. [82] is recommended to provide an initial regression rate estimate. Note, however, that all current analytical approaches to estimate hybrid fuel regression rates (including the approach described in Ref. [82]) should be treated as approximate only and, as such, should be verified via hot fire test as soon as possible.

6.5.1 Variability

All hybrid propulsion engineers have likely found themselves in the position of trying to design a new hybrid rocket using published regression rate data. A particular challenge in doing this is deciphering which regression rate data to use for a given propellant combination, given the typically large scatter in reported regression rates between motors. Fig. 6.6 provides an example of the scatter in reported test data for the combustion of HTPB with oxygen. There can be multiple reasons for the scatter seen in hybrid regression rate data for what should be the same propellant combination. This includes different geometric scales, injection schemes, data reduction techniques, using a_o across a different range of O/F, etc. Suppose tests are conducted at very high or very low oxidizer mass fluxes. In that case, the regression rate can become pressure dependent (see Section 3.2.4 of Chapter 3 for more discussion of this), and any data collected in this regime will not be applicable to hybrids operated at more typical conditions. Whenever there

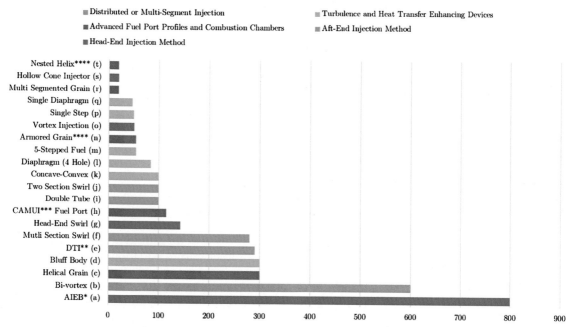

FIGURE 6.7 Experimentally observed percentage increase in regression rate compared to a baseline reference case for various changes to the rocket configuration (including the oxidizer injection scheme). Figure is based on Ref. [101]. Note: *Axial Injection End Burning, **Distributed Tube Injector, ***CAscaded MUltistage Impinging-jet, ****Categorization different from Ref. [101]. References: a. [114], b. [118], c. [239], d. [236], e. [240], f. [241], g. [237], h. [242], i. [230], j. [229], k. [243], l. [235], m. [244], n. [245], o. [246], p. [123], q. [247], r. [248], s. [249], t. [234].

is metal loading in the fuel grain, radiative (pressure) effects will also need to be included [18]. Even without additives, there can also be issues with the fuel formulation not being identical despite having the same name. This is a particular issue for Hydroxyl Terminated PolyButadiene (HTPB), as discussed in Ref. [238]. Ref. [101] provides a great summary of various techniques which can be employed to increase the fuel regression rate without changing the propellant combination. It is the authors' experience that increasing fuel regression rate is not always desirable, depending on the application under consideration, especially since many techniques to increase fuel regression rate can lead to large fuel sliver fractions or introduce combustion instability challenges. However, Ref. [101] serves as a great summary of the potential impact that oxidizer injection and motor configuration can

have on reported regression rates. Fig. 6.7, which is generated based on Ref. [101], shows the percentage increase in regression rate that has been observed experimentally from changes to the oxidizer injection schemes, fuel grain geometry, or combustion chamber configuration etc. It is worth noting that this increase in regression rate may be specific to a given test configuration and thus should be treated as a more qualitative indication of the approximate order of magnitude change in regression rate that might be seen for different rocket configurations.

To evaluate which test data should be used to predict the performance of a new design, it is recommended that the propulsion engineer considers the scale, range of oxidizer mass fluxes, and range of O/F over which the new design will operate and endeavor to match this to reported test data as much as possible. If extrapolation is required beyond the reported range of test data, then care should be taken to evaluate how the regression rate parameters were determined. The authors have found that determining regression rate coefficients via the space-time averaging technique does not extrapolate as well to new test conditions as well as determining the coefficients that minimize the expected error in burn time [9] or in the mass of fuel burned [203]. If care is taken to either avoid extrapolation or to extrapolate carefully, then using published regression rate data can be extremely helpful in trading propellant combinations and performing a first-pass hybrid design. This section is written to ensure that the propulsion engineer understands the potential pitfalls associated with this approach, but they also need to be aware that the application of Marxman's regression rate law to predict the performance of new hybrid rocket motors has been used successfully for many programs.

Unfortunately, the predicted motor ballistics for a new motor is very sensitive to the regression rate exponent, n. Small variations in this value can notably change the predicted rocket behavior in terms of burn time, thrust, O/F etc. Wherever possible, a hot fire test campaign is recommended to validate a new motor design. Sub-scale tests are often used initially to predict motor performance, but for the reasons discussed in Section 6.3.8, a full-scale test is also recommended as soon as it is practicable to do so.

6.5.2 Regression rate coefficient unit conversion

The units of a and a_o are unusual in that they are a function of the exponent n and also of the exponent m (if m is non-zero). Consider the two functional forms of the regression rate law for hybrid rockets, Eq. (6.1) and (6.2). Evaluation of the dimensions of these equations gives the units of a and a_o per Eqs. (6.28) and (6.29), re-

spectively.

$$[a] = \frac{Length^{1+2n+m}Time^{n-1}}{Mass^n} \quad (6.28)$$

$$[a_o] = \frac{Length^{1+2n}Time^{n-1}}{Mass^n} \quad (6.29)$$

Many sources have reported the data in different units, such as Refs. [82] and [93]. This is done to make the resulting value easy to understand. The regression rate is often on the order of a millimeter per second, so it is easier to think in terms of 1 mm/s instead of 10^{-3} m/s. However, since the regression rate equation is not dimensionless, and the values of a or a_o and n are used in other design equations, using a mix of units requires a great deal of attention. To avoid checking units in all the design equations, the authors prefer to work exclusively in SI units. Converting the reported regression rate coefficients so that they apply to SI units should be straightforward but is a common source of error for propulsion engineers new to the task. For this reason, a quick example of converting units of a_o is provided here.

Consider the example where the regression rate data is presented such that $\dot{r}$ is in units of mm/s, and the oxidizer mass flux, G_o, is in units of g/cm^2. This selection of units (including two different units for length) is used regularly in the hybrid rocket community. For example, Ref. [82] gives a wonderful summary of regression rate coefficients a_o and n for the combustion of various fuels with oxygen from various other experiments. The empirical constants listed in this reference for the combustion of PolyEthylene (PE) wax fuel (Marcus 200) with GOx are $a_o = 0.188$ and $n = 0.781$. Converting the units of these constants begins by recognizing that the exponent, n, is assumed to be dimensionless and so does not change. The first step is to use a single-length scale: the regression rate is converted into the same units as the mass flux (cm). Since a_o initially gave the regression rate in terms of mm, this requires that it must be divided by 10 to give the regression rate in terms of cm, i.e., $a_o = 0.0188$. Now, Eq. (6.29) can be used to convert from a length scale in cm to a length scale in meters, and from a mass scale in grams to a mass scale in kilograms, see Eq. (6.30). The value of a_o in SI units is found to be $a_o|_{S.I.} = 3.11 \times 10^{-5}$. Note that in this example, there is no time unit conversion since seconds is consistently used as the unit of time.

$$[a_o]_{new} = [a_o]_{old}$$

$$\times \frac{\left(\dfrac{Length_{new}}{Length_{old}}\right)^{1+2n} \left(\dfrac{Time_{new}}{Time_{old}}\right)^{n-1}}{\left(\dfrac{Mass_{new}}{Mass_{old}}\right)^{n}}$$

$$a_o|_{S.I.} = \frac{0.188}{10} \frac{\left(\dfrac{1}{100}\right)^{1+2n}}{\left(\dfrac{1}{1000}\right)^{n}}$$

$$a_o|_{S.I.} = 3.11 \times 10^{-5}$$

$$(6.30)$$

If there is any doubt whatsoever about the conversion process, then it is recommended that the propulsion engineer sanity-check the conversion by inputting mass flux values in the original units and ensuring that the same regression rate is calculated for that same mass flux in the new units. This simple check can save a lot of heartache that will result from designing a hybrid rocket using the wrong regression rate parameters.

7

Hybrid design

7.1 Introduction

The goal of this chapter is to outline an approach that can be used to conduct a "first-pass" hybrid propulsion system design and assist in design trade studies. The focus here is on the top-level geometric design and mass estimate of the system, not on power requirements or detailed component, or internal motor design, which will need to come later. The first section of this chapter focuses on providing an overview of the design process. This process makes use of much of the material in the rest of the book and the propulsion engineer will be pointed to various sections throughout it. In working through this process, the challenges associated with designing these propulsion systems are elucidated, given the current Technology Readiness Level (TRL) of hybrid rocket motors and the limited availability of test data for these systems. Wherever possible, guidance on best practices is given. Propulsion engineers are cautioned not to set up the code outlined here and then run it blindly. Think through the results, and ask colleagues and friends for their experience and advice. The design will only be as good as what you put into it: garbage in, garbage out.

This chapter intends to include all the equations needed to design a hybrid rocket motor. Some have been presented previously, and the subsections here will both reference where they came from and include them again for convenience. At the end of each subsection, the new parameters that should have been determined will be listed. An example of how this process can be used to size a system is given in Chapter 10.

7.2 Overview

When starting a propulsion system design, the propulsion engineer typically knows the total spacecraft mass (m_i) and the necessary ΔV (see Chapter 2 for more information on these variables). This is a reasonable assumption as they are often set by the mission objectives and/or launch vehicle. However, it should be noted that an estimate for the propulsion system mass would have been needed to calculate the total spacecraft mass. Fig. 7.1 shows the steps in this iterative design process. The equations presented in this chapter generally assume a single, cylindrical port fuel grain. More complex hybrid fuel grain geometries were introduced in Fig. 1.5 of Chapter 1. However, modification of the design equations would be necessary for these more complex variations.

There are two major uses for this preliminary design code. The first is to compare a hybrid system to an existing design (e.g., a monopropellant

117

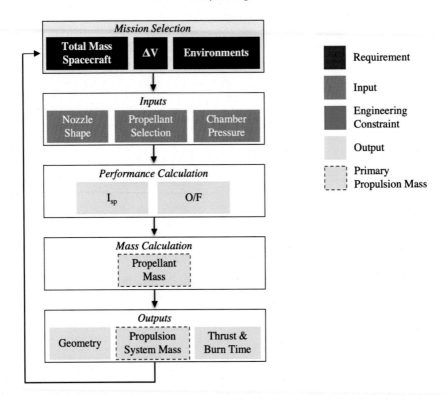

FIGURE 7.1   Simplified flow of the hybrid rocket motor design process.

CubeSat propulsion system, a solid upper-stage motor, or an orbit insertion motor). The system masses and capability of the baseline design can usually be found in published information, such as [250]. Any savings in mass in the hybrid propulsion system over the baseline would then translate into increased payload mass or increased ΔV, depending on which is more useful in the application. Since a number of the variables are known in this case, the process does not need to be iterative.

The second option is to use this method for a brand-new design. In this case, an initial estimate for the propulsion system wet mass would be needed to start the process. This can be done using an assumed mass fraction (see Chapter 2). An example in Chapter 10 discusses the poten-

tial issues with making this estimation; however, it is a necessary starting point.

The main mission requirements: total mass and ΔV are not the only inputs into the design process. The environments to which the propulsion system is being exposed are critical to propellant selection. The design process generally begins with choosing a fuel and oxidizer combination that satisfies the application specific requirements for environments/storability, safety, system heritage, performance, and potentially also compatibility.

To evaluate the performance of the selected fuel and oxidizer, the chamber pressure and nozzle area ratio must first be selected. These values, along with the thermochemical properties of the propellants, are inputs to a chemical equilibrium solver, which determines the final composition of the mixture after combustion and the associated performance in terms of I_{sp} and C^*. The chemical equilibrium solver will give an ideal performance; therefore a knockdown factor, or efficiency, is used to predict real performance. In this case, several efficiencies are required. The ideal specific impulse is attenuated (multiplied) by the combustion (C^*) efficiency, η_{C^*}, and nozzle efficiency, η_n. These will be discussed in greater detail in Sections 7.5.2 and 7.4.4 for η_{C^*} and η_n, respectively.

The real performance can then be used to calculate the mass of propellant required to deliver the ΔV to the spacecraft with the specified initial wet mass via the rocket equation, see Section 7.5.4. The propellant mass can then be divided into the mass of the oxidizer and the mass of fuel using the ideal O/F ratio. This is used to ascertain the required mass of usable fuel and oxidizer. For this discussion, the standard hybrid configuration (solid fuel and liquid or gaseous oxidizer) is assumed. A reverse hybrid could still be designed with only the need to flip the words fuel for oxidizer in the approach. The unburned portion of the fuel grain, or sliver fraction, should be added to the total mass of the fuel. The residual and hold-up for the oxidizer also need to be added to the oxidizer mass. Once the total mass of fuel and oxidizer are known (the usable and unusable portions) the volume of each then also be determined. Packaging constraints and motor ballistics are then considered to design the combustion chamber and oxidizer tank(s). There are a few different ways to design the combustion chamber depending on what is driving the design (e.g., a target thrust, fuel grain length, or end-of-life mass flux). Once the oxidizer tank(s) and combustion chamber are designed and the (optional) pressurization system is sized, a first-order estimate of the primary propulsion system mass can be generated. If this first-order propulsion system mass is inconsistent with the assumed initial spacecraft mass, then the design process in Fig. 7.1 must be iterated upon until convergence is achieved. The

rest of this chapter walks through this process in more detail. The simplified flow, introduced in Fig. 7.1, is intended to allow the propulsion engineer to keep track of the overall process and key steps as the chapter dives into the details.

7.2.1 Masses

A key aspect of any initial hybrid rocket design is evaluation of the primary masses, including:

- Propellant mass: fuel mass and oxidizer mass, including both usable and unusable mass for each propellant.
- Pressurant mass: pressurant gas mass, if applicable.
- Tank masses: oxidizer and, if applicable, the pressurant tank mass. These tank masses include the added mass for bosses, mounting accommodation, and propellant management devices, if applicable.
- Hybrid motor dry mass: at a minimum, this will include the structural shell of the combustion chamber, structural attachments and the nozzle mass. A more detailed design should also include any insulation, mixing device mass, and the mass associated with oxidizer injection.
- Feed system mass: mass for oxidizer and pressurant feed system components. This includes the mass of any pumps, if applicable.
- Attitude control system: mass of components, tanks, and propellant if required for attitude control.
- Igniter mass: mass of the overall ignition system, which could include additional components, tanks, and propellant if required for ignition.
- Thrust vector control system: mass of components, tanks, and propellant if required for thrust vector control.

Secondary system masses not sized here include all cabling/harnessing mass, thermal hardware mass, mass of the power system, and structural support mass. These masses may affect the trade results for certain point designs and so should be considered as the design matures. Ref. [250] is a good reference for the propulsion engineer looking to include an estimate of these other masses. When trading a hybrid rocket design with another propulsion system design, care must be taken to ensure that the same categories of mass are being considered for each system.

7.3 Mission selection

7.3.1 Spacecraft mass and ΔV

The overall requirement for a propulsion system design is that it is able to deliver a specified amount of ΔV to a spacecraft with an initial total wet mass, m_i. It is assumed when starting a propulsion system design that the propulsion engineer knows the total spacecraft mass, m_i, and the necessary ΔV for the application/mission under consideration. This is a reasonable assumption as they are often set by the mission objectives and/or launch vehicle. Chapter 2 provides more information on these variables and, in particular, gives examples of how to estimate ΔV requirements for some example missions.

Knowns: ΔV and m_i

7.3.2 Environments

The mission environments, particularly the thermal and dynamic environments (shock and vibration), can drive the design by limiting the selection of propellants. For example, cryogenic propellants are challenging for deep space missions because they would need to be maintained at very low temperatures for a long period of time. Similarly, neat paraffin, which is brittle, may have more challenges in applications with high shock loading. Therefore the propellant selection for each application requires more than a simple performance (I_{sp} or C^*) optimization.

The mass impact of maintaining an environment where the propulsion system can survive should be captured as part of the design process.

Temperature

The operating and storage (or nonoperating) temperature requirements may limit propellant selection and should be included early in the propulsion system design process. Thermal control systems, in the form of heaters, insulators, and thermal isolation, may be used to control the propellant temperature, and the associated mass and power requirements must be accounted for in the system design. Therefore the propellant selection that minimizes the need for thermal control can often trade favorably.

The upper temperature limit for the mission will generally drive the design pressure, or Maximum Expected Operating Pressure (MEOP) for propellant and pressurant feed systems. Earth-based launch systems typically allow for a greater range of propellants to be used since ground-based thermal control systems can be used on the launch pad, avoiding significant mass and/or power penalties for the launch vehicle.

The temperature ranges may vary over mission phases. Equatorial launches can often be quite warm and can drive the upper-temperature limits for the mission. Cruise towards outer planets can become cold, depending on the amount of heat the avionics transfer into the spacecraft. Planetary exploration can range from hot and hostile (e.g., Venus) to cold, with huge diurnal temperature variations (e.g., Mars). The space environment near the Earth, including heating in Earth orbit, is described in [250]. Luckily for most propulsion engineers, our good friends in thermal engineering often help bound the temperatures needed for survival.

There are many other scenarios where a propulsion system may be required to operate over a range of temperatures, which can often limit or even drive the propellant selection. In the absence of thermal control systems, care must be taken to ensure that the nonoperating temperature range of the application of interest will not detrimentally affect the fuel or oxidizer. Temperature requirements (glass transition, melting/vaporization, softening, etc.) for common hybrid fuels are discussed in Chapter 5.

The oxidizer selection is similarly constrained by temperature. Care must be taken to ensure that the oxidizer will not freeze or dissociate during the mission. The designer should also check the vapor pressure of the oxidizer at the maximum and minimum temperatures. This will ensure that there is sufficient pressurant in the low-temperature case and that the maximum temperature does not detrimentally drive the tank masses to contain high pressures.

Cryogenic oxidizers (e.g., LOx) must be maintained either at or below their boiling point, or the tank should be vented to allow for the evaporating gas to escape. Boil off of the oxidizer is expected and must be accounted for in the design or loading operations [251]. These propellants require special consideration during tank design as the oxidizer tanks will actually shrink when loaded with a cryogen.

Propulsion system design with cryogenic oxidizer requires a transient analysis of heat transferred into the system upon ascent. Aeroheating causes stratification in cryogenic propellants, which should be modeled for accurate performance information. The propulsion engineer is directed to Reference [252] for information on how this is done. Ref. [251] is also a good reference for the design and analysis of cryogenic systems.

Knowns: Allowable temperature (storage and operation).

Dynamics

Structural loads can be applied during ground operations, due to pyrotechnic devices on board, or self-induced during ascent. High shock environments may impact fuel selection. Liquefying fuels such as paraffin can be brittle

without strength additives and therefore are not suitable for all missions.

Knowns: Dynamic environment, mainly affects fuel selection.

7.4 Inputs

The propellants (fuel and oxidizer), the combustion chamber pressure, the nozzle shape, and the pressurization mechanism for the oxidizer must be selected prior to design calculations.

7.4.1 Propellant selection

A number of factors contribute to propellant selection, including performance, density, flight heritage, and mission environments, including potential long-term storage requirements. The propellant selection will impact nearly every other aspect of the propulsion system design.

Chapter 5 provides an overview of possible fuel and oxidizer options, including some discussion of their potential benefits and challenges for various applications and flight heritage in other systems. The performance of common propellant combinations at a single design pressure and area ratio is provided both in Fig. 5.8 and Table 5.5 of Chapter 5. The properties of the propellants are provided in Tables 5.2 and 5.3.

Once the propellants have been selected, the ambient, storage and operating conditions (specifically temperature and pressure) are used to determine fuel and oxidizer densities. While seemingly obvious, the allowable temperatures for the fuel and oxidizer must be checked to ensure the propellant can be stored or operated at the selected conditions (i.e., it does not freeze, including during blow-down conditions of a pressurant gas).

Assuming the propellant combination is not novel, regression rate parameters, a_o and n, needed to solve Eq. (7.1) can be assumed. See, for example, Table 6.2. These parameters are key to evaluating the motor ballistics and will be used in many of the design equations in the following sections. Note that accurate determination of regression rate parameters can be challenging and dependent on the test conditions and data reduction method used, see Chapter 6. Regression rate data does not exist for all propellant combinations. It is possible to estimate the regression rate parameters for the initial design. For example, the behavior of paraffin with N_2O_4 was originally estimated using the regression rate parameters for paraffin with N_2O [226]. Validation through testing will be required and the tests should be conducted at a relevant scale whenever it is practical to do so.

$$\dot{r}(t) = a_o G_o^n(t) \qquad (7.1)$$

Throughout this chapter, the simplifying assumption that the regression rate depends only on the oxidizer mass flux is made. Most published regression rate data and design papers make the same assumption [183,253–255]. However, the propulsion engineer is cautioned that this can introduce error in the predicted regression rate, particularly at low O/F. Regression rate error can impact the achieved O/F, thrust, I_{sp}, and C^*. Design correction to account for the new regression rate may require changes to the fuel grain geometry, oxidizer mass flow rate, or even to the propellant selection.

At this point, the use of any additives should be determined, since their presence changes the performance. For example, adding aluminum to paraffin reduces the optimal O/F. It can also enhance stability and may decrease nozzle erosion (as long as it does not mechanically enhance it, see Section 7.10.4). The addition of a hypergol might simplify ignition at the cost of performance. This is something that can be iterated on during the design processes if necessary.

Knowns: Fuel, Oxidizer, ρ_F, ρ_O, a_o, and n for the propellant combination.

7.4.2 Pressurization mechanism

The selection of the pressurization mechanism has a first-order impact on the architecture. There are four main options:

1. No pressurization (blow-down system)
2. Using a self-pressurizing oxidizer (N_2O or, to a lesser extent, LOx, would have had to have been selected as the oxidizer)
3. An external pressurization system
4. A pump

Applications that value simplicity over all else often choose the first or second option. A number of sounding rockets have used nitrous oxide for the operational simplicity self-pressurizing oxidizers lend (one fewer tank and an associated reduction in plumbing). However, the propulsion engineer is discouraged from thinking that modeling the self-pressurizing oxidizer's behavior will be simple. Most high-performance options will need an external pressurization system or a pump. With the advances in electrical pumps that have materialized in the last decade or so, those options are looking even more attractive.

There are also many options for an external pressurization system. A stored gas pressurization system is essentially a tank containing pressurant in the gas phase, e.g., Helium or Nitrogen. A gas generator or pressurant stored as a liquid (sometimes inside a cryogenic oxidizer tank) are also possibilities. The authors suggest Ref. [252] for further reading on alternatives.

Knowns: Pressurization mechanism, masses will be determined later.

7.4.3 Chamber pressure

The selection of the motor chamber pressure is generally not simple in the early phases of the design process. An increase in combustion chamber pressure will generally slightly increase the ideal specific impulse of the motor, see Fig. 7.2. Figs. 7.2 and 7.3 provide examples of the

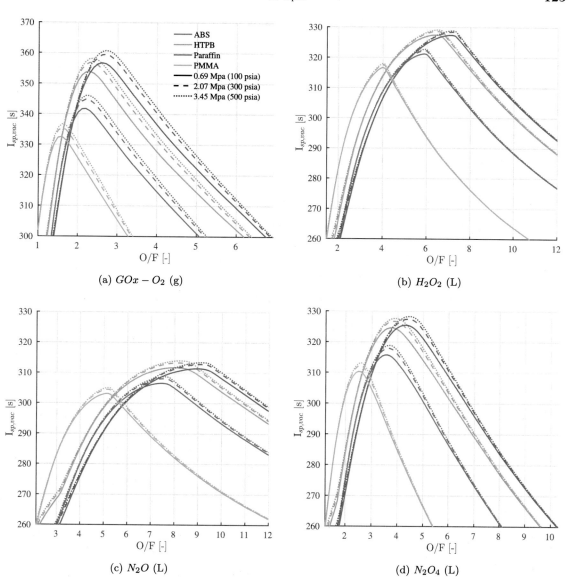

FIGURE 7.2 Ideal vacuum specific impulse versus O/F ratio and chamber pressure for selected hybrid propellant combinations. The legend in (a) applies to all figures. Note that the axis ranges are not identical across the four figures in order to show the pressure effect more clearly. The results shown assume equilibrium combustion with fuel properties, per Table 5.2 of Chapter 5. They are determined using CEA [74].

effect of chamber pressure on performance (I_{sp} and C^*, respectively) for a selection of common propellant combinations.

All the performance data in Fig. 7.2 and Fig. 7.3 is calculated using the following assumptions:

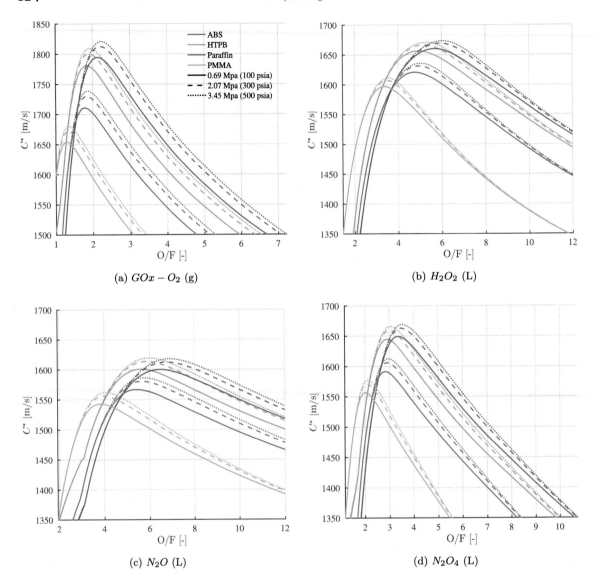

FIGURE 7.3 Ideal C^* versus O/F ratio and chamber pressure for selected hybrid propellant combinations. The legend in (a) applies to all figures. Note that the axis ranges are not identical across the four figures in order to show the pressure effect more clearly. The results shown assume equilibrium combustion with fuel properties per Table 5.2 of Chapter 5. They are determined using CEA [74].

- Shifting equilibrium combustion
- Propellant temperature: 293.15 K
- Nozzle area ratio: 40:1
- Fuel properties per Table 5.2 of Chapter 5.

An increased combustion chamber pressure also leads to a reduction in the size and mass of the nozzle. This can be seen by evaluating Eq. (7.2) (which is Eq. (3.64), solved for the throat area) and recognizing that at a given propellant mass flow rate, $\dot{m}$, and temperature, T, the design nozzle throat area, A_{th}, is inversely proportional to the chamber pressure, P_c. The smaller nozzle throat area leads to a proportional reduction in the exit area for the same area ratio, and therefore also to a reduction in nozzle length. Thus increasing the chamber pressure can be beneficial for mass and length-constrained systems, see Eq. (7.2).

$$A_{th} = \frac{\dot{m}_p C^*_{ideal} \eta_{C^*}}{P_c C_d} \qquad (7.2)$$

Note that C_d in Eq. (7.2) is a function of geometry between the combustion chamber and nozzle throat. As an initial approximation, $C_d = 1$ can be assumed. η_{C^*} is the combustion efficiency discussed in Section 7.5.2.

The above discussion implies that increasing chamber pressure only has positive effects, reducing the system mass and volume. However, this is not necessarily true. Increasing the chamber pressure may result in an increase chamber structural mass (if the chamber wall thickness is above the minimum material thickness). Since the standard for stable combustion is generally pressure variations within $+/-5\%$ of the mean, a pressure drop into the combustion chamber large enough to ensure positive upstream pressure with that amount of pressure fluctuation is necessary. Therefore for typical operating pressures of hybrid rockets, a margin of 15–20% is needed between the combustion chamber pressure and the oxidizer tank pressure at the end of the liquid burn phase [2]. Therefore increasing the chamber pressure may also increase the required feed system pressure and thus will increase the MEOP for the oxidizer and pressurant feed lines and tanks. This, in turn, will likely increase the tank, feed line, and pressurization system mass. Increasing chamber pressure also serves to increase radiative heat transfer in the combustion chamber, thereby increasing the required mass of insulation and potentially decreasing the system performance due to an in-crease in nozzle erosion. Thus the design chamber pressure is a nuanced decision that can be difficult to quantify without a dedicated development test program or representative erosion data.

Chamber pressures in the range of 1–6.9 MPa (150–1000 psia) are typical for hybrid rocket motors (e.g., [32]). For simplicity, select an initial chamber pressure in the middle of that range to start the design process and then iterate on this value if required to reduce the potential erosion (a relationship between erosion/ablation rate and pressure would be needed to capture this behavior), or improve system packaging. Note for nitrous oxide: the chamber pressure should be below saturation pressure of N_2O so the droplets vaporize across the injector instead of resaturating [47].

Knowns: Target P_c.

7.4.4 Nozzle area ratio, shape and efficiency

Most rocket nozzles have a converging (subsonic) section, a throat (sonic), then a diverging (supersonic) section. These are also known as De Laval nozzles. The shape of the subsonic portion is not critical because anything that is relatively smooth will have minimal energy losses. Therefore the focus of nozzle design is on optimizing the shape of the diverging section of a nozzle. Since the flow velocity is supersonic, there is greater potential for losses [2]. Conical and bell nozzles are the most common nozzles used.

The nozzle area ratio, $\mathcal{R}$, is given by Eq. (7.3). Here, A_e is the cross-sectional area of the nozzle exit, A_{th} is that of the nozzle throat, r_e is the nozzle exit radius, and r_{th} is the nozzle throat radius.

$$\mathcal{R} = \frac{A_e}{A_{th}} = \frac{r_e^2}{r_{th}^2} \qquad (7.3)$$

All that is needed now is the area ratio, as the throat area will be calculated in the following sections. Area ratios between about 3–25 are used between sea level and 10 km, and ratios between 40–100 can be used for high altitude/in-space applications, though higher ratios (up to 400) have been used [6].

At this point in the design process, the most useful information is the nozzle efficiency, η_n. There are many types of losses in a nozzle as introduced in Ref. [6]. The real nozzle efficiency should take as many of these into account as possible. The divergence losses are the most widely known; however, if they are the only thing accounted for in the design, the propulsion engineer is likely to overpredict performance. The various efficiencies can be multiplied together to produce an overall nozzle efficiency to apply to the design [223].

1. Divergence losses, $\eta_{n,d}$: the flow exits the nozzle at an angle (see Eq. (7.6) for conical nozzles and Fig. 7.5 for bell nozzles).
2. Low nozzle contraction areas or a small chamber/port cross-sectional area compared to the nozzle throat causes pressure losses that reduce performance. Ref. [6] gives a performance loss of an estimated 0.31% for a chamber-to-throat area ratio of 3.5:1 and up to 1.34% for a 1:1 ratio.
3. Boundary layer/wall friction losses (usually less than 1% performance loss [6]).
4. Two-phase flow losses, $\eta_{n,2P}$. Very small particles ($D < 0.005$ mm) do not impact performance. A small mass fraction (less than 6%) of small particles ($D < 0.01$ mm) corresponds to a reduction in performance of $\approx 2\%$ ($\eta_{n,2P} \approx 0.98$) . Larger particles or larger mass fractions of particles could see much larger losses, e.g., larger solid particles or liquid droplets can lead to a 5% [6] - 10%[24] reduction in performance ($\eta_{n,2P} = 0.9$–0.95). High area ratio nozzles operating at high altitude/space can see condensation or even freezing (e.g., H_2O snow). However, the effect on performance from condensation is small unless a large amount of snow is produced.

5. Chemical reactions within the nozzle. Ref. [6] reports this to be about a 0.5% reduction in performance. However, hydrocarbons are typical hybrid fuels, and the reaction of CO_2 formation is much slower than of other common products, e.g., H_2O. This can account for a substantial decrease in nozzle efficiency [155] when compared to other propellants (e.g., H_2 and O_2).
6. Nozzle throat erosion. Erosion of the nozzle throat diameter in small nozzles can reduce the nozzle area ratio and, in all nozzles, can introduce an uneven throat surface with increased boundary layer losses. Ref. [6] found that erosion of the nozzle throat diameter by 1-6% would drive a decrease in I_{sp} of up to 0.7%.
7. Incomplete mixing or combustion.
8. Real gas properties. Refs. [6] and [131] reported this as reducing η_n by 0.2–0.7%, and 0.4%, respectively.

Conical nozzles

Conical nozzles are simple and are used when manufacturing costs outweigh performance requirements or if large amounts of metallic additives are being used in the fuel. Conical nozzles are defined by their half angle, θ_e, see Fig. 7.4. The standard has become a 15-degree conical nozzle, $\theta_e = 15°$, since it is a good compromise between mass, length, and performance. The length of the expanding part of a conical nozzle is given by Eq. (7.4).

$$L_{n,cone} \approx \frac{r_e - r_{th}}{tan(\theta_e)} \tag{7.4}$$

In a real nozzle contour, a circular inlet radius, $r_{th,inlet}$, of 0.5–1.5 r_{th} leads into the conical portion of the nozzle, see Fig. 7.4, increasing the nozzle length [2]. The nozzle length is more accurately given by Eq. (7.5).

$$L_n = \frac{r_{th}(\sqrt{\mathcal{R}} - 1) + r_{th,inlet}(sec(\theta_e) - 1)}{tan(\theta_e)} \tag{7.5}$$

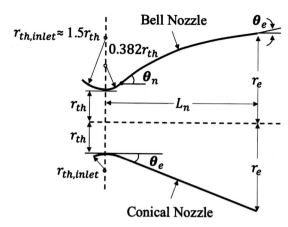

FIGURE 7.4 Comparison of a parabolic approximation of a Bell Nozzle (Rao Nozzle) and a Conical Nozzle, adapted from Ref. [2].

There is a reduction in nozzle efficiency in conical nozzles because not all of the exit flow is in the axial direction, the divergence loss introduced above. For a 15-degree nozzle, this corresponds to $\eta_{n,d} = 0.983$ (for the flow direction correction only). This can be found from Eq. (7.6), from Ref. [2], and is plotted in Fig. 7.5. However, as described in Section 7.4.4, there are many other contributing factors to the nozzle efficiency.

$$\eta_{n,d} = \frac{1}{2} \left(1 + cos(\theta_e)\right) \qquad (7.6)$$

Here, θ_e is the angle between the axial direction of thrust and the exit angle of the nozzle. This is equal to the half angle in a conical nozzle, see Fig. 7.4. See Ref. [24] for a derivation of this equation.

Bell nozzle

Bell nozzles have a large angle expansion section directly behind the throat (θ_n is 20° to 40°) [2]. The flow is expanded rapidly up until an inflection point at which the expansion angle is at a maximum. The expansion continues at a decreasing rate after the inflection point. The angle at the exit plane is small, usually less than 10°, to minimize the divergence losses discussed above.

The difference between the inflection and exit angles is called the turn-back angle.

The method of characteristics is used to design these nozzles. The propulsion engineer is pointed to Ref. [256] for an explanation of this numerical technique. Rao nozzles make a parabolic approximation for the bell expansion [257] that is also used for their design.

At this point in the propulsion system design process, the most useful information is the length of the nozzle. Bell nozzles can be shorter than conical nozzles, which helps with packaging constraints and reduces mass. Bell nozzles from 75–85% length have been shown to be nearly as efficient as full-length conical nozzles with the same area ratio [6]. Performance begins dropping beyond that point (see Fig. 7.5), so 80% nozzle length is most often adopted. The length of a bell nozzle can be calculated using the percentage reduction times L_n from Eq. (7.5).

Fig. 7.5 compares the divergence losses of the two nozzle types. The advantages of the bell nozzle come at an increase in manufacturing cost. However, modern manufacturing advances, such as 3D printing, have minimized this concern.

Note on bell nozzles for fuels with metallic additives: The bell shape is quite efficient at turning gases. However, unburned solid particles, such as Al_2O_3, have been observed to continue in a straight line and hit the nozzle wall at high velocity. This has caused appreciable damage (e.g., erosion of the nozzle wall). Therefore inflection and turn-back angles are typically much smaller for metalized propellants than for pure liquids. Borrowing from solid rocket propulsion, bell nozzles with metalized propellants generally have an inflection angle in the range of 20–26° and a turn-back angle of 10–15°. While liquids can be up to double those values: 27–50° for inflection angle and 15–30° for turn-back angle. It should be noted that the performance benefit of the bell nozzle will be somewhat reduced for a hybrid with solid particles entrained in the exhaust [6].

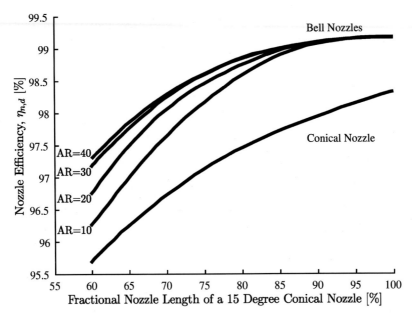

FIGURE 7.5 Divergence losses for conical and bell nozzles as a function of the fractional nozzle length compared to a 15-degree conical nozzle. Note that AR is the nozzle area ratio (elsewhere referred to as $\mathcal{R}$). The figure is made using data from Ref. [2].

Knowns: Initial selection of the nozzle area ratio, $\mathcal{R}$, and a prediction for η_n.

7.5 Performance calculation

The combustion chamber pressure, P_c, nozzle area ratio, $\mathcal{R}$, and the thermochemical properties of the propellants are needed to evaluate performance in terms of C^*, I_{sp}, and I_d. A chemical equilibrium solver is used to determine the final composition of the mixture after combustion by finding the composition that minimizes the Gibbs free energy of the mixture. The details of these calculations are provided in Chapter 4. In this chapter, a one-dimensional solver: Chemical Equilibrium with Applications (CEA) [74] is used to perform all such calculations and to help determine the target O/F. The selection of O/F resulting in maximum specific impulse, I_{sp}, can optimize the total mass of the final design

for some systems [258] and thus will be used as the design criteria here. To determine this O/F, the ideal specific impulse, $I_{sp,ideal}$, is evaluated across the range of oxidizer-to-fuel mass ratios (O/F) for the propellant combination, with the O/F corresponding to the maximum ideal specific impulse selected for use for the remainder of the design.

Note that there can be instances were nonideal O/F ratios are desirable, such as when volume constraints outweigh performance. The total system mass is also not *guaranteed* to minimize at the peak I_{sp}, and it may instead optimize at the O/F that maximizes density impulse, I_d, or C^*, or potentially even at some other less intuitive value. Other considerations, such as wall cooling, stability, or even rocket geometry, may drive the design O/F away from the value for peak I_{sp} in order to minimize mass. The average O/F is chosen based on the requirements of the design, and this value is used in the chemical equilibrium calculations. As a starting point, it is recommended to use the O/F that maximizes $I_{sp,ideal}$, but the propulsion engineer should feel empowered to explore other O/F ratios for their design.

Knowns: Target average O/F

7.5.1 Chemical equilibrium solver

The propulsion engineer now has enough information to calculate the chemical equilibrium conditions at the design O/F. A chemical equilibrium solver should be used to determine the final composition of the mixture at that design point as well as the ideal performance, $I_{sp,ideal}$, and the properties of the combustion products at multiple stations, including ratio of specific heats (γ), temperatures, and densities.

Inputs: Target average O/F, nozzle area ratio ($\mathcal{R}$), chamber pressure (P_c), propellant selection, and initial propellant temperature

Knowns: The outputs of the chemical equilibrium solver include the ideal performance, such as $I_{sp,ideal}$. Also, the composition of the combustion products at multiple stations, including ratio of specific heats (γ), temperatures, and densities. Finally, the composition of the exhaust, e.g., concentration of oxidizing species, is also a result of this analysis.

7.5.2 Combustion efficiency

The chemical equilibrium solver will output an ideal performance. If the equilibrium is allowed to progress along the length of the nozzle (shifting equilibrium), the ideal performance is typically somewhat optimistic. If equilibrium is frozen at the combustion chamber or nozzle throat, the ideal performance is somewhat conservative. Further discussion of this is provided in Section 4.9 of Chapter 4. The real performance of the rocket will also vary from the ideal performance based on how well the motor is optimized. The reported C* efficiency, η_{C*}, will always be relative to an ideal C* that will differ depending on whether frozen or shifting equi-

librium is assumed. The assumption of shifting equilibrium will be assumed for all ideal calculations presented in this book, and thus the reported C* efficiencies, η_{C*}, will also be relative to ideal values calculated with shifting equilibrium. Eq. (7.7) shows how to calculate $I_{sp,real}$ using an assumed η_{C*}.

$$I_{sp,real} = I_{sp,ideal}\eta_{C*}\eta_n \qquad (7.7)$$

The authors have found a challenging but achievable value for the C* efficiency to be 95% ($\eta_{C*} = 0.95$). This is consistent with the achieved ground results of Ref. [177]. It is not recommended that this value be used for all designs since a reasonable amount of optimization would be necessary to achieve that performance (e.g., enhanced mixing or residence time). Ref. [131] compared the performance of hydrocarbon/N_2O motors with (84 tests) and without (44 tests) mixing devices and found those with a mixing device had consistently higher performance (by about 10%) for the short duration tests analyzed. However, a trade of the mass required to achieve high efficiency vs the lower dry mass but lower performance alternative should be completed. The size of the motor can also play a role in this trade. Small hybrids often report η_{C*} in the range of 0.75–0.90. Finally, the amount of data available for a design similar in size, thrust, injector configuration, etc., should affect how aggressive one can be in initial performance assumptions.

Knowns: Assumed or target η_{C*} and $I_{sp,real}$. These values will eventually need to be determined from testing.

7.5.3 Nozzle exit pressure and thrust coefficient

The nozzle exit pressure, P_e, and the ideal thrust coefficient, C_F, are outputs of most chemical equilibrium solvers. They can also be calculated manually from the chamber pressure, P_c, the ratio of specific heats, γ, and nozzle area ratio, $\mathcal{R}$. The equations to calculate these values are provided here so that the propulsion engineer has a better understanding of them, but it is also acceptable (and generally more accurate) to take the values for P_e and C_F directly from the chemical equilibrium code output.

Eq. (7.8) can be solved numerically for the exit pressure, P_e. This equation assumes one-dimensional, frozen, isentropic flow and would need 3D corrections and corrections for γ to be completely accurate.

$$\left(\frac{\gamma+1}{2}\right)^{\frac{1}{\gamma-1}}\left(\frac{P_e}{P_c}\right)^{\frac{1}{\gamma}}\sqrt{\frac{\gamma+1}{\gamma-1}\left[1-\left(\frac{P_e}{P_c}\right)^{\frac{\gamma-1}{\gamma}}\right]}$$
$$=\frac{1}{\mathcal{R}} \qquad (7.8)$$

The nozzle exit pressure is important, particularly relative to atmospheric pressure, P_a, as together they dictate how the flow will behave at the exit of the nozzle:

- Under expanded: $P_e > P_a$. Expansion waves will occur at the nozzle exit, with a normal shock downstream.
- Perfectly expanded: $P_e = P_a$. Ideal case with mostly axial flow.
- Over expanded: $P_e < P_a$. Oblique shocks will occur at the nozzle exit. Potential for separated flow within the nozzle when highly over expanded.

Most nozzles can not change the exit pressure to match the altitude through which they are flying (aerospikes are the counterexample of this). When the design exit pressure is higher or lower than the ambient pressure, the performance is reduced (typically on the order of 1–5% [6]). This can also decrease thrust substantially as will be shown by the contribution of the right most term of Eq. (7.9).

Next, the ideal thrust coefficient can be obtained from Eq. (2.4) or in terms of γ, Eq. (2.6). It should be noted that C_F is maximized when $P_e = P_a$, perfectly expanded. Eq. (2.6), has been rewritten below as Eq. (7.9), where the total pressure in the combustion chamber has been estimated to equal P_c (this is equivalent to assuming that velocities in the combustion chamber are low). This equation also assumes one dimensional, frozen, isentropic flow of a calorically perfect gas.

$$C_F = \left\{\left(\frac{2\gamma^2}{\gamma-1}\right)\left(\frac{2}{\gamma+1}\right)^{\frac{\gamma+1}{\gamma-1}}\left[1-\left(\frac{P_e}{P_c}\right)^{\frac{\gamma-1}{\gamma}}\right]\right\}^{\frac{1}{2}}$$
$$+\left(\frac{P_e}{P_c}-\frac{P_a}{P_c}\right)\mathcal{R} \qquad (7.9)$$

Knowns: The nozzle exit pressure, P_e, and thrust coefficient, C_F

7.5.4 Propellant mass calculation

The real performance can then be used to calculate the mass of propellant required to achieve the mission objective: delivering the required ΔV to the spacecraft with the specified total initial wet mass. This is done through the rocket equation, originally presented as Eq. (2.27), but rewritten here as Eq. (7.10).

$$m_p = m_i \left(1 - e^{\frac{-\Delta V}{I_{sp,real}\, g_0}}\right) \qquad (7.10)$$

Here m_p is the usable propellant mass, m_i is the initial or total spacecraft wet mass, $I_{sp,real}$ is the ideal specific impulse of the motor multiplied by the c^* and nozzle efficiencies (see Eq. (7.7)), and g_0 is the standard gravitational acceleration, 9.81 m/s^2. It is critical to remember that m_p is only the usable propellant. The residuals need to be added prior to sizing the motor case. This will be discussed in Section 7.9.

The usable masses of the fuel and oxidizer can be found from the O/F ratio.

$$m_F = \frac{m_p}{1 + O/F} \qquad (7.11)$$

$$m_O = \frac{O/F}{1 + O/F} m_p \qquad (7.12)$$

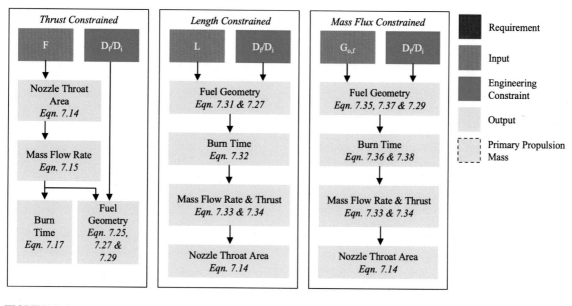

FIGURE 7.6 Alternative approaches to calculate hybrid rocket ballistics.

Knowns: The usable propellant, fuel, and oxidizer masses, m_p, m_F, and m_O, respectively.

7.6 Motor ballistics, packaging and outputs

At this point, the propulsion engineer is ready to start modeling the ballistics and sizing the motor. The simplified regression rate law (Eq. (3.48) from Chapter 3 and Eq. (7.1) in Section 7.4.1 of this chapter) is used and is included again for convenience, as Eq. (7.13). Recall that the derivation of this equation makes use of the simplification that the regression rate depends only on the oxidizer mass flux.

$$\dot{r} = a_o G_o^n \qquad (7.13)$$

7.6.1 Design process options

Packaging constraints and motor ballistics are used to design the combustion chamber, but there are actually multiple valid ways to design the hybrid rocket at this point. Depending on which parameter is driving the design (e.g., target thrust, fuel grain length, or end-of-life mass flux), a slightly different process is followed to determine the fuel grain geometry, burn time, and oxidizer mass flow rate. Fig. 7.6 lays out the design process for three of the possible driving design constraints. All of these processes use the same underlying set of equations to ensure that the regression rate law of Eq. (7.13) is always applied consistently, and conservation of mass is not violated. Examination of Fig. 7.6 shows that although the same overall set of equations is used, they are rearranged as required to solve for the various parameters. Other design processes, and different driving constraints, may also be used as long as the equations are applied consistently. It is very important that regardless of the design process used, the checks of Section 7.8 are always applied to ensure that the design is reasonable.

The discussion throughout this section assumes that the design is constrained to achieve a target thrust. If the propulsion engineer would like to design for a rocket that is instead constrained by fuel length or minimum oxidizer mass flux, then the process of Fig. 7.6 should be followed.

7.6.2 Target thrust

A hybrid rocket design might be thrust-constrained for various reasons. Some hybrid rockets have a target thrust defined so that they can be a drop-in replacement for an existing propulsion system or so that they meet a minimum thrust requirement for a mission. A hybrid rocket may also have a maximum thrust defined so that it has low thrust to ensure it is controllable from an attitude control perspective. Regardless of the reason for designing an average target thrust, it is assumed at this point that the propulsion engineer knows the target thrust of their rocket.

Note that some general rules of thumb can be applied early in the design process if thrust requirements are likely to drive the design. As an example, if high thrust is needed, then liquefying (aka high regression rate) fuels should be favored over classical fuels and/or nontraditional designs should be considered, such as swirl injection hybrids or multiport fuel grains. If low thrust is desired, e.g., because of acceleration limits on the payload, then slower-burning fuels or perhaps an end-burning hybrid design should be favored in the design trade. Thinking through such thrust considerations early in the design phase can save time calculating performance and iterating on unsuitable designs.

Knowns: Target thrust, F.

7.6.3 Nozzle throat area

The nozzle throat area, A_{th}, can be determined using the design thrust, F and the thrust coefficient, C_F, see Eq. (7.14). The discharge coefficient, C_d in this equation (and Eq. (7.15)) is a function of geometry between the combustion

chamber and nozzle throat. As an initial approximation, $C_d = 1$ can be assumed.

$$A_{th} = \frac{F}{C_F P_c \eta_n C_d} \qquad (7.14)$$

The average propellant mass flow rate, $\dot{m}_p$, can be determined from Eq. (7.15) or (7.16).

$$\dot{m}_p = \frac{P_c C_d A_{th}}{\eta_{C^*} C^*} \qquad (7.15)$$

$$\dot{m}_p = \frac{F}{I_{sp} g_0 \eta_{C^*} \eta_n} \qquad (7.16)$$

Knowns: Throat area, A_{th}, propellant mass flow rate $\dot{m}_p$

7.6.4 Burn time

The burn time, t_b, can be calculated easily based on the propellant mass flow rate and total usable propellant mass, see Eq. (7.17).

$$t_b = \frac{m_p}{\dot{m}_p} \qquad (7.17)$$

The oxidizer mass flow rate, $\dot{m}_O$, which is assumed to be constant, can then also be calculated easily via Eq. (7.18).

$$\dot{m}_O = \frac{m_O}{t_b} \qquad (7.18)$$

Knowns: Burn time, t_b, oxidizer mass flow rate $\dot{m}_O$

7.6.5 Fuel grain geometry

A hybrid rocket motor requires a specific fuel grain geometry to achieve the desired performance similarly to a solid rocket motor. The port diameter, D_p, will set the oxidizer mass flux, which changes with time based on the regression rate equation, see Eq. (7.19).

$$\frac{dD_p}{dt} = 2\dot{r} = 2a_o G_o^n \qquad (7.19)$$

Where, D_p is the diameter of the port. Eq. (7.20) can be found by substituting in for the oxidizer mass flux, assuming a known fuel grain geometry. In this case, a single, cylindrical port is used.

$$\frac{dD_p}{dt} = \frac{2^{2n+1}a_o}{\pi^n}\frac{\dot{m}_O^n}{D_p^{2n}} \tag{7.20}$$

If the oxidizer mass flow rate is constant, a further simplification can be made, and Eq. (7.21) can be used to calculate the instantaneous port diameter. Note that the empirical constants a_o and n need to be in SI units. It is easy to make errors in equations using the regression rate coefficients if the units are not consistent, see Section 6.5.2 of Chapter 6 for a discussion of unit conversion for these coefficients.

$$D_p(t) = \left[D_i^{2n+1} + \frac{(2n+1)(2^{2n+1})a_o}{\pi^n}\dot{m}_O^n t\right]^{\frac{1}{2n+1}} \tag{7.21}$$

Eq. (7.21) cannot be solved until the fuel grain geometry is determined (D_i must be known). The fuel grain geometry is a function of both structural integrity and rocket ballistics.

Fuel grain structural integrity

The fuel grain also needs to withstand structural loads. For a center-perforated fuel grain, this generally materializes in the ratio of initial to final diameters (or radii) of the fuel grain, see Fig. 7.7.

Chamber pressure is not the only structural load; thermal changes can produce internal stress (grain cool down during casting and thermal cycling), self-induced loads during flight, slump (gravity), shock, and aerodynamic loads should also be considered. These can be more or less severe, depending on the application. A finite element analysis will eventually be necessary. Ref. [259] gives a good overview of this process. However, simple calculations focused on pressure loading (or another unique driving load) are appropriate for this point in the design.

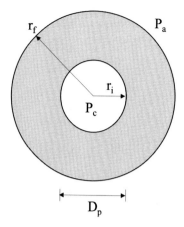

FIGURE 7.7   Fuel Grain Geometry.

The fuel grain can be treated as thick wall cylinder to determine the stresses. The radial and hoop stresses, σ_r and σ_θ, respectively, are inversely proportional to the radius of the fuel grain and given by Eqs. (7.22), also known as Lamé's equations. The calculated stresses would then need to be compared against the failure criteria for the material of interest.

$$\sigma_r = \frac{r_i^2 P_c - r_f^2 P_a}{r_f^2 - r_i^2} - \frac{(P_c - P_a)r_i^2 r_f^2}{(r_f^2 - r_i^2)r^2} \tag{7.22}$$

$$\sigma_\theta = \frac{r_i^2 P_c - r_f^2 P_a}{r_f^2 - r_i^2} + \frac{(P_c - P_a)r_i^2 r_f^2}{(r_f^2 - r_i^2)r^2} \tag{7.23}$$

Higher ratios of the inner radius to the final or outer radius lead to increased stresses, particularly at the initial fuel grain surface.

Another consideration in evaluating the fuel grain geometry is the amount of volume taken up by fuel, also known as the volumetric loading of the fuel, V_L. It is given by Eq. (7.24). For a cylindrical port, as is assumed here, this is also dependent on the port diameter ratio, $\frac{D_f}{D_i}$.

$$V_L = 1 - \left(\frac{D_i}{D_f}\right)^2 \tag{7.24}$$

Often a port diameter ratio, $\frac{D_f}{D_i}$, is selected that balances between the desire for high volumetric loading, V_L, (high $\frac{D_f}{D_i}$) and the structural limitations of the fuel grain (low $\frac{D_f}{D_i}$). For high regression rate fuels, such as paraffin and PE wax, a reasonable final to initial fuel grain radius ratio is 2–3. This number can be increased by using stronger materials or adding strength additives to the fuel grain.

Once the ratio of final to initial port diameters is selected, the initial port diameter, D_i, can be found from Eq. (7.25). Again, this equation assumes a constant oxidizer mass flow rate, $\dot{m}_O$.

$$\frac{D_f}{D_i} = \left(1 + \frac{(2n+1)2^{2n+1}a_o \dot{m}_O^n t_b}{D_i^{2n+1}\pi^n} \right)^{1/(2n+1)}$$
(7.25)

The final port diameter can be determined by Eq. (7.26) or more simply via Eq. (7.27).

$$D_f = \left\{ \left[\frac{(2n+1)2^{2n+1}a_o}{\pi^n} \right] \times \frac{\dot{m}_O^n t_b}{1 - (D_i/D_f)^{2n+1}} \right\}^{1/(2n+1)}$$
(7.26)

$$D_f = \left(\frac{D_f}{D_i} \right) D_i$$
(7.27)

Alternatively, in terms of the volumetric loading, V_L and performance (specific impulse, $I_{sp,real}$, and total impulse, I_{real}) the final port diameter can be determined from Eq. (7.28):

$$D_f = \left\{ \left[\frac{(2n+1)2^{2n+1}a_o}{(I_{sp,real}g_0\pi)^n} \right] \right.$$
$$\left. \times \frac{\left(\frac{O/F}{1+O/F} \right)^n I_{real}^n t_b^{(1-n)}}{1 - (1 - V_L)^{(2n+1)/2}} \right\}^{1/(2n+1)}$$
(7.28)

In this equation, I_{real} is the delivered total impulse and $I_{sp,real}$ is the real specific impulse (see Eq. (7.7)).

Fuel grain length

Once the initial and final port diameters are known, the total mass of fuel, m_F, sets the fuel grain length, L_F, see Eq. (7.29). In general, a minimum length-to-diameter ratio, L/D, is desired from packing considerations. However, there can be a desired length-to-diameter ratio range to ensure adequate mixing/performance. This check is described in Section 7.8.3.

$$L_F = \frac{4m_F}{\pi \rho_F (D_f^2 - D_i^2)}$$
(7.29)

Or in terms of the performance and volumetric loading, Eq. (7.30) gives:

$$L_F = \frac{4}{\pi \rho_F V_L} \frac{I_{real}}{I_{sp,real}g_0} \frac{1}{(1 + O/F)D_f^2}$$
(7.30)

The fuel grain diameter over time, $D_p(t)$, can also be determined at this point using Eq. (7.21).

Knowns: The port initial diameter, D_i, final diameter, D_f, port diameter over time, $D_p(t)$, fuel grain length, L_F, and volumetric loading of the fuel, V_L.

7.6.6 Additional motor ballistics equations

If either of the alternative design approaches of Fig. 7.6 are taken, then variations of the equations throughout this section (Section 7.6) must be used. These can be derived using the equations already presented but, for convenience, are included here in the form needed. For brevity, the discussion of the various terms and design decisions is not repeated. The discussion throughout Section 7.6, as well as the discussion in the following section (Section 7.8), still applies.

Equations with fuel length constrained

Some missions are very constrained in terms of packaging, and so the fuel grain length, L_F, becomes the driving constraint for the rocket. Thus the fuel grain length, L_F, is an input to this design approach. The ratio of fuel grain diameters, D_f/D_i, is also assumed to be known (see the discussion in Section 7.6.5).

The initial fuel grain diameter, D_i, can be calculated via Eq. (7.31).

$$D_i = \sqrt{\frac{4m_F}{\pi \rho_F L_F \left[\left(\frac{D_f}{D_i}\right)^2 - 1\right]}} \tag{7.31}$$

The final port diameter, D_f, can then be calculated using Eq. (7.27). The fuel grain geometry is then fully defined.

The burn time, t_b, can be determined from the fuel grain geometry and the oxidizer mass, m_O, using Eq. (7.32). Note that the equation assumes accurate knowledge of the regression rate parameters and a constant oxidizer mass flow rate. It essentially calculates the time it takes for the port to regress from the initial to final diameters (D_i and D_f, respectively).

$$t_b = \left[\frac{\left(D_f^{2n+1} - D_i^{2n+1}\right)\pi^n}{a_o(2n+1)2^{2n+1}m_O^n}\right]^{\frac{1}{1-n}} \tag{7.32}$$

The oxidizer mass flow rate, $\dot{m}_O$, is calculated via Eq. (7.18). The average overall propellant flow rate is then given by Eq. (7.33). The average thrust is given by Eq. (7.34), and the nozzle throat area can be calculated via Eq. (7.14). In these equations, C^*, C_F, and I_{sp} are the ideal values.

$$\dot{m}_p = \frac{m_p}{t_b} \tag{7.33}$$

$$F = \dot{m}_p C^* \eta_{C^*} C_F \eta_n = I_{sp}\dot{m}_p g_0 \eta_{C^*}\eta_n \tag{7.34}$$

The instantaneous oxidizer mass flux, fuel mass flow rate, chamber pressure, and thrust are now able to be calculated as functions of time using the equations in the following section, Section 7.7.

Equations with min. mass flux constrained

In missions that require low thrust, the minimum mass flux to ensure stable operation may be the driving constraint. Thus it is assumed that the input here is a minimum oxidizer mass flux, $G_o|_{min}$. For a constant oxidizer mass flow rate, the minimum oxidizer mass flow rate occurs at the end of the burn, $G_o|_{min} = G_{o,f}$. The end of burn mass flux, $G_{o,f}$, is, therefore, an input to this design process.

The ratio of fuel grain diameters, D_f/D_i, is also assumed to be known (see the discussion in Section 7.6.5). The final fuel port diameter can be calculated in terms of the total oxidizer mass, m_O, the ratio of fuel grain diameters, D_f/D_i, and the mass flux at the end of the burn, $G_{o,f}$, see Eq. (7.35).

$$D_f = \left\{\frac{(2n+1)8a_o m_O G_{o,f}^{n-1}}{\pi\left[1 - \left(\frac{D_i}{D_f}\right)^{2n+1}\right]}\right\}^{1/3} \tag{7.35}$$

The oxidizer mass flow rate, $\dot{m}_O$, is then able to be calculated via Eq. (7.36).

$$\dot{m}_O = \frac{\pi G_{o,f} D_f^2}{4} \tag{7.36}$$

The initial fuel diameter, D_i, can be calculated via Eq. (7.37).

$$D_i = \frac{D_f}{\left(\frac{D_f}{D_i}\right)} \tag{7.37}$$

The fuel grain length, L_F, is then able to be determined from Eq. (7.29). At this point, the fuel grain geometry is fully defined.

The burn time, t_b, can be calculated using Eq. (7.38).

$$t_b = \frac{m_O}{\dot{m}_O} \tag{7.38}$$

The average overall propellant flow rate is then given by Eq. (7.33). The average thrust is given by Eq. (7.34), and the nozzle throat area can be calculated via Eq. (7.14). In these equations, C^*, C_F, and I_{sp} are the ideal values.

The instantaneous oxidizer mass flux, fuel mass flow rate, chamber pressure, and thrust are now able to be calculated as functions of time using the equations in the following section, Section 7.7.

7.7 Performance predictions

Now that the fuel grain dimensions (D_i, D_f, and L_F), oxidizer mass flow rate ($\dot{m}_O$), instantaneous port diameter ($D_p(t)$), and nozzle throat area (A_{th}) are known, other important parameters can all be found as functions of time, including the regression rate, $\dot{r}(t)$, the fuel mass flow rate, $\dot{m}_F(t)$, the oxidizer to fuel ratio, $O/F(t)$, the chamber pressure $P_c(t)$ and the thrust, $F(t)$. The oxidizer mass flux is Eq. (3.49), rewritten here as Eq. (7.39) in terms of the instantaneous port diameter ($D_p(t)$). The oxidizer mass flow rate is known (assumed to be constant).

$$G_o(t) = \frac{4\dot{m}_O}{\pi D_p^2} \qquad (7.39)$$

Since the regression rate constants a_o and n are known, the instantaneous regression rate can also be solved for using Eq. (7.1).

The fuel mass flow rate, $\dot{m}_F$, can also be calculated using the regression rate and port diameter, see Eq. (7.40).

$$\dot{m}_F(t) = \rho_F \pi D_p L_F \dot{r} \qquad (7.40)$$

The chamber pressure can also be modeled as a function of time, see Eq. (7.41).

$$P_c(t) = \frac{(\dot{m}_O + \dot{m}_F)C^*\eta_{C^*}}{A_{th}C_d} \qquad (7.41)$$

Recall that C_d in Eq. (7.41) is the discharge coefficient between the combustion chamber and

nozzle throat. As an initial approximation $C_d = 1$ can be assumed. Note also that C^* in Eq. (7.41) is a function of $O/F(t)$ and, as such, is also a function of time. For simplicity, C^* can often be treated as a constant in early design if n is close to 0.5 or for short burns, see the discussion in Section 7.8.6. The nozzle throat area, A_{th} is also treated as a constant in Eq. (7.41). This is not true if there is likely to be significant throat erosion during the burn but is a reasonable assumption for a first pass at the design.

The instantaneous O/F ratio can be determined by recalling that it is equal to the mass flow rate of the oxidizer divided by the mass flow rate of the fuel. The latter is given by Eq. (7.40). The functional form of the regression rate, Eq. (3.48), along with the oxidizer mass flux (Eq. (7.39)) is used to get Eq. (7.42).

$$O/F(t) = \frac{\dot{m}_O}{\dot{m}_F} = \frac{\dot{m}_O^{1-n} D_p^{2n-1}}{4^n \pi^{1-n} \rho_F L_F a_o} \qquad (7.42)$$

Knowns: instantaneous oxidizer mass flux, $G_o(t)$, regression rate, $\dot{r}(t)$, fuel mass flow rate, $\dot{m}_F(t)$, chamber pressure, $P_c(t)$, and $O/F(t)$.

Transient thrust

A target thrust is used in the design process shown here, see Section 7.6.2; however, the transient rocket thrust is dictated by motor ballistics. It can be calculated as an output of the design process, see Eq. (7.43). The I_{sp} in Eq. (7.43) is a function of $O/F(t)$ and, as such, is also a function of time (just like C^*). This should be accounted for if a significant O/F shift is expected during the burn but, for simplicity, I_{sp} can often be treated as a constant in early design, particularly if n is close to 0.5 or for short burns, see the discussion in Section 7.8.6.

$$F(t) = (\dot{m}_O + \dot{m}_F)C^*\eta_{C^*}C_F\eta_n = I_{sp}\dot{m}_p g_0 \eta_{C^*}\eta_n \qquad (7.43)$$

In Eq. (7.43), C^* and I_{sp} are the ideal values predicted from the chemical equilibrium solver. The combustion efficiency, η_{c^*}, reduces the performance to account for incomplete combustion and thermal losses. The nozzle efficiency, η_n, is being applied conservatively since, in this form, it applies to both the momentum and pressure terms. However, the lack of data on the effect of boundary layer shear on performance makes this conservatism necessary.

Knowns: thrust as a function of time $F(t)$.

7.8 Design assessment

At this point, the rough rocket design is pretty much done. The major design choices have been made, and the propulsion engineer can iterate on it easily. However, the design needs to be evaluated to ensure there was not an error in the inputs or a fundamental design flaw that could cause the propulsion system to perform poorly/unstably or not be manufacturable. The highlighted areas below are not hard constraints but are meant to point out the areas beyond which, the design may begin to run into issues. Assuming that the design passes these checks, then the system mass can be evaluated. If the design does not pass these checks then the design needs to be updated (e.g., via a different propellant selection, target thrust, target P_c, etc.).

7.8.1 Separation in the nozzle

Separated flow within the nozzle can, at best, reduce performance (essentially to not take advantage of the entire nozzle expansion area, while having to pay for the mass of the entire nozzle). At worst, an asymmetry in the separated flow can cause control or stability issues in the vehicle and/or damage the nozzle. An empirical rule of thumb is that the flow will not separate for a nozzle with an exit pressure that is one-third of the ambient pressure [24]. If the expected nozzle pressure, P_e, is low enough that flow separation is likely, then the propulsion engineer should start the design over with a smaller nozzle area ratio, $\mathcal{R}$.

7.8.2 Operating mass flux range

The exact stability limits for oxidizer mass flux range are still actively being debated in research. The regression rate will develop a dependency on pressure at the lower and upper ranges, as depicted in Fig. 3.2 of Chapter 3 and Fig. 6.1 of Chapter 6. It is also important to understand how well the regression rate coefficients selected for the propellant combination represent the range of oxidizer mass fluxes in the design. This is discussed further in Chapter 6.

Instabilities such as chuffing (low mass flux) and blowout (high mass flux) would also be expected at the extremes of mass flux. The oxidizer mass flux of a hybrid propulsion system decreases over time since it is inversely proportional to the cross-sectional area of the fuel port, which increases with burn time. Therefore the final oxidizer mass flux in the motor should be evaluated for low-end stability. The authors have typically limited their designs to greater than 50 kg/m²s. However, the stable motor operation has been reported at as low as 0.43 kg/m²s [147]. Note that stability does not mean free of pressure dependency. On the other end of the spectrum, the Peregrine motor [260] ran mass fluxes as high as 1200 kg/m²s and struggled with flame-holding instabilities. An upper limit of 650 kg/m²s is suggested for single port paraffin/N_2O motors [260].

7.8.3 Fuel grain L/D

A reasonable fuel grain length-to-diameter ratio (L/D) allows for adequate mixing and fuel utilization. For a single port, axial injection hybrid without additional mixing enhancements, a rule of thumb is to use an $L/D_f \approx 6$ [238]. If L/D_f is significantly less than 6 then the design may need to be re-evaluated, particularly if the characteristic length, L^*, is also short.

The characteristic chamber length, L^* is defined by Eq. (7.44), where V_C is the volume of the combustion chamber and A_{th} is again the cross-sectional area of the throat (see Chapter 2, Section 2.2.9 for a discussion of the characteristic length). It is a way to describe/visualize the resonance time in the combustion chamber. Regression rate enhancements, such as those described in Chapter 3 Section 3.5 can increase the resonance time. While increased resonance time may be helpful, the extent of mixing achieved is what actually increases performance. Looking at L^* at this early phase of the design can give the propulsion engineer an idea of whether regression rate enhancements or mixing devices may be necessary to achieve the desired performance.

$$L^* = \Psi_C / A_{th} \qquad (7.44)$$

7.8.4 Mach number in fuel port

The ratio of port diameter, D_p, to nozzle throat diameter, D_{th}, is used to ensure that the Mach number at the end of the fuel port is low throughout the burn. Flow will be choked at the nozzle throat during nominal operation; therefore the Mach number in the fuel port can be estimated using Eq. (7.45). This equation is only approximate as it ignores heat addition due to combustion in the post combustion chamber and assumes isentropic flow between the fuel grain port and the nozzle throat.

$$\frac{D_p}{D_{th}} = \left[\left(\frac{\gamma+1}{2} \right)^{\frac{\gamma+1}{2(\gamma-1)}} \right.$$

$$\left. \times \frac{M}{\left(1 + \frac{\gamma-1}{2} M^2 \right)^{\frac{(\gamma+1)}{2(\gamma-1)}}} \right]^{-1/2} \qquad (7.45)$$

M in Eq. (7.45) is the Mach number at the end of the fuel port. Recall that the outputs of the chemical equilibrium calculation will have included a value for the ratio of specific heats

in the combustion chamber, γ. A good rule of thumb is $M \lesssim 0.3$, which typically corresponds to a diameter ratio, D_p/D_{th} of at least 1.5. D_p/D_{th} is a minimum at the beginning of the burn, so, generally, the port Mach number only needs to be checked at that time (i.e., when $D_p = D_i$).

7.8.5 Maximum vehicle acceleration

The maximum acceleration of the vehicle can be a design constraint, depending on the requirements of the payload. Per Ref. [238], if the propulsion system is being used on a spacecraft with deployed solar arrays, the maximum acceleration should be 1 g (9.8 m/s²). A maximum acceleration of 3 g (29.4 m/s²) for spacecraft without solar arrays is also recommended. Launch vehicles can typically sustain much larger accelerations of 10–15 g. Human rating limits are surprisingly high, with sustained accelerations in the range of 3–10 g, transient operations up to 19–22 g, and emergencies as high as 29–38 g [261]. The thrust over time, taking real performance into account, can be found from Eq. (7.43). The acceleration is this thrust, $F(t)$, divided by the spacecraft wet mass at that time, see Eq. (7.46).

$$a(t) = \frac{F}{m_i - \int_0^t \dot{m}_p dt} \qquad (7.46)$$

7.8.6 O/F shift

As discussed in Section 3.4.3 of Chapter 3, some O/F shift over the burn time is expected for all hybrid rockets, except for the special case where the regression rate exponent, n, equals 0.5. The reduction in performance due to O/F shift is typically too low to even measure it in test, but it is more significant if the exponent n is far from 0.5.

As a sanity check on the design, it is good practice to evaluate the expected range of O/F that will be encountered across the full mission.

This can be evaluated early in the design phase via:

- A quick assessment of the O/F at BOL, Eq. (7.47), and EOL, Eq. (7.48). These can be compared qualitatively to the design O/F to ensure that the expected variation in performance is not large.

$$O/F|_{BOL} = \frac{\dot{m}_O}{\dot{m}_F}\bigg|_{BOL} = \frac{r_i^{2n-1}\dot{m}_O^{1-n}}{2a_o\rho_F\pi^{1-n}L_F} \tag{7.47}$$

$$O/F|_{EOL} = \frac{\dot{m}_O}{\dot{m}_F}\bigg|_{EOL} = \frac{r_f^{2n-1}\dot{m}_O^{1-n}}{2a_o\rho_F\pi^{1-n}L_F} \tag{7.48}$$

- An assessment of O/F evolution during the burn using the result of Eq. (7.42) evaluated at every time step for the total mission burn duration. If this approach is used, then the mean specific impulse, $\bar{I}_{sp}$ can be assessed relative to the design specific impulse, I_{sp}. If desired, the reduction in performance can be accounted for in another design iteration using the mean specific impulse $\bar{I}_{sp}$. In general, though, the impact of O/F shift on performance is less than the uncertainty in the assumed combustion efficiency, η_{C^*}, so it is typically ignored at this point.

7.8.7 Injector pressure drop

The oxidizer tank pressure must be sufficiently above the combustion chamber pressure to prevent feed system coupled instabilities, see Chapter 8, Section 8.4.2 for a discussion of this type of instability. Gaseous oxidizers can achieve this by choking the flow at (or near) the injector. Note that supersonic oxidizer in the combustion chamber can have its own issues. Chamber pressure variations are used as a means to evaluate stability, with $\pm 5\%$ commonly used as a target. Therefore up to $\pm 5\%$ uncertainty can be expected in the chamber pressure, and a larger margin on the pressure drop

across the injector is necessary. Liquid rockets use an injector pressure drop in the range of 15–20% of the chamber pressure to ensure stability [2]. If a pressurization system is used, the end-of-life pressure in the oxidizer tank should be the regulated pressure value.

The mass flow rate of a liquid through an injector is given by Eq. (7.49).

$$\dot{m}_{inj} = \sum_{N_{inj}} C_d A_{inj}\sqrt{2\rho_O\Delta P} \tag{7.49}$$

Here, N_{inj} is the number of injector elements (or holes), C_d is the discharge coefficient for the injector element, A_{inj} is the area of a single injector element, ρ_O is the density of the oxidizer, and ΔP is the pressure drop across the injector. The discharge coefficient for specific injector designs are given in Chapter 9, Table 9.4. In the final design, $\dot{m}_{inj}$ needs to be equal to the oxidizer mass flow rate $\dot{m}_O$.

The Weber number is a dimensionless number that is used to understand the likelihood of atomization across an injector. It is the ratio of the fluid's inertia to surface tension. The Weber number as applied to an injector is given in Eq. (7.50).

$$We = \frac{\rho_O V^2 D_{inj}}{\sigma_O} \tag{7.50}$$

Here, V is the fluid velocity, D_{inj} is the hydraulic diameter of the injector element, and σ_O is the surface tension of the oxidizer. The Weber number should be quite large (Ref. [24] suggests $We = 10^5 - 10^6$ are easily achievable) in order to break the flow into many tiny drops and facilitate combustion.

7.9 Residuals

Assuming that the rocket ballistics all look good based on the recommended checks of the previous section, then the propulsion engineer can start to move into determining the mass of the propulsion system. The residual mass of propellant must be accounted for when evaluating the system mass (both in terms of added propellant mass and in terms of scaling up the system to account for the additional volume required).

The usable propellant, or propellant that will be converted into the necessary ΔV, was calculated in Section 7.5.4. However, practically, there will always be some amount of oxidizer that cannot be expelled from the tank or left in the lines. This is called "residual and hold up" and can account for 1–2% of the total oxidizer unless the oxidizer is self-pressurizing [262].

The fuel residual or "sliver fraction" is dependent on the grain design and whether the grain is restrained. As the fuel grain approaches burnout, it is possible for chunks to sluff off and entrain in the flow in such large pieces that they cannot be fully burned. The propulsive efficiency is the highest at the end of the burn, further increasing the desire to minimize a sliver fraction. The authors found success achieving low sliver fractions for single port hybrids at SmallSat scale (see Ref. [10]) and medium scale [48]. Targets for sliver fraction can be around 3% for single port fuel grains. Multiport fuel grains generally have much higher sliver fractions, >12%, though high-strength multiport grains have endeavored to reduce the fuel residuals to that of the single port values [38]. The amount of insulator used in the combustion chamber can be increased to allow for some uneven burning in the fuel grain. For example, while the axial variation in the fuel grain is often ignored for simplicity, it is observed practically. Therefore adding slightly more insulation will enable the fuel grain to be burned longer while protecting the structural portion of the combustion chamber.

The uncertainty in the mixture ratio also leads to increased residuals. For hybrid rockets, this can be caused by large error bars in the regression rate parameters, variations in the oxidizer mass flow rate (due to temperature/pressure/density variations, a poorly calibrated venturi in the oxidizer line, etc.), an unexpected combustion efficiency, manufacturing tolerances in the fuel grain, or the O/F shift if $n \neq 0$. Additional fuel and/or oxidizer needs to be carried to cover this potential gap. Hybrid rockets typically have more flexibility to carry additional oxidizer, but in this case, care needs to be taken to ensure that the rocket will not run out of fuel and insulation prematurely.

The propellant masses that the tanks and combustion chamber are being sized for need to account for these uncertainties. Therefore the propellant masses used in the following sections are actually a factor times m_O and m_F, with an aggressive (for flight) choice being 1.03 or more. Note that to achieve this, the fuel sliver fraction on the ground would need to be below approximately 2% (to give additional margin for mixture ratio uncertainty).

Accounting for this residual for oxidizer is as simple as increasing the total mass of oxidizer and sizing the tanks accordingly.

For the fuel, it is recommended that the additional residual mass only be added as increased diameter beyond D_f, rather than also increasing the fuel grain length. In this case, the outer diameter of the fuel, D_o, can be calculated from Eq. (7.51). Here $m_{F,sliver}$ is the fuel residual mass, $m_{F,sliver} \simeq 0.03 m_{F,usable}$.

$$D_o = \sqrt{\frac{4 m_{F,sliver}}{\rho_F \pi L_F} + D_f^2} \qquad (7.51)$$

Knowns: fuel grain outer diameter accounting for sliver fraction, D_o.

7.10 Combustion chamber and nozzle mass

7.10.1 Combustion chamber mass

The combustion chamber mass is broken up into two main parts. The structural mass, which is responsible for the pressure and external loads. Second, the insulation, which protects the main structure from the combustion gases, will be calculated in the following section, Section 7.10.2.

The combustion chamber is generally a cylindrical pressure vessel with hemispherical or ellipsoidal end caps. The former can be expensive to manufacture and adds length to the design (bad for packaging, good for mixing and L^*). The length of the cylindrical section must be at least the fuel grain length, plus any additional length for mixing devices and/or pre/post combustion chamber for increased mixing/residence time. Assuming the pressure loads are driving, the equations in Sections 7.11.1 or 7.11.2 are acceptable for this design. The combustion chamber thickness should be checked to ensure that it is at least the minimum material thickness discussed in Section 7.11.1. The propulsion engineer needs to be careful of this as hybrids often operate at low pressure (relative to other propulsion systems) and therefore can have wall thicknesses driven by machinability limits rather than pressure loads. Chapter 10 provides an example of estimating the combustion chamber mass.

The combustion chamber maximum expected operating pressure (MEOP) should be evaluated for the worst-case combination of operating conditions; i.e. it should be calculated accounting for the highest possible propellant mass flow rate (highest fuel regression rate and oxidizer mass flow rate) and the smallest possible as-built nozzle throat [24]. A safety factor, k_s, is applied to the MEOP to determine the design burst pressure of the combustion chamber (this is essentially the same approach as for tanks though different safety factors may be used). The safety factor for combustion chambers are generally $k_s = 1.25$ for robotic spacecraft/rockets and $k_s = 1.4$ for human rated rockets [24]. Significantly larger safety factor values are often used for ground test.

7.10.2 Insulation

Insulators are often used in hybrid motors to protect (minimize) combustion chamber structural materials. In the absence of insulation, the motor case must be designed to withstand highly oxidizing gases and temperatures above 3000 K. Materials such as columbium are capable of retaining structural integrity under these conditions but are expensive and heavy. Even so, for designs where performance is of greater concern than cost, these materials can be traded against the total mass of the combustion chamber and insulator mass for an insulated design.

Insulation mass should be determined by considering the ideal expected conditions inside the combustion chamber and the total burn time. Insulators can generally be divided into two categories: ablative and non-ablative insulators. Both classes of insulators will generally experience some erosion during combustion. However, the ablative insulators will erode more quickly and will therefore generally only have a thin thermal layer at the surface of the material. This prevents significant heat transfer beyond the insulator and into the case of the motor or surrounding materials. An example of a non-ablative insulator that is typically used for ground testing is graphite. Graphite is often used in the post combustion chamber and as a nozzle throat. Graphite is chemically stable, and is able to withstand high temperatures for extended periods of time without significant erosion. However, graphite will absorb large amounts of heat, which is then transferred to the surrounding materials. This is generally not an issue for short-duration ground tests with a nitrogen or noble gas purge or a large thermal mass to conduct away the heat. However, they can be problematic for longer tests or flight-like scenarios with thin walls and without a purge. The time scales associated with heat transfer through graphite are generally sufficiently slow that peak heating of the surrounding materials occurs after test completion. The phenomena of heat transfer beyond the end of the burn is known as "soak back."

The insulator is selected so that the insulator density, ρ_{ins}, and the insulator ablation/regression rate are known or estimated. The thickness of insulating material is based on the amount of time for which the structural material needs to be protected, t_{ins}. The combustion chamber is usually split into three parts to size the insulation:

1. The precombustion chamber, which needs to be protected for $t_{ins} = t_b + t_m$, where t_m is a margin on the burn time.
2. The area behind the fuel grain, which needs to be protected for $t_{ins} \approx t_m$, or slightly longer if there is some m exponent dependence.
3. The post combustion chamber, which needs to be protected for $t_{ins} = t_b + t_m$ and usually from harsher conditions than the precombustion chamber.

It is evident that knowledge of the burn time is necessary before sizing the insulation mass, thankfully, that is known from the motor ballistics, Eq. (7.17). The burn time margin, t_m, needs to be sufficient such that heat will not penetrate the insulator to adversely affect the structure across the range of burn possibilities. A reasonable assumption for t_{margin} lies in the range of 10-15% of the burn time, or at least 3 s. However, this will need to be refined based on the particular design (e.g. how fast the valves open, how well the regression rate is known, what the fuel utilization target is, etc.).

Examples of representative erosion rates for some insulation materials (like carbon cloth phenolic and silica cloth phenolic) are provided in Table 9.6 of Chapter 9. Note that these erosion rates are provided for guidance only since erosion rates can vary significantly depending on the combustion chamber pressure and the propellant combustion products, specifically the oxidizing species. This will be covered in more depth in this chapter in Section 7.10.4 and is discussed in Section 9.8.2 of Chapter 9.

To obtain an initial estimate of insulation mass, the surface area can be estimated as the exposed area of the chamber, and its thickness can be calculated using Eq. (7.52). This approach can also be adapted to size the minimum required thickness of any mixing devices inside the combustion chamber. For mixing devices, the full burn time should generally be considered, i.e., $t_{ins} = t_b + t_{margin}$. As an initial conservative approach, the thickness calculated from Eq. (7.52) should be added to the minimum required structural thickness of any mixing devices. This should help ensure that the mass is not underpredicted since the devices generally should be structurally sound throughout the burn (to avoid plugging the nozzle).

$$th_{ins} = t_{ins}\dot{r}_{ins} \qquad (7.52)$$

Here, th_{ins} is the insulator thickness, $\dot{r}_{ins}$ is the ablation rate of the insulator at representative conditions, and t_{ins} is the amount of time that the insulator is designed to survive.

The mass of the insulator, m_{ins} is then easily calculated using the motor geometry and insulator density, ρ_{ins}. For a cylindrical post combustion chamber with internal radius r_{cc} and length L_{pcc}, the post combustion chamber insulation mass (for the cylindrical section) is given by Eq. (7.53).

$$m_{ins}|_{cyl} = \rho_{ins}\pi L_{pcc}\left[r_{cc}^2 - (r_{cc} - th_{ins})^2\right] \quad (7.53)$$

The insulation mass for a hemispherical post combustion chamber is given by Eq. (7.54).

$$m_{ins}|_{hemisphere} = \frac{4}{6}\rho_{ins}\pi\left[r_{cc}^3 - (r_{cc} - th_{ins})^3\right] \qquad (7.54)$$

The insulation mass for an elliptical post combustion chamber is given by Eq. (7.55), where r_{cc} is the semi-major axis, and r_b is the semi-minor axis of the ellipse.

$m_{ins}|_{ellipse}$

$$= \frac{4}{6}\rho_{ins}\pi \left[r_{cc}^2 r_b - (r_{cc} - th_{ins})^2 (r_b - th_{ins}) \right]$$

(7.55)

Note that Eqs. (7.54) and (7.55) are conservative in that they assume complete insulation without accounting for the nozzle throat or insulation mass in the converging section that might already be included in the nozzle mass.

In general, the fore end of the motor will be a more benign environment than the aft end of the motor. Thus the insulation in the fore-end can be sized for a reduced percentage of the total burn time, and in some cases, may not be required at all, as recirculation of oxidizer being injected into the combustion chamber protects it from the hot combustion gases. The ability to withstand the oxidizing environment and high temperature of steel can be improved with a simple coating, such as Room Temperature Vulcanizing (RTV) silicone coating [83] or a surface zirconium oxide layer [263]. RTV is often used as a cheap means of ablatively protecting materials in hybrid motors, especially, with complex surface geometries since it can be painted on, but is difficult to apply in thick layers. This is a good time to remind the propulsion engineer to read the safety data sheets for any insulator selected. RTV has plenty of health hazards associated with it.

Knowns: insulation mass, m_{ins}

7.10.3 Nozzle mass

There are several ways to do an initial sizing for a nozzle. The first would be to design a 15-degree conical nozzle (or Rao nozzle) and calculate the hoop stress at each point along the length to determine the thickness required for the structural material. Then an insulating layer would be applied to that based on the burn time. A throat insert would have to be added since it has appreciable mass as well. Finally, the nozzle will need to be connected to the rest of

the combustion chamber, so a factor similar to that for tanks (15–30%) would be applied. However, Ref. [1] introduces an empirical equation for rocket nozzles, captured here as Eq. (7.56), that is much simpler and is reasonable for this point in the design process. Eq. (7.56) estimates the nozzle mass, m_n, as a function of several parameters

$$m_n = 2.56 \times 10^{-5} \left[\frac{(m_p C^*)^{1.2} \mathcal{R}^{0.3}}{\left(\frac{P_c}{10^6} \right)^{0.8} t_b^{0.6} (tan\theta_e)^{0.4}} \right]^{0.917}$$

(7.56)

where, m_n is the mass of the nozzle [kg], m_p is the propellant mass [kg], $\mathcal{R}$ is the nozzle area ratio, P_c is the combustion chamber pressure [Pa], t_b is the burn time [s], C^* is the characteristic velocity [m/s], and θ_e is the nozzle half angle. This relationship was developed using both space motors and strap on boosters (without TVC) for solid rockets. Ref. [1] found this to be accurate to within about 20% for the data set used. Solid rockets typically have a less harsh nozzle environments than hybrids. Therefore this equation will almost certainly underestimate the mass, and it may be desirable to place a model uncertainty factor on top of the calculated value. The challenges surrounding multiple burns (e.g., soak back) were not considered here and it is critical to note that Eq. (7.56) will not scale well to small scales (SmallSat/CubeSat scale). A more ground-up approach would be needed to estimate mass for this type of nonstandard rocket scale.

Knowns: nozzle mass, m_n

7.10.4 Nozzle erosion

If the nozzle is being sized using a ground-up approach, then nozzle erosion must be considered. Hybrid rockets commonly employ ablative materials for the nozzle throat, just like solid rockets. Nozzle throat erosion, while still an area of active research, has been found to depend on several fundamental parameters, several of which will be discussed here to help the propulsion engineer make educated decisions.

- The concentration of oxidizing species in the exhaust gas is probably the most important parameter for nozzle erosion. The concentration depends on the propellant combination and the O/F ratio. Nozzle erosion is usually at a maximum near stoichiometric conditions (e.g., where the adiabatic flame temperature is maximum and it is desirable to operate the motor) and at high O/F where the nozzle material will be exposed to more oxidizer-rich products. The O/F and therefore the nozzle erosion can vary over the burn time. The addition of aluminum to the fuel grain has been shown to decrease nozzle erosion by up to 45% [264]. This is thought to mainly be due to the fact that aluminum forms aluminum oxide and decreases the concentration of oxidizing species in the exhaust gases. It can sometimes also create a protective layer over the nozzle throat. However, the particles have also been known to mechanically damage the nozzle.
- Chamber pressure also directly impacts the nozzle throat erosion since it increases heat transfer to the throat. Ref. [265] found a nearly linear increase in throat erosion rate with pressure for their simulations of HTPB with N_2O_4.
- The adiabatic flame temperature drives the wall temperature at the nozzle throat and therefore nozzle erosion.
- Real performance in testing can be different from simple predictions. The observed nozzle erosion usually doesn't start immediately and can increase over time [131], [264]. Therefore it can be difficult to design for nozzle throat erosion without dedicated test data with a flight weight design in a representative environment.

While this may sound daunting, and the detailed design will likely be, a first-order estimate is within reach. Common nozzle materials are given in Chapter 9, Section 9.8.2, along with some estimates for nozzle erosion. Graphite is a common choice, and the propulsion engineer will note that the range of regression rates given is quite large (see the discussion in the previous paragraph about the dependency on pressure and propellant combination). A constant regression rate is a good place to start at this point in the design process, e.g., a modeled regression rate for the selected propellant combination [265] or application of any empirical data available.

Knowns: estimate for the nozzle throat regression rate

7.11 Tank masses

Tanks can dominate the propulsion system dry mass. Tank sizing applies to the combustion chamber, oxidizer tanks, and pressurant tanks (if applicable). The oxidizer and pressurant tanks are selected to be either monolithic metallic tanks or Composite Overwrapped Pressure Vessels (COPVs). Three methods of approaching tank mass estimation are presented. The first introduces tank design based on the mechanical properties of a metal tank. The second method uses empirical scaling laws for Composite Overwrap Pressure Vessels (COPVs). Finally, a survey of tanks with publicly available information was completed. Note that the tank mass estimates do not include any insulation or thermal hardware mass. These masses must be included if thermal control of the tanks is required.

Recall that the required mass of oxidizer is generally increased to account for hold-up oxidizer in the feed lines, residual oxidizer in the tanks, as well as for potential off-nominal O/F during the burn. The tanks must be sized for this increased oxidizer mass. A minimum oxidizer tank ullage of 5% is also recommended as a starting point for all liquid oxidizers.

TABLE 7.1 Common combustion chamber materials. General material properties are provided for reference only, material certifications should be used for fabrication. Adapted from Table 5–15 of Ref. [1].

	Density (ρ) [kg/m^3]	Ultimate Tensile Strength (F_{tu}) [MPa]	Yield Strength [MPa]	CTE [μm/m-C]	Efficiency $\left(\frac{F_{tu}}{\rho g_0}\right)$ [km]	Cost
Aluminum (6061-T6)	2700	310	276	24	12	Low
Titanium (Ti-6Al-4V)	4480	1170	1100	8 to 10	27	High
Steel 304	8000	640	235	17	8	Low
Graphite (Carbon Fiber) [1]	1550	895		−0.6 [266]	59	High

7.11.1 Metal tanks

The masses of metallic tanks for propulsion systems are typically calculated using structural calculations for a thin-walled pressure vessel. This makes the assumption that the driving load is the storage or combustion pressure as opposed to external loads. The calculated wall thickness is checked against minimum thicknesses for manufacturability. If the pressure loads require less than the minimum wall thickness, the calculated thickness is replaced by this value. This is usually good enough for a first pass at the design. However, the external loads (primary, coupled, shock, etc.) will need to be evaluated to see if the tank or combustion chamber needs to be stiffened. The material must be chosen based on both strength and compatibility in the case of the oxidizer tank. The geometry can then be selected by packaging constraints or driven by the fuel grain geometry for the combustion chamber.

The wall thickness of a spherical tank is given by Eq. (7.57).

$$th_{sphere} = \frac{k_s P r_{sphere}}{2\sigma_y} \qquad (7.57)$$

Where σ_y is the yield stress of the tank material, P is the Maximum Expected Operating Pressure (MEOP) of the tank, and r is the tank radius. The yield stress, σ_y, for common tank materials can be found in Table 7.1. A safety factor, k_s, is used for all structural calculations. See Ref. [7] for recommended values. The safety fac-

tor multiplied by the MEOP is essentially the design burst pressure, P_b of the tank, $P_b = k_s P$.

If the required thickness of the tanks is calculated to be less than the minimum material thickness for machinability, then the tank thickness is set to this minimum material thickness. This minimum thickness varies by material. The propulsion engineer can talk to local machinists or search available materials to get a good feel for this value, but 0.762 mm (0.03 inch) will be used for all structures presented in this chapter.

The thickness of the cylindrical section will be twice that of the hemispherical section and is given by Eq. (7.58). Do not forget to check this against the minimum thickness for the material.

$$th_{cyl} = \frac{k_s P r_{cyl}}{\sigma_y} \qquad (7.58)$$

A 2:1 ellipsoid is most commonly used for propulsion tank end caps (the radius of the tank is twice the length of the end cap). The stress concentration factor, K_{sc}, for an ellipsoid is given by Eq. (7.59).

$$K_{sc,ellipsoid} = \frac{1}{6}\left[2+\left(\frac{r_{tank}}{r_b}\right)^2\right] \qquad (7.59)$$

Where r_{tank} is the semi-major axis, and r_b is the semi minor axis. Then Eq. (7.60) can be used to determine if the endcap is expected to yield.

$$\frac{k_s K_{sc,ellipsoid} P r}{th_{ell}} < \sigma_y \qquad (7.60)$$

Again, check the calculated tank thickness against the minimum thickness for machinability. The propulsion engineer has been reminded several times about this because hybrids can operate at low pressures and often need to consider this. If the minimum material thickness is larger than the required thickness for the pressure loads, then it should be used to estimate the tank mass.

The mass of the metallic oxidizer and pressurant tanks can now be determined based on the density of the selected materials, ρ_{tank}, along with the external geometry and wall thickness of the tanks. The mass of a cylindrical tank is given by Eq. (7.61) where m_{endcap} can be determined via either Eq. (7.64) or (7.65) for a hemispherical or ellipsoidal end cap, respectively. The mass of a spherical tank is given by Eq. (7.62).

$$m_t|_{cyl,total} = m_t|_{cyl} + 2m_{endcap} \quad (7.61)$$

$$m_t|_{sphere,total} = 2m_t|_{hemisphere} \quad (7.62)$$

$$m_t|_{cyl} = \rho_{tank}\pi L_{tank,cyl}\left[r_{tank}^2 - \left(r_{tank} - th_{cyl}\right)^2\right] \quad (7.63)$$

Here, $L_{tank,cyl}$ and r_{tank} are the length and radius, respectively, of the cylindrical section of the tank. Note that the volume inside the tank needs to account for the tank thickness.

The mass for a hemispherical endcap is given by Eq. (7.64).

$$m_t|_{hemisphere} \quad (7.64)$$
$$= \frac{2}{3}\rho_{tank}\pi\left[r_{tank}^3 - \left(r_{tank} - th_{sphere}\right)^3\right]$$

The mass for an elliptical endcap is given by Eq. (7.65).

$$m_t|_{ellipse} \quad (7.65)$$
$$= \frac{2}{3}\rho_{tank}\pi\left[r_{tank}^2 r_b - (r_{tank} - th_{ell})^2(r_b - th_{ell})\right]$$

The tank mass calculated to this point does not make accommodations for boss(es), mounts/skirts, and welds. Standard tank designs often

add 15–30% for this value. Then this could be refined as the design is matured.

Liquid oxidizer tanks may need a diaphragm or other propellant management device (PMD). This can be a substantial increase to the tank mass. The propulsion engineer is directed to Section 9.3.1 of Chapter 9 to learn more about PMDs and to Section 7.11.3 of this chapter to understand how a PMD can affect the tank mass.

7.11.2 Empirical tank scaling law and composite overwrapped pressure vessels (COPVs)

The tank mass can also be calculated using an empirical scaling law, Eq. (7.66). This equation can be used for both metallic tanks and for Composite Overwrapped Pressure Vessels (COPVs). It also applies to both tanks and to combustion chambers [1,24]. COPVs are tanks made up of a liner and composite material. The liner is constructed of a material compatible with the oxidizer and pressurant and prevents leaks through the composite material. COPV tank mass estimation is most often estimated in this phase of the design using the empirical scaling law given by Eq. (7.66) rather than a bottoms-up approach like that described in the previous section for metallic tanks due to the complexities of the composite lay-up.

$$\frac{P_b V}{g_0 m_t} = (PV/W) \quad (7.66)$$

In Eq. (7.66), P_b is the burst pressure for the tank [Pa], V is the tank volume [m^3], m_t is the tank mass [kg], g_0 is Earth gravity, 9.81 [m/s^2], and (PV/W) is a constant for the tank material and has units of length, in this case, meters. As mentioned earlier, the tank burst pressure is the MEOP multiplied by a factor of safety. Typical values of (PV/W) for metal combustion chambers and tanks are from 7620 to 15,240 m [24]. Recommended values of (PV/W) for composite combustion chambers and tanks (COPVs)

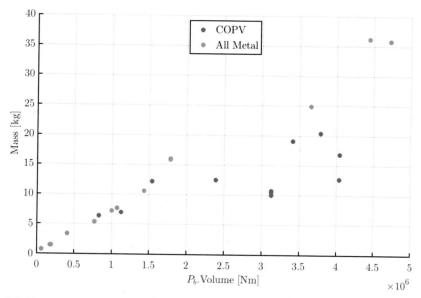

FIGURE 7.8 COPV and metallic pressurant tank masses versus tank design burst pressure times the available volume. Figure is produced using data from Ref. [267].

are 18,000–35,000 m; these were evaluated based on the data set presented in the following section. They generally agree with existing flight tank masses except that they can underpredict the mass of small COPVs, essentially very small COPV masses can approach the mass of metallic tanks. It is recommended that the propulsion engineers compare their COPV values to Fig. 7.8 and Fig. 7.9 to ensure that they are reasonable.

7.11.3 COTS tanks

A bottoms-up approach, as described in Section 7.11.1, will produce reasonably accurate metallic tank masses. However, the scaling provided in Eq. (7.66) will give an approximate tank mass estimate that can be used for the purposes of first-pass system designs and trades. Many tank vendors provide information on their existing flight tanks online. If existing tanks closely match the tank design constraints of the mission, then the published mass can be used directly since it will be more accurate than any calcula-

tion. For missions that are cost-constrained more than performance constrained, it may be desirable to use heritage tanks even if they are not perfectly suited for the mission.

Real tank masses will differ from the predicted values depending on their geometry, mounting, material, and the existence of a diaphragm or other Propellant Management Device (PMD). A comparison of COPV tank mass and metallic tank mass for pressurant tanks is provided in Fig. 7.8. Tank mass is plotted versus the burst pressure times the total internal volume. These tanks are all produced by the same tank vendor (Northrop Grumman) and use titanium metal (both for the tank liner in COPVs and for the metallic tanks). This figure shows that while similar at low pressure and small scale, COPV tanks tend to be more mass efficient than metallic tanks as the design pressure and/or tank volume increases.

Fig. 7.9 shows how COPV mass scales to larger tanks. Data is included from two different

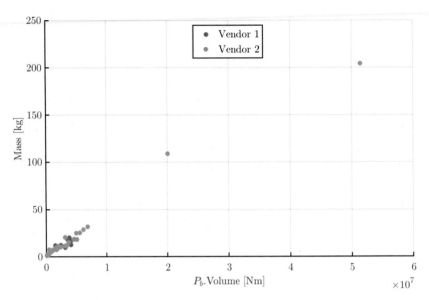

FIGURE 7.9 COPV tank masses for two tank different vendors versus tank design burst pressure times the available volume. Figure is produced using data from Refs. [268] and [267].

vendors (Northrop Grumman and Arde), and it includes different material liners.

Metallic tanks can have significant mass variation, particularly when comparing liquid propellant tanks to pressurant tanks. Fig. 7.10 is included here to show how the presence of PMDs or diaphragms tends to increase the mass of the tank. Note that Fig. 7.10 had to use MEOP instead of the tank burst pressure as the design burst pressure was not available publicly for the tanks with PMDs.

Knowns: masses of the oxidizer tank, pressurant tank and combustion chamber structural mass

7.11.4 Other tank considerations

Geometry changes due to pressure

The tanks will actually expand in diameter, D (Eq. (7.68)), and length, L_{tank} (Eq. (7.67)), when pressure is applied. The amount of expansion depends on the material properties of the tank, namely the Elastic modulus, E, and Poisson's

ratio, ν. This is important when designing the support structure for the propulsion system, for example.

$$\Delta L_{tank} = \frac{P L_{tank} D}{4 E th}(1 - 2\nu) \qquad (7.67)$$

$$\Delta D = \frac{P D^2}{4 E th}\left(1 - \frac{\nu}{2}\right) \qquad (7.68)$$

Tanks for cryogenic propellants

An oxidizer tank shrinks when loaded with cryogenic oxidizer per Eq. (7.69), where α_L is the coefficient of linear expansion for a metal tank [252].

$$V_{cold} = V_{warm}\left[1 - 3\alpha_L(T_{warm} - T_{cold})\right] \qquad (7.69)$$

7.12 Pressurization subsystem

As introduced in Section 7.4.2, there are multiple options for pressurizing the oxidizer. The most common is to utilize a tank of stored gas. It is also possible to operate a hybrid rocket with

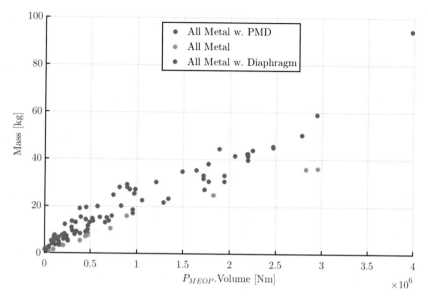

FIGURE 7.10 Metallic tank masses versus tank MEOP times the available volume for pressurant tanks and liquid tanks with PMDs or diaphragms. Figure is produced using data from Refs. [267], [269], and [270].

the oxidizer in blow-down mode, in this case, the oxidizer tank pressure and oxidizer flow rate will vary (decrease) over the course of the burn. A special use case of a "self-pressurizing," high vapor pressure oxidizer is discussed in Section 7.12.2. More often, pressurant gas is used to keep the oxidizer tank pressure approximately constant throughout the burn. This can be achieved by controlling the pressurant gas flow via either a mechanical regulator or active pressure control. The latter is colloquially referred to as "bang-bang" because a number of solenoid valves are opened/closed to keep the pressure within the desired tolerance.

7.12.1 Pressure-fed systems (stored gas)

Systems that use gas pressurization are called pressure-fed systems or stored gas pressurization systems. These regulated pressurization systems are the focus of this subsection.

The oxidizer tank design pressure is input to these calculations and needs to be selected at this point. The mass of gas necessary to maintain the tank pressure can be conservatively calculated assuming that all of the liquid oxidizer is expelled from the oxidizer tank [1]. This is conservative since, as discussed in Section 7.9, there will always be some liquid left in the tank. Conservation of mass of the pressurant gas in both the pressurant tank(s) and oxidizer tank(s) between the Beginning Of Life (BOL) and End Of Life (EOL) of the mission allows the pressurant tank to then be sized, see Eq. (7.70). Here m_{press} is the total mass of pressurant gas (the sum of the pressurant gas mass in all pressurant and oxidizer tanks), ρ_{press} is the density of the pressurant in the pressurant tank, $\rho_{press,ullage}$ is the density of the pressurant in the oxidizer tank ullage, V_{ullage} is the total ullage volume (the sum of the ullage volume in each oxidizer tank), and V_{press} is the total pressurant tank volume (the sum of the volume in each pressurant tank). Note that $\rho_{press,ullage}$ is the density of only the pressurant gas in the ullage, instead of just the total ullage density, as the ullage will likely also contain some oxidizer vapor (the amount of which will depend on the vapor pressure of the oxidizer).

$$m_{press,BOL} = m_{press,EOL}$$

$$\left[\rho_{press}V_{press}\right]_{BOL} + \left[\rho_{press,ullage}V_{ullage}\right]_{BOL}$$
$$= \left[\rho_{press}V_{press}\right]_{EOL} + \left[\rho_{press,ullage}V_{ullage}\right]_{EOL} \tag{7.70}$$

The propellant tank ullage will change (increase) over the course of the mission as oxidizer is expelled, but to first order the pressurant tank volume will not change, thus Eq. (7.70) can be rewritten as Eq. (7.71) and rearranged to solve for total pressurant tank volume, V_{press}, per Eq. (7.72).

$$\rho_{press,BOL}V_{press} + \left[\rho_{press,ullage}V_{ullage}\right]_{BOL}$$
$$= \rho_{press,EOL}V_{press} + \left[\rho_{press,ullage}V_{ullage}\right]_{EOL} \tag{7.71}$$

$$V_{press} = \frac{\left[\rho_{press,ullage}V_{ullage}\right]_{EOL} - \left[\rho_{press,ullage}V_{ullage}\right]_{BOL}}{\rho_{press,BOL} - \rho_{press,EOL}} \tag{7.72}$$

Examination of Eq. (7.72) makes it clear that to determine the amount of pressurant gas needed, the BOL and EOL pressurant gas densities must be known in the oxidizer ullage and in the pressurant tank. These are a function of temperature and pressure and can be approximated with the ideal gas law (see Eq. (4.3) in Chapter 4), or more accurately determined using a real gas equation of state (see Section 4.4 of Chapter 4), or the NIST real gas properties [271]. It is straightforward to determine the BOL densities as the BOL temperature and pressure are typically known a priori. Calculation of the pressurant gas density at the pressurant tank MEOP and maximum design temperature give the BOL density of the gas, $\rho_{press,BOL}$ (see Section 7.3.2). The pressurant tank pressure at BOL is then the pressure that corresponds to $\rho_{press,BOL}$ at BOL temperature (which is generally different from the maximum design temperature). The ullage density, $\rho_{press,ullage,BOL}$, can be calculated from the design pressure of the oxidizer tank at BOL

(or the density at the launch oxidizer tank pressure and launch temperature if the tank is not initially at the design pressure/temperature). Note that for oxidizers with high vapor pressure (such as NTO), the contribution of oxidizer vapor should be accounted for when determining the mass of pressurant in the ullage. In the BOL this can be as simple as saying that the pressurant density is a function of the propellant tank temperature and the propellant tank pressure minus the oxidizer vapor pressure at that temperature.

The EOL conditions are more challenging to determine and depend on the burn duration(s), frequency, and the existence (or not) of tank heaters and thermal isolation. Consider the gas being expelled out of the pressurant tank(s), the two bounding cases for the EOL temperature and pressure in the pressurant tank are isothermal or adiabatic blowdown of the tank [73]. If the oxidizer is expelled slowly (or via a series of short burns over a long period of time) and heat is available to heat the pressurant tanks, then the blowdown can be considered isothermal. This essentially says that the heat available from the rest of the spacecraft will equalize any cooling provided by the expansion of the pressurant gas. If the oxidizer is expelled very quickly so that there is not sufficient time for heat to be transferred back into the pressurant gas, the blowdown is considered adiabatic [1]. The adiabatic assumption is typically overly conservative, even for a single, high-thrust burn. The intermediate case, which generally represents a more realistic scenario between the two other extremes, is for the blowdown to be considered polytropic. For any of these cases, the conditions in the tank are governed by Eq. (7.73) [272]. Here, γ_p is the polytropic exponent, see Fig. 7.11. For isothermal blowdown, $\gamma_p = 1$, for adiabatic blowdown, $\gamma_p = \gamma$ (recall that γ is the ratio of specific heats, see Chapter 4 for details). The polytropic exponent for pressurant gas blowdown that is neither adiabatic nor isothermal will be some value between these two extremes.

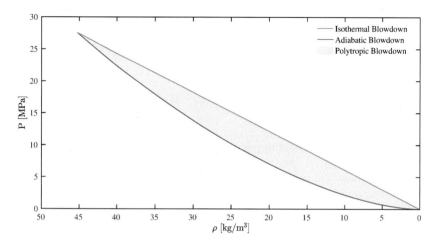

FIGURE 7.11   Pressure versus decreasing density for Helium gas with isothermal, adiabatic, and polytropic blowdown.

A full thermal model or test is generally required to accurately determine γ_p, but in the early design phase, an approximate value is usually estimated, with larger values being more conservative.

$$\frac{P_{press}}{(\rho_{press})^{\gamma_p}}\bigg|_{BOL} = \frac{P_{press}}{(\rho_{press})^{\gamma_p}}\bigg|_{EOL} \qquad (7.73)$$

The EOL pressure in the pressurant tanks is often constrained to be some minimum value depending on the system design. For obvious reasons, this pressure cannot be less than the oxidizer tank pressure, and, generally, some additional pressure margin is required above this to account for pressure drop in the feed line between the pressurant tank(s) and oxidizer tank(s). Thus the EOL pressure can be set as a design value, and the EOL density in the pressurant tank, $\rho_{press,EOL}$, can then be calculated via Eq. (7.73).

The EOL oxidizer tank pressure is simply the regulated design pressure, but the accurate determination of the ullage temperature requires a full thermodynamic model. Again, if the use of oxidizer is slow, and heat is available to heat the oxidizer tanks, then the ullage temperature can be treated as constant (i.e. it is the BOL tempera-

ture) [73]. If the propellant is used quickly, then the ullage temperature at EOL will typically be less than the BOL temperature. The pressurant tank sizing can be looked at for a few estimates of EOL ullage temperature based on prior experience with similar missions, or a thermodynamic model, like the one discussed in Ref. [252] or [273], can be developed. Assuming an EOL temperature allows the ullage density at EOL, $\rho_{press,ullage,EOL}$, to be calculated. Eq. (7.72) can then be used to determine the total pressurant tank volume needed as well as the pressurant gas mass. The number and shape of the pressurant tank(s) is a design variable that is often restricted by packaging requirements. Increasing the number of pressurant tanks is generally less mass efficient but can help with packaging and generally increases heat transfer to the pressurant gas during blowdown. This decreases γ_p, which helps reduce the mass of pressurant gas required.

Following the calculation of the required pressurant mass along with the corresponding volumes of the pressurant tank(s), the mass of these tanks is then determined using the approach discussed in Section 7.11. The pressurant tank feed system components should also be considered (see Chapter 9 for a discussion of possible components) and, if possible, estimates of the component mass should be included in the propulsion system dry mass, see Section 7.13.1 of this chapter.

7.12.2 Self-pressurizing oxidizers

High vapor pressure oxidizers are often referred to as "self-pressurizing" in the hybrid rocket community. The most commonly used self-pressurizing oxidizer is nitrous oxide; however, LOx can also be used at lower pressures. Initially the oxidizer will be at equilibrium at the beginning of life temperature and pressure, with some amount of ullage that may or may not use a second gas (such as helium or oxygen) to "supercharge" the pressure above the oxidizer vapor pressure. See the discussion on N_2O safety for why this is recommended for nitrous oxide in Chapter 11, Section 11.5.2. As liquid oxidizer is expelled from the tank, the oxidizer tank pressure drops, and when it gets below the vapor pressure of the oxidizer, some liquid oxidizer will vaporize. Generally the vaporization rate of the oxidizer is too slow to maintain a constant pressure in the oxidizer tank. The ensuing variation in oxidizer tank pressure will lead to a variation in the oxidizer flow rate into the hybrid rocket. Thus a variable flow rate of oxidizer will need to be accounted for in the ballistic model of the hybrid rocket.

Determining what the oxidizer tank pressure will be over time analytically (which is key to accurately predicting the performance and thrust of the rocket) has been the subject of much research, see, for example, [274], [176], [275], [276]. However, the relative simplicity (in implementation, not performance modeling) of this type of pressurization scheme has led to it being favored for sounding rocket applications, e.g., [71], [277] and [278].

An oxidizer tank is modeled as two parts: vapor and liquid, see Fig. 7.12. Ref. [279] explains that an initial transient phase begins with

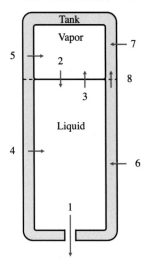

FIGURE 7.12 Simple tank model for self-pressurizing oxidizers, adapted from Ref. [176]. 1 shows the mass transfer of liquid oxidizer out of the tank. 2 is the heat and mass transfer via condensation, diffusion, and convection from the vapor to liquid, and 3 is via boiling, evaporation, diffusion, and convection from the liquid to vapor. 4-7 indicate heat transfer between regions. 8 indicates heat transfer via conduction and mass transfer via a moving boundary.

the ullage becoming subcooled. The homogeneous condensation begins, and the homogeneous two-phase mixture expands. This decreases the system pressure, which superheats the liquid, and heterogeneous nucleation begins on the tank walls. These bubbles break off and rise until they reach the interface with the free surface. It takes an appreciable time to superheat the liquid and then for vapor to reach the surface through boiling. This lag corresponds to the drop and recovery in pressure that is usually observed with self-pressurizing oxidizer flow.

Ref. [279] noted that very small changes in the initial state of the oxidizer tank, specifically the temperature, fill level and bubble population, produced substantial changes in the pressure, and visualization data.

7.12.3 Pumped systems

The use of pumps allows the oxidizer to be stored at a lower pressure, thereby reducing the associated tank mass and increasing safety. In terms of mass, pumped systems tend to trade well over gas-fed pressurization systems for high thrust applications. However, the mass savings of pumped systems for large hybrid rockets are at the cost of significantly increased system complexity. Modern electric pumps are now also making this option possible for smaller hybrid propulsion systems. The trade between a turbopump system and gas pressurization is explored more in Chapter 10, Section 10.6.

The design of pumped systems is a rich area, with the potential to get lost in a lot of technical detail and physics. Detailed design should delve into this extensive topic by consulting references such as [280], [2] or [24]. However, for the initial design approach adopted in this chapter, a more parametric first-order approach is used. The focus in this section is on mass estimation for a gas turbine-driven centrifugal pump, together referred to as a turbopump. If the turbopump dimensions are needed, they can also be estimated in a similar high-level manner to the presented mass calculation using the approaches of Ref. [6] or [1].

The mass of turbomachinery for a centrifugal turbine-driven pump, m_{pump} [kg], generally scales well with torque in the main pump shaft per Eq. (7.74) [1]. This simple equation somewhat masks the need for a slightly deeper look at the turbopump design that is actually required in order to solve for the pump torque and thereby determine the mass. However, this section will give the propulsion engineer a brief, high-level glance into the design process.

$$m_{pump} = \alpha_{pump}\tau_{pump}^{\beta_{pump}} \quad (7.74)$$

In Eq. (7.74), $\alpha_{pump} = 1.3$ to 2.6 and $\beta_{pump} = 0.6$ to 0.667 are empirical constants. For initial design, it is common to use $\alpha_{pump} = 1.5$ and

$\beta_{pump} = 0.6$. The pump shaft torque, τ_{pump} [Nm], is calculated from Eq. (7.75).

$$\tau_{pump} = \frac{\mathfrak{P}_{pump}}{N_{pump}} \quad (7.75)$$

$\mathfrak{P}_{pump}$ and N_{pump} are the pump power (in W) and rotation rate (in rad/s), respectively. The pump power is a function of the pressure change required to be generated by the pump, ΔP_{pump} [Pa], the volumetric flow rate of oxidizer through the pump, $\frac{\dot{m}_O}{\rho_O}$ [m^3/s], and the pump efficiency, η_{pump} [-], per Eq. (7.76).

$$\mathfrak{P}_{pump} = \frac{\Delta P_{pump}\dot{m}_O}{\rho_O \eta_{pump}} \quad (7.76)$$

To determine the efficiency and rotation rate of the pump the propulsion engineer now needs to consider an important pump parameter: the pump stage-specific speed, N_s [rad.m$^{0.75}$/s$^{1.5}$]. This number can be derived from dimensional analysis of key pump parameters, see Ref. [280] for the derivation, to generate Eq. (7.77).

$$N_s = \frac{N_{pump}\sqrt{\frac{\dot{m}_O}{\rho_O}}}{\left(\frac{H_p}{n_{stages}}\right)^{0.75}} \quad (7.77)$$

Here, $\dot{m}_O$ [kg/s] and ρ_O [kg/m^3] are the mass flow rate and density of the oxidizer, respectively. N_{pump} [rad/s] is the rotation rate of the pump and will be calculated in the next paragraph. H_p [m] is the head rise, given by Eq. (7.81) and n_{stages} [-] is the number of stages, given by Eq. (7.82).

Pump efficiency is a function of stage-specific speed and can be read off of Fig. 7.13. This is based on experimental data that should be representative enough for preliminary design, however, it should be noted that the maximum pump efficiency can be slightly higher than this figure, up to around 90% [6].

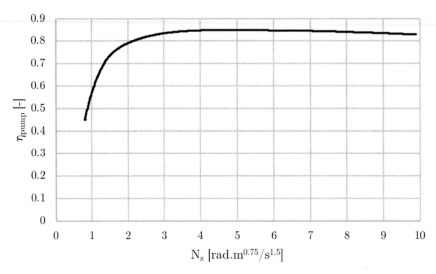

FIGURE 7.13 Pump efficiency versus stage specific speed. Figure is generated using data originally from Rocketdyne, taken from Ref. [1].

The rotation rate of the pump can be calculated by considering the maximum rate to avoid cavitation, see Eq. (7.78), and the limit on the stage-specific speed, see Eq. (7.79) (a rearranged form of Eq. (7.77)). Per Ref. [1], the stage-specific speed limitation typically also ensures that the pump rotation rates are within the safe operational limits of the bearing and turbine. The rotation rate of the pump, N_{pump}, is the lesser value of Eq. (7.78) and Eq. (7.79). If a booster pump is used, then the cavitation limit is no longer a concern, and Eq. (7.79) can be used directly.

$$N_{pump} = \frac{V_{SS}\,(NPSH)^{0.75}}{\sqrt{\dfrac{\dot{m}_O}{\rho_O}}} \quad (7.78)$$

$$N_{pump} = \frac{N_s \left(\dfrac{H_p}{n_{stages}}\right)^{0.75}}{\sqrt{\dfrac{\dot{m}_O}{\rho_O}}} \quad (7.79)$$

In Eq. (7.78), V_{SS} refers to the suction specific speed in the units of rad m$^{0.75}$/s$^{1.5}$. For cryogenic oxidizers, a reasonable suction specific speed limit is $V_{SS} = 90$ rad m$^{0.75}$/s$^{1.5}$; for other oxidizers, a reasonable suction specific speed limit is $V_{SS} = 70$ rad m$^{0.75}$/s$^{1.5}$ [1].

The Net Positive Suction Head (NPSH) [m] for the pump is calculated by Eq. (7.80). It is a measure of the margin to avoid cavitation within the pump [1]. The critical NPSH is dependent on the pump design and is often determined experimentally. The critical NPSH is defined in practice to generate a 2–3% head-generation loss. Cavitation much beyond this range creates visibly unsteady flow and can actually damage the hardware [2].

$$NPSH = \frac{P_{inlet} - P_{vapor}}{g_0 \rho_O} \quad (7.80)$$

The stage-specific speed, N_s of Eq. (7.79) is chosen to increase pump efficiency without requiring a large inducer diameter. For initial design, it is common to use a stage-specific speed of $N_s = 3.0$ rad.m$^{0.75}$/s$^{1.5}$ [1]. From Fig. 7.13, $N_s = 3.0$ rad.m$^{0.75}$/s$^{1.5}$ corresponds to a pump efficiency of $\eta_{pump} = 83.5\%$.

If Eq. (7.78) is used to size the pump (i.e., if it produced a lesser value for N_{pump} than

Eq. (7.79), and there is no booster pump), then Eq. (7.77) must be used to calculate N_s using this new value for N_{pump} from Eq. (7.78). This new value of N_s should then be used to determine the pump efficiency via Fig. 7.13.

Note that it is common to express the required pressure change across the pump as a head rise, H_p [m], as in Eq. (7.77) and (7.79). The head rise is the height a column of propellant would rise (in units of length) under pressure as given by Eq. (7.81).

$$H_p = \frac{\Delta P_{pump}}{g_0 \rho_O} \qquad (7.81)$$

The number of stages, n_{stages}, is determined from Eq. (7.82), which simply states that there is a limit of maximum increase in head pressure per stage of a centrifugal pump, $\Delta P_{stage,lim}$. For initial pump design with typical hybrid rocket oxidizers, it is reasonable to limit this to $\Delta P_{stage,lim} = 47 \times 10^6$ Pa [1].

$$n_{stages} \geqslant \frac{\Delta P_{pump}}{\Delta P_{stage,lim}} \simeq \frac{\Delta P_{pump}}{47 \times 10^6} \qquad (7.82)$$

7.13 Other masses

7.13.1 Component masses

Feed system components are discussed in Chapter 9. For a flight system, the number and mass of components must be minimized. Sketching out the plumbing and instrumentation diagram (P& ID), also called a schematic, is quite helpful here, as that will identify the number of valves, transducers, and other components. Fig. 7.14 shows several published design concept schematics. They illustrate the number and type of valves for a potential Cube-Sat propulsion system and Mars Ascent Vehicle (MAV) design. Not including the Reaction Control System (RCS) hardware, tanks or hybrid rocket combustion chamber, these designs have 9 and 29 components, respectively, including valves, regulators, pressure transducers, test ports, etc. These components are required for safety, to load propellant as well as test and operate the motors. The complexity is quite different between the two, with the major differentiator being the phase of the oxidizer. The CubeSat concept uses gaseous oxygen, while the notional MAV uses liquid MON with a stored gas pressurization system.

Unfortunately, determining the total component mass is not as straightforward as multiplying the number of components times an average mass. Component mass is generally a function of the design pressure, flow rate, materials, response time, etc. However, as a propulsion engineer gains experience, it is possible to put together a database of candidate component masses to pull from. Many vendors have catalogs or data sheets with publicly available information online. Table 7.2 has mass ranges of common components. Regulators and main flow valves are highly dependent on the design (e.g., flow rates) and cannot easily be summarized. For example, a SmallSat design might use a 0.025 kg main oxidizer valve [282], while a launch vehicle could use a 2 kg (or heavier) main oxidizer valve. A linear fit of thruster valves ranging from 10 to 2780 N from several references ([283], [284], [285]) gives an empirical equation for thruster valve mass, $m_{v,thruster}$ in kg, as a function of thrust, F in N, see Eq. (7.83).

$$m_{v,thruster} = 8.0 \times 10^{-4} F + 0.266 \qquad (7.83)$$

TABLE 7.2 Masses for common components.

Component	Mass [kg]	Reference
Fill and Drain Valve	$0.070 - 0.155$	[286], [287]
Fill and Drain with Cap	$0.090 - 0.209$	[286], [287]
Pressure Transducer	< 0.28	[288], [289]
Pyro Valve	< 0.16	[287]
Check Valve	$0.02 - 0.170$	[290]
Gas Filter (Etched Disk)	0.025	[291]
Liquid Filter (Etched Disk)	$0.160 - 1.0$	[291]
High Pressure Latch Valve	$0.340 - 0.800$	[292]
Low Pressure Latch Valve	$0.168 - 0.340$	[293]

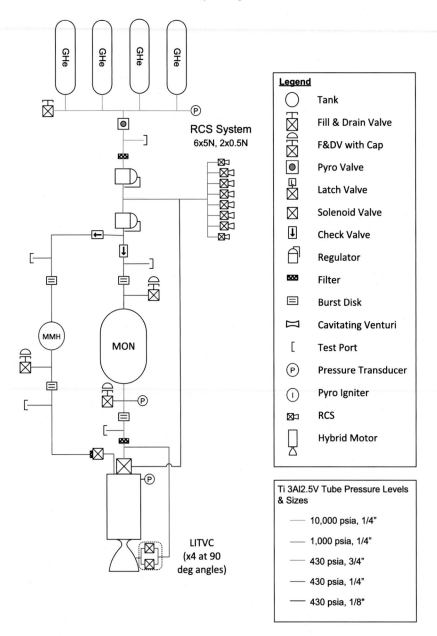

FIGURE 7.14 **Left:** Schematic for a potential CubeSat propulsion system [281] and **Right:** Mars Ascent Vehicle [48].

Chapter 9 gives a more in depth discussion of possible flight components, and the factors that should be considered as part of component selection.

7.13.2 Tubing mass

For preliminary calculations, it is typical to make some assumptions on tubing mass (e.g., assign a percentage of the total mass). To increase fidelity, the size of the tubing can be found based on pressure of the fluid and the maximum velocity desired within the tube, see Eq. (7.84). The minimum tubing size is calculated for a specified max fluid velocity and constant density. For gaseous oxygen, $V_{max} = 30$ m/s.

$$D_{tube} \geqslant \sqrt{\frac{4\dot{m}_O}{V_{max}\rho_O\pi}} \qquad (7.84)$$

The minimum diameter, D_{tube}, and the maximum expected operating pressure (MEOP), size the tubing required for the feed lines. Once a minimum diameter is calculated, the tube size should be chosen from the nearest (larger) commercially available option (see Tables 9.2 and 9.3 in Chapter 9). Note, that this table lists the tube outer diameter, OD, and wall thickness, th_w, and D_{tube} from Eq. (7.84) refers to the inner diameter ($D_{tube} = OD - 2th_w$). Also, it is easiest not to choose the *next* larger available size, but the closest larger size that matches the inlets/outlets of the components that will make up the feed system. In the United States, common component sizes are 1/4, 3/8, 1/2, 3/4 inch, etc. This will save the propulsion engineer many tube diameter transitions, which generally add feed line length and increase the build complexity. More details on feed line analysis are included in Chapter 8.

Using the outer diameter, OD, and wall thickness, th_w, from Table 9.2 or 9.3, Eq. (7.85) gives the mass of tubing as a function of tube length.

$$m_{tube} = \pi\rho_{tube}\left[\left(\frac{OD}{2}\right)^2 - \left(\frac{OD}{2} - th_w\right)^2\right]L_{tube} \qquad (7.85)$$

7.13.3 Igniter mass

The mass of the igniter is commonly ignored at this point, since it is small compared to the rest of the system mass. However, if the design is also small, e.g., a CubeSat motor, that will not be the case. The solid rocket industry uses an empirical correlation based on the amount of free volume in the combustion chamber, V_{free}, see Eq. (7.86) from Ref. [24]. In this case, the free volume is the volume of the combustion chamber minus the volume of the propellant and any insulation (or mixing devices), e.g., $V_{free} = V_C - V_p - V_{ins}$.

$$m_{igniter}[lbs] = 0.0664V_{free}^{0.571} \qquad (7.86)$$

Here, V_{free} is in cubic inches and $m_{igniter}$ is in pounds. Alternatives to Eq. (7.86) include adding a known mass or assuming the igniter requires a percentage of the flow rate (e.g., 0.5%) for several seconds. Another option is just to add the mass for a standard ignition system, e.g., a pyrogen. This is highly dependent on the ignition scheme. If monopropellant ignition using a catalyst is selected, there will be appreciable mass that needs to be modeled. However, on the other hand, hypergolic additives may add very minimal mass but may need to be accounted for in the performance calculation. See Section 9.9 of Chapter 9 for a description of ignition options.

7.13.4 Safety considerations

Safety requirements should be considered early. Additions such as a safe and arm device or Flight Termination System (FTS) often add appreciable mass to the system. Safety considerations are discussed in Section 9.12 of Chapter 9.

7.14 Mass margin

Two types of mass margins are applied to the dry masses in a propulsion system. The first depends on the maturity of the design, called Mass Growth Allowance or MGA. Ref. [294] defines standards for spacecraft subsystems, including propulsion. A Commercial Off the Shelf (COTS) part could have an MGA as low as 3%, while a novel hybrid motor is more likely to have at least 25-30% MGA. Then system margin is applied to the entire dry mass. This value is bought down over the evolution of the project: the earlier in the project life cycle, the higher percentage is held. At the conceptual design phase, a system margin of 15% is desirable. The mass margin should be added to the propulsion system mass estimate before trading options since options with lower design maturity/flight heritage will carry a higher MGA penalty.

7.15 Mass evaluation

The propulsion engineer should evaluate the total propulsion system mass by adding up the propellant and dry mass of the system. Ensure that the mass margins of the previous section are accounted for, as it is important to account for the expected mass growth of hardware with low heritage/Technology Readiness Level (TRL). At this point in the design process, the propulsion engineer should have a good idea of the propulsion system mass (wet and dry mass), the expected rocket ballistics, and the overall propulsion system configuration and dimensions. If the overall propulsion system mass is not consistent with the assumed total wet spacecraft mass, m_i, then the design will need to be iterated on with the new m_i or lower payload mass to ensure that the ΔV requirement is able to be met.

7.16 Other considerations

7.16.1 Variable oxidizer mass flow rate

The equations throughout this chapter generally assume a constant oxidizer mass flow rate. The summary of design equations provided in Appendix C lists which equations use this assumption. If the propulsion engineer wishes to design a rocket with a variable oxidizer mass flow rate, then it is recommended that an average $\dot{m}_O$ initially be used for a first pass of the design to evaluate fuel grain geometry, etc. The transient ballistics and performance with the variable oxidizer mass flow rate, $\dot{m}_O(t)$, should then be calculated using the approach laid out in Section 3.6.1 of Chapter 3.

7.16.2 Vacuum versus ground performance

While perhaps not imminently necessary at this point in the design process, the propulsion engineer should be aware that the thrust of a rocket on the ground is not equal to its thrust at vacuum conditions. Equations for the real vacuum and ground thrust levels can be written as Eqs. (7.87) and (7.88), respectively.

$$F|_{vac,real} = \eta_{c^*}\eta_n \left(\dot{m}_p g_0 \, I_{sp}|_{vac} \right) \quad (7.87)$$

$$F|_{ground,real} = \eta_{c^*}\eta_n \left(\dot{m}_p g_0 \, I_{sp}|_{vac} - P_a A_e \right) \quad (7.88)$$

Or,

$$F|_{ground,real} = \eta_{c^*}\eta_n \left(\dot{m}_p g_0 \, I_{sp}|_{ground} \right) \quad (7.89)$$

Where:

$$I_{sp}|_{ground} = I_{sp}|_{vac} - \frac{P_a A_e}{\dot{m}_p g_0} \quad (7.90)$$

Note that the ideal vacuum-specific impulse, $I_{sp}|_{vac}$, may not be the same for ground testing as vacuum testing. In addition to the pressure term due to testing at a higher ambient pressure, a lower expansion nozzle is often used to ensure there is no flow separation in the nozzle (see Section 7.8.1). Similarly, the nozzle efficiency, η_n, is likely to be different depending on the shape of the truncated ground nozzle versus the final flight shape.

7.16.3 Volume constraints and impulse density

If the spacecraft is volume constrained more than mass constrained, then the design O/F can

TABLE 7.3 Masses of spacecraft subsystems by type of spacecraft. LEO refers to Low Earth Orbit, HEO refers to High Earth Orbit and planetary refers to interplanetary spacecraft. Data from [250].

Spacecraft Subsystem	Percentage of Total (wet) Mass [%]			
	No propulsion	LEO	HEO	Planetary
Payload	41%	24%	19%	7%
Structure and Mechanisms	20%	21%	14%	12%
Thermal	2%	2%	2%	3%
Power (inc. harness)	19%	17%	10%	10%
Telecom	2%	2%	2%	3%
Avionics	5%	4%	2%	2%
Attitude Determination and Control	8%	5%	3%	3%
Propulsion	0%	2%	4%	6%
Other (including balance mass)	3%	2%	2%	1%
Propellant	0%	21%	42%	52%

be selected such that impulse density, rather than specific impulse, is optimized. For some propellant combinations, optimization of the O/F for peak impulse density may also lead to a minimum overall propulsion system mass. This is because tank and combustion chamber structural masses tend to be a function of volume, rather than propellant mass. However, this may require running at off-peak conditions for greater density or modification of the propellants selected for performance optimization at an alternative O/F.

Impulse density, I_d, is defined per Eq. (2.22) of Chapter 2 and is simply the specific impulse, I_{sp}, multiplied by the propellant density, ρ_p. The average density of a given propellant combination, ρ_p, must therefore be calculated to determine the density impulse. It can be calculated via Eq. (2.23).

7.16.4 Power

Propulsion system design choices have a substantial impact on the power required by the system. The biggest driver is the survival and operational heating required for the propellant combination. Other operational necessities (e.g., ignition, valve operation, and duty cycle) also contribute to the power budget. This is being

brought up because power needs during flight translate directly into mass via batteries.

A Power Equipment List (PEL) of all the components in the propulsion system that consume power should be produced. The power draw in each possible state (e.g., off, idle, average, transient/peak) is assigned. The time spent in each state can then be used to determine the maximum power and energy needs.

7.16.5 Masses of other subsystems

When a propulsion engineer looks at a spacecraft, they often only see the propulsion system. However, in truth, the success of any mission relies on all the subsystems working together. The mass of the propulsion system can be large in the case of a launch vehicle or quite small for spacecraft. Ref. [250] gathered data on average spacecraft subsystem dry masses that helps the propulsion engineer understand the other system drivers and how propulsion fits into the larger mission design. This data is presented here as a percentage of the dry mass in Table 7.3.

Booster scale motors have slightly different allocations. Table 7.4 shows the percentage breakdown of subsystem masses for a booster motor study. The intent of this study was to use two larger boosters or eight smaller ones as a re-

TABLE 7.4 Subsystem Masses used in a trade of Hybrid Booster Motors (HBM). Compares Liquid Injection and Flexseal Nozzles for Thrust Vector Control. Adapted from Table 3-3 in Ref. [371].

Description	Solid Rocket Motor (Flexseal)	4.57 m HBM (LITVC)	4.57 m HBM (Flexseal)	2.44 m HBM (LITVC)	2.44 m HBM (Flexseal)
Nosecap and Frustum	9.7%	10.6%	10.5%	9.9%	9.6%
Forward Skirt	14.8%	4.6%	4.6%	8.4%	8.1%
Aft Skirt	25.7%	30.6%	30.4%	29.2%	28.5%
Hold Down Posts	6.3%	6.9%	6.8%	0.0%	0.0%
Attach Structure	3.5%	0.0%	0.0%	0.0%	0.0%
Systems Tunnel	1.7%	2.2%	2.2%	6.2%	6.1%
Separation Ring	0.4%	0.0%	0.0%	0.0%	0.0%
Thermal Protection	4.2%	4.3%	4.3%	7.1%	6.9%
Paint and Sealants	34.4%	38.2%	37.9%	82.9%	80.7%
Deceleration System	22.1%	21.0%	20.9%	15.9%	15.5%
Separation Motors	3.1%	3.4%	3.4%	3.3%	3.2%
Batteries/ Generator/ Conductor	0.1%	0.3%	0.3%	1.0%	1.0%
Electrical/ Instruments	2.3%	2.0%	2.0%	6.5%	6.4%
Avionics and Flight Controls	0.0%	0.1%	0.1%	0.2%	0.2%
TVC Control System	5.4%	5.2%	5.9%	2.4%	4.9%
Range Safety and Abort	0.3%	0.4%	0.4%	1.2%	1.2%
Subsystem Mass	100.0%	92.0%	92.1%	92.2%	92.4%
Interstage Structure	0.0%	8.0%	7.9%	7.8%	7.6%
Total Subsystem Mass	100.0%	100.0%	100.0%	100.0%	100.0%

placement for an Advanced Solid Rocket Motor (Shuttle Boosters). The actual masses for these two options are given in Table 10.13 of Chapter 10.

Launch vehicles are described by their payload mass fraction, Γ. Recall that this is the percentage of the total wet mass made up by the payload, as defined by Eq. (2.47). This is typically less than 5% for LEO but could be much less than 1% for planetary missions. Efficient launch vehicles have dry masses of about 10%.

8

Hybrid design challenges

8.1 Introduction

This chapter takes the next step in hybrid rocket design, providing the detail required to move from initial, paper designs/trades to a more robust, detailed design. Specific challenges are discussed here, such as calculating feed system pressure drops, estimating water hammer pressure surges, evaluating/avoiding combustion instabilities, and propellant book-keeping for extended-duration flight missions.

The calculations that follow assume that the reader is familiar with both incompressible and compressible fluid flow. The most common equations used by propulsion engineers are laid out. However, if the reader has never used such equations previously, then background reading is recommended to become familiar with the theory. There are numerous books that provide a detailed description of fluid flows, such as Refs. [142], [295], [296].

Inputs include the oxidizer, design oxidizer mass flow rates, and a desired range of combustion chamber pressures. A single feed line may be used to provide a range of oxidizer mass flow rates, though the upper limit of this range will be restricted by maximum feed line pressure, typically limited by the selected components. Flow velocities can also limit the operating range of a feed line since high velocities should be avoided for most oxidizers. The approaches discussed below can be used to size and estimate pressure drops in the feed line. All calculations should be verified via cold-flow tests of the feed line prior to attempting a hot fire.

8.2 Incompressible feed line analysis

This section discusses the approach for calculating pressure drops in low Mach number and incompressible oxidizer feed lines, i.e., when the oxidizer density, ρ, is assumed to be constant.

8.2.1 Pressure losses

Changes in pressure in the feed line occur as a result of 1) area changes, 2) friction, 3) bends within the feed line, and 4) head pressure due to changes in feed line height in the presence of gravity. Valves in the feed line typically produce pressure losses via the first three mechanisms. Pressure changes from the various mechanisms can be added together to produce the total change in pressure.

Flow within the feed line is assumed to be turbulent. Unfortunately, the presence of turbulence means that a purely theoretical approach cannot be used. Instead, first-order physics and dimensional analysis will be used to correlate

161

experimental data. This analysis starts by looking at the modified Bernoulli equation for incompressible flow with losses between two locations, 1 and N, in a feed line:

$$\Delta\text{Pressure} = -\Delta\text{Kinetic Energy} + \text{Friction Losses} + \text{Minor Losses} - \Delta\text{Gravitational Energy}$$

$$P_1 - P_N = -\frac{1}{2}\rho\left(V_1^2 - V_N^2\right) \quad +\frac{1}{2}\rho\int_1^N \frac{V^2 f}{D_h}dx + \frac{1}{2}\rho\sum_i K_i V^2 \quad -\rho g\left(h_1 - h_N\right) \tag{8.1}$$

Here P is the static pressure, V is the average fluid velocity at a given x location, where x is the location along the feed line, g is the body acceleration, typically gravity, D_h is the hydraulic diameter, which is simply the internal tube diameter for circular tubes, f is the friction factor, and K_i is the minor loss coefficient.

The first and last terms on the right side of Eq. (8.1) should be familiar to the reader from the Bernoulli equation for incompressible flow. The middle two terms may be unfamiliar, so some time is spent here discussing their origin. Note that these first three terms on the right side of Eq. (8.1) capture the change in flow conditions from area changes, friction, bends, and valves.

As mentioned above, evaluating pressure losses due to frictional effects in a turbulent flow requires a mix of empirical, experimental data and an understanding of the physics of the flow. Consider the dimensional analysis of this flow: the change in pressure of the fluid is observed to depend on the tube diameter, D, the surface roughness of the tube, ϵ, the length of tube, l, the flow velocity, V, the fluid viscosity, μ, and the fluid density, ρ. This can be written as:

$$\Delta P|_{friction} = f^n\left(D, \epsilon, l, V, \mu, \rho\right) \tag{8.2}$$

Here f^n is a yet to be determined function. There are seven physical variables (ΔP, D, ϵ, l, V, μ, ρ) and three physical dimensions (length, mass, time). Thus from Buckingham Pi Theorem, four functional groups are expected. Select three repeating variables to be ρ, V, D, which allows Eq. (8.2) to be rewritten in dimensionless form:

$$\frac{\Delta P|_{friction}}{\rho V^2} = f^n\left(\frac{\epsilon}{D}, \frac{\mu}{\rho V D}, \frac{l}{D}\right) \tag{8.3}$$

A factor of two is also applied to the left side of Eq. (8.3) by convention [296] to make the denominator of the kinetic energy per unit volume of flow. The pressure loss has been observed to be directly proportional to the tube length. Also, the dimensionless form of the viscosity, μ, in Eq. (8.3) is $\frac{1}{Re_D}$, where Re_D is the Reynolds number using the internal diameter per Eq. (8.5). Eq. (8.3) can be updated to become:

$$\frac{\Delta P|_{friction}}{\frac{1}{2}\rho V^2} = \frac{l}{D}f^n\left(\frac{\epsilon}{D}, Re_D\right) \tag{8.4}$$

$$Re_D = \frac{\rho V D}{\mu} = \frac{\dot{m}D}{A\mu} \tag{8.5}$$

Comparison of Eq. (8.4) with Eq. (8.1) makes it apparent that the friction factor, f, can now be defined as the function in Eq. (8.4). It is dependent on the roughness of the tube relative to the tube diameter, $\frac{\epsilon}{D}$, as well as the Reynolds number of the flow, Re_D. f is generally determined from experimental data. It can be read from figures, such as Fig. 8.1, determined from the implicit Colebrook-White Equation, Eq. (8.6), or estimated from explicit approximations to the Colebrook-White Equation, such as that shown in Eq. (8.7), known as the Zigrang-Sylvester approximation [297].

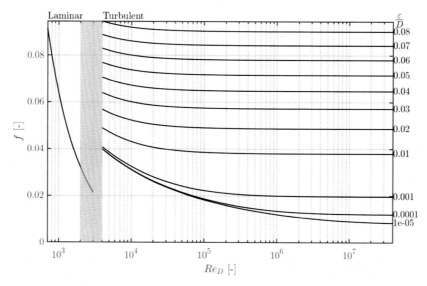

FIGURE 8.1 Friction factor, f, versus Reynolds number, Re_D, and surface roughness, $\frac{\epsilon}{D}$. The gray region is the transition region from laminar to turbulent flow. The friction factor in this region is not well defined as a result of the uncertainty in the transition Reynolds number.

$$\left.\frac{1}{\sqrt{f}}\right|_{turbulent} = -2log_{10}\left(\frac{\epsilon}{3.7D} + \frac{2.51}{Re_D\sqrt{f}}\right) \tag{8.6}$$

$$f|_{turbulent,approx} =$$
$$\frac{0.25}{\left(log_{10}\left(\frac{\epsilon}{3.7D} - \frac{5.02}{Re_D}log_{10}\left(\frac{\epsilon}{3.7D} + \frac{13}{Re_D}\right)\right)\right)^2} \tag{8.7}$$

Again, the Reynolds number here is defined using the internal pipe diameter, per Eq. (8.5).

Note that this friction factor, called f here, is equivalent to the friction factor that is occasionally defined as $4f$ in other sources such as Ref. [252]. If other sources are being used to determine the friction factor, f, then care should be taken by the reader to be sure that a factor of 4 is not being applied mistakenly.

The friction factor f, used in this text is also related to the Coefficient of Friction, C_f, of Eq. (8.45) for compressible flow, according to

Eq. (8.8).

$$f = 4C_f \tag{8.8}$$

The discussion so far has been focused on turbulent flow, but the friction factor can also be calculated for laminar flow, see Eq. (8.9) for flow in a circular pipe. It is difficult to predict where exactly transition from laminar to turbulent flow will occur, but, in general, flow is assumed to be turbulent if the Reynolds number, Re_D, is greater than 4000.

$$f|_{laminar} = \frac{64}{Re_D} \tag{8.9}$$

8.2.2 Minor losses

The minor feed line losses of Eq. (8.1) are used to broadly capture a number of common sources of pressure drop in the feed line. The general definition of the minor loss factor, K, is given by Eq. (8.10). In order to calculate the pressure loss from the loss factors (K) the relevant velocity must be used. This can be ambigous at times, especially in the case of area changes. Therefore, it will be defined for each loss factor below.

$$K = \frac{\Delta P}{\frac{1}{2}\rho V^2} \tag{8.10}$$

The approach for calculating minor losses in a feed line again makes use of empirical data and is applied to pressure losses from:

• A rapid expanding area change, per Fig. 8.2, Eq. (8.11) [298]. This pressure loss factor $K_{expansion}$, depends on the area ratio. When calculating the ΔP, the flow velocity in the small tube, A_1 should be used.

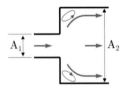

FIGURE 8.2   Schematic of sudden expansion resulting in a minor loss.

$$K_{expansion,1} = \left(1 - \frac{A_1}{A_2}\right)^2 \tag{8.11}$$

• A rapid contracting area change, per Fig. 8.3, Eq. (8.12) [295]. This pressure loss factor, $K_{contraction}$, is also based on the area ratio and the flow velocity in the small tube, A_2 should be used to calculate the pressure drop.

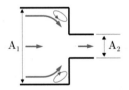

FIGURE 8.3   Schematic of sudden contraction resulting in a minor loss.

$$K_{contraction,2} \approx 0.42\left(1 - \frac{A_2}{A_1}\right) \tag{8.12}$$

• Bends in the feed line. See Eq. (8.13) [299] for a smooth 90-degree bend in a round tube, per

Fig. 8.4 and Eq. (8.14) [299], and for a smooth 45-degree bend in a round tube. These equations are valid when $Re_D\left(\frac{r}{D}\right) > 364$. Note that K_{90deg} and K_{45deg} in Eqs. (8.13) and (8.14) include the frictional losses within the bend as well as the minor loss.

FIGURE 8.4   Schematic of the smooth 90-degree bend resulting in a minor loss.

$$K_{90deg} \approx 0.388\left(0.95 + 4.42\left(\frac{r}{D}\right)^{-1.96}\right)$$
$$\times \left(\frac{r}{D}\right)^{0.84} Re_D^{-0.17} \tag{8.13}$$

$$K_{45deg} \approx 0.194\left(1 + 5.13\left(\frac{r}{D}\right)^{-1.47}\right)$$
$$\times \left(\frac{r}{D}\right)^{0.84} Re_D^{-0.17} \tag{8.14}$$

• Flow junctions such as a"T" with line diverging flow per Fig. 8.5, or converging flow per Fig. 8.6.

• The combination of losses that occur across valves and other feed line components. For this final item, most suppliers provide a measured discharge coefficient in the form of C_d or C_V, for their component. It is recommended that the equation provided by the supplier should be used to determine the pressure drop for a given discharge coefficient to be consistent. If this is not available, then the loss coefficient for C_d or C_V can generally be calculated from Eq. (8.15) or (8.17), respectively. Note that these equations calculate the loss factor, K, in terms of the velocity in the

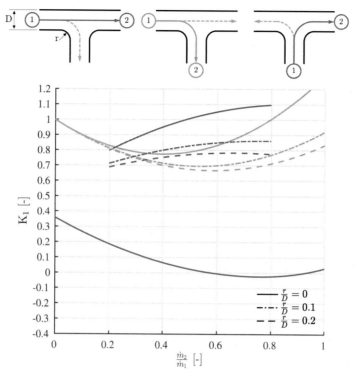

FIGURE 8.5 Minor loss factor, K_1, for T flow junction with diverging flow. The reported minor loss factor is calculated in terms of the velocity of the combined flow before diverging, i.e., at location 1 in the T. The figure is generated assuming that all three branches of the T have equal diameters. Note that the loss factor through the run does not depend on the bend radius to inner tube diameter ratio, $\frac{r}{D}$. The figure is generated using equations from Ref. [300].

line, not the velocity inside the valve.

$$K = \frac{A_{line}^2}{C_{d,valve}^2 A_{valve}^2} \qquad (8.15)$$

Eq. (8.15) assumes that the discharge coefficient, $C_{d,valve}$, is defined per Eq. (8.16).

$$C_{d,valve} = \frac{\dot{m}}{A_{valve}\sqrt{2\rho\Delta P}} \qquad (8.16)$$

$$K = \frac{2A_{line}^2}{\rho C_{V,valve}^2} \qquad (8.17)$$

Eq. (8.17) assumes that the valve flow coefficient, $C_{V,valve}$, is defined per Eq. (8.18).

$$C_{V,valve} = \frac{\dot{m}}{\rho\sqrt{\Delta P}} \qquad (8.18)$$

Sometimes suppliers will give a pressure drop across a component at a known mass flow rate. This can then be converted to a loss coefficient, K, per the general minor loss equation, Eq. (8.10). Note that care must be taken if cavitation is likely to occur across a component. In this case, only data at representative flow conditions can be used to determine K.

It is important to remember that minor loss coefficients are always calculated in terms of a velocity in the system (e.g., the flow velocity upstream, downstream, or at the smallest cross section of an impediment can all be used), and the

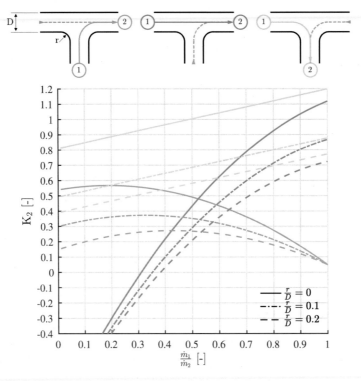

FIGURE 8.6 Minor loss factor, K_2, for T flow junction with converging flow. The reported minor loss factor is calculated in terms of the velocity of the combined flow after converging, i.e., at location 2 in the T. The figure is generated assuming that all three branches of the T have equal diameters. The figure is generated using equations from Ref. [300].

location at which this velocity is calculated will therefore change the value of K. Thus when using these minor loss coefficients, care must be taken to be consistent in using the same definition of velocity in order to produce accurate results; i.e. if a loss coefficient is calculated using the upstream line velocity, then the velocity upstream of that loss must always be used to calculate the pressure drop with that K. The minor loss factor can be converted to be in terms of other velocities if desired, by simply multiplying the loss factor by the dynamic head ratio; i.e. if the loss factor is calculated in terms of the velocity at point 1, then the loss factor calculated in terms of the velocity at point 2 can be calculated using the head ratio between these points as the

correction factor, see Eq. (8.19).

$$K_2 = K_1 \frac{V_1^2}{V_2^2} \qquad (8.19)$$

All of the minor loss coefficients, the known initial conditions in the feed line, and the feed line lengths/area changes can then be used to evaluate the static pressure and mass flow rate through the feed line. An example of this is provided for incompressible flow (see Section 8.2.5).

8.2.3 Cavitating venturi

Cavitating venturis are often used in liquid systems in a similar manner to choked orifices in gas systems. When used correctly, cavitating venturis allow the propulsion engineer to control the flow of liquid through a feed line using only the upstream pressure. The flow rate through the venturi will be independent of the pressure downstream of the venturi as long as the downstream pressure is not increased to the point at which cavitation would cease. The rule of thumb to maintain cavitation in the venturi is to ensure that the pressure downstream of the venturi is less than 85% of the pressure upstream [272]. This 85% pressure recovery is equal to the sum of the inlet-to-throat and throat-to-exit minor loss coefficients ($K_{IN} \simeq 0.05$ and $K_{OUT} \simeq 0.15$ where both loss coefficients are defined using the fluid velocity at the throat [298]). This rule of thumb is a little different for high vapor pressure oxidizers; in this case, the critical pressure ratio to ensure cavitation must account for vapor pressure, see Eq. (8.20).

$$\frac{[P - P_{vapor}]_{OUT}}{[P - P_{vapor}]_{IN}} \leqslant \sim 0.85 \quad (8.20)$$

Once cavitation is assured, the mass flow rate of the venturi can then be calculated from Eq. (8.21), with a typical discharge coefficient, C_d, of 0.93 [272]. This discharge coefficient is a typical value but be aware that some venturis may have discharge coefficients as high as nearly one. Note that the pressure, P, in Eq. (8.21) is the static pressure immediately upstream of the venturi.

$$\dot{m} = C_{d,venturi} A_{venturi} \sqrt{2\rho \left(P - P_{vapor}\right)} \quad (8.21)$$

Take care applying these equations to a cryogenic fluid. Directly consult a cryogenic flow text, such as Ref. [251], for accurate flow rate calculations with two phase flow of a cryogenic fluid.

8.2.4 Water hammer

Water hammer is a phenomenon in which a section of feed line experiences a localized, transient increase in pressure due to a rapid change

in momentum from the propellant flow. This can occur when a system is first being primed or when the oxidizer flow is suddenly stopped due to the rapid closure of a valve. The equations presented in the previous section can be used to determine mass flow rates, and thereby estimate the first-order increase in pressure from this phenomena, per Eq. (8.22). Eq. (8.22) is derived in Appendix A.

$$\Delta P = \rho V a = \frac{\dot{m}a}{A} \quad (8.22)$$

Where ρ is the propellant density, V is the velocity of the propellant at the location where momentum is halted, A is the pipe cross-sectional area at the location where momentum is halted, and a is known as the speed of sound, or celerity, of the fluid. a is given by Eq. (8.23).

$$a = \frac{a_{HL}}{\sqrt{1 + \frac{K_{prop}D}{th_w E}}} \quad (8.23)$$

Here, a_{HL} is the hard liquid speed of sound (Eq. (8.24)), K_{prop} is the bulk modulus of elasticity of the propellant, E is the modulus of elasticity of the tubing, th_w is the thickness of the tubing, and D is the diameter of the tubing. Note that the only difference in a_{HL} and a is that the latter accounts for the adjustment to the speed of sound as a result of elastic deformation of the tubing.

$$a_{HL} = \sqrt{\frac{K_{prop}}{\rho}} \quad (8.24)$$

Eq. (8.22) holds when a valve closes rapidly, i.e. when the time taken for the pressure wave to travel from the valve to the inlet and back to the valve exceeds the time taken to close the valve, ($2L/a \geqslant t_{close}$). If the valve closes more slowly than this threshold, the pressure surge is reduced and must be evaluated with a transient solution using the method of characteristics, see Chapter 7 of Ref. [252].

Note that Eq. (8.22) does not consider a potential increase in peak pressure due to constructive resonance from pressure surges in multiple line branches. Again, method of characteristics can be used for a more rigorous approach to calculate the expected transient pressure rise. The method of characteristics should also be used if the designer needs to account for vaporization of the propellant during priming, as the presence of gas in the line also reduces the pressure surge predicted by Eq. (8.22). The reader is directed to the wonderful transient treatment of system priming in Ref. [252].

Eq. (8.22) will give a reasonable prediction of water hammer peak pressure for oxidizers with low vapor pressure and a single main feed line. Eq. (8.22) is conservative except when there is likely to be line reflections from multiple feed line branches. During system priming, the fluid mass flow rate will be dictated by the change in pressure between the tank and the downstream line pressure, as well as the impediments to the flow per Eq. (8.1). The designer will need to solve this equation to evaluate the expected pressure surge from water hammer. To do this, it is helpful to first re-write Eq. (8.1) to explicitly describe the pressure drop(s) in terms of mass flow rate.

First, recall that mass flow rate is given by Eq. (8.25).

$$\dot{m} = \rho V A \qquad (8.25)$$

Eq. (8.25) is substituted into Eq. (8.1) to generate Eq. (8.26). Note that the last term was been converted from an integral to an summation. This practical form enables the evaluation of discrete sections of line. The number of segments, i, and the line lengths, L_i, correspond to primed segments and primed line lengths only. This equation also assumes that when priming lines the downstream pressure is approximately at vacuum, i.e., $P_N = 0$. For systems that are primed with initially evacuated lines, the initial back pressure, P_N, is approximately the propellant vapor pressure (since the liquid front flashes

to vapor), and if $P_{vapor} \ll P_{tank}$, then the back pressure may be taken as zero. If the propellant has high vapor pressure relative to the tank pressure, or if the lines are not initially evacuated, then P_N should be included in Eq. (8.26).

$$P_{tank} = -\rho g \left(h_1 - h_N \right)$$
$$+ \frac{\dot{m}^2}{2\rho} \left[\frac{1}{A_N^2} - \frac{1}{A_1^2} + \sum_i \frac{K_i}{A_i^2} + \sum_i \frac{f_i L_i}{D_i A_i^2} \right]$$
$$\qquad (8.26)$$

As discussed in the previous section, many component manufacturers will report a discharge coefficient and equivalent area, $C_d A$, rather than a loss factor, K, the conversion is typically straightforward via Eq. (8.27) (this is equivalent to Eq. (8.15), but here K is kept in terms of the valve velocity for convenience; any equation will produce the correct answer as long as the correct A is always used with each K).

$$\frac{K_i}{A_i^2} = \frac{1}{\left(C_{d,i} A_i \right)^2} \qquad (8.27)$$

Note that some components, such as filters, will typically have laminar flow, rather than turbulent flow, through them. The pressure drop may be determined experimentally as a function of mass flow rate for these components. Then this relationship may be explicitly used in Eq. (8.26). The friction factor, f, is a function of the Reynolds number and therefore will change with mass flow rate as described previously. Thus solution of Eq. (8.26) will require iteration. Once the mass flow rate has been determined, the velocity for evaluating the water hammer pressure surge of Eq. (8.22) can be calculated easily from continuity via Eq. (8.25).

8.2.5 Incompressible flow calculation example

This example applies the equations and coefficients presented in the preceding sections to ensure that the propulsion engineer understands how to apply them to the analysis of a

TABLE 8.1 Internal line diameters and lengths for the schematic shown in Fig. 8.7. Note: * Line segment 6 does not include a downstream length for friction calculations as the minor loss factor for the 90-degree bend already accounts for friction losses in that segment.

Location	Internal Diameter		Downstream Line Length	
	[m]	[inch]	[m]	[inch]
1	1.092E-02	0.430	1.270	50
2	1.092E-02	0.430	-	-
3	8.103E-03	0.319	1.016	40
4	8.103E-03	0.319	-	-
5	8.103E-03	0.319	1.524	60
6*	8.103E-03	0.319	-	-
7	8.103E-03	0.319	0.254	10
8	8.103E-03	0.319	-	-
9	1.588E-03	0.063	-	-

feed system. Consider the liquid flow through a tube from a tank, with friction, an area change, a valve, a bend in the feed line, and a cavitating venturi as shown in Fig. 8.7. As a reminder, it is assumed for this example that the flow is incompressible.

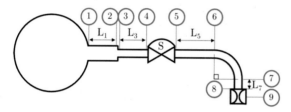

FIGURE 8.7 Example schematic for calculating pressure drop with incompressible flow. The schematic is illustrative only and not to scale. The locations correspond to: 1. Tank outlet. 2. Upstream of area change but with the area change loss factor accounted for. 3. Downstream of area change. 4. Upstream of valve. 5. Downstream of valve. 6. Upstream of the 90-degree bend. 7. Downstream of the 90-degree bend. 8. Upstream of cavitating venturi. 9. At cavitating venturi.

For this example, assume that Nitrogen Tetroxide (NTO), N_2O_4, is the fluid in the tank stored at 2.758 MPa (400 psia) and 25°C. Assume also that the tank is maintained at a constant pressure (though the upstream pressure regulation system is not shown or analyzed here since this section is dedicated to the analysis of

the incompressible liquid feed lines). NTO has a density, ρ, of 1445 kg/m^3, a viscosity, μ, of 0.396×10^{-5} Pa s, and a vapor pressure, P_{vapor}, of 120 kPa (17.4 psia), per Table 5.3 of Chapter 5. Let us assume that given the feed line geometry/components of Fig. 8.7, we wish to calculate the mass flow rate of NTO as well as the static pressure at each location in the feed line. Such calculations are common for propulsion engineers, though the approach here could easily be modified to instead size a venturi to deliver a specified mass flow rate (also a common calculation). For simplicity, it is assumed that the feed line is at the same height, so gravitational forces can be neglected, and any variations in fluid properties (such as viscosity, μ, or vapor pressure, P_{vapor}) are assumed to be negligible.

The internal line diameters and lengths for this example are provided in Table 8.1 with the locations defined per Fig. 8.7. The bend radius of segment 6 is 0.127 m (5 inches). It is assumed that the feed lines are mechanically polished, and as such, a surface roughness, ϵ, of 2.540 $\times 10^{-3}$ mm (1.0 $\times 10^{-4}$ inches) is used when calculating the friction factor, f. The venturi area is as provided in Table 8.1, and the discharge coefficient, $C_{d,venturi}$, is assumed to be 0.93 [272].

To start these calculations, first estimate the approximate mass flow rate of NTO in the feed

TABLE 8.2 Feed line conditions calculated using an initial mass flow rate estimate of 0.1607 kg/s. Note: * Friction factors are only calculated for segments of line where they are needed for friction calculations. Line segment 6 does not include a friction calculation as the minor loss factor for the 90-degree bend already accounts for friction losses in that segment.

Location	V	Re_D	f	P	
	[m/s]	[-]	[-]	[Pa]	[psia]
1	1.187	4.731E+06	0.0144	2.757E+06	399.85
2	1.187	4.731E+06	-	2.755E+06	399.61
3	2.157	6.378E+06	0.0152	2.752E+06	399.17
4	2.157	6.378E+06	-	2.746E+06	398.24
5	2.157	6.378E+06	0.0152	2.523E+06	365.93
6*	2.157	6.378E+06	-	2.513E+06	364.54
7	2.157	6.378E+06	0.0152	2.512E+06	364.41
8	2.157	6.378E+06	-	2.511E+06	364.17

line. To do this, the pressure losses in the feed line are initially ignored (or alternatively, they could have been estimated as a percentage of the tank pressure). Then using Eq. (8.21), the mass flow rate through the venturi can then be estimated to be 0.1607 kg/s.

This mass flow rate estimate can now be used to approximate the velocity of the fluid, V_i in each segment, i, of the feed line from continuity per Eq. (8.28). Here, $A_{int,i}$ is the internal cross-section area of the feed line at location i.

$$V_i = \frac{\dot{m}}{\rho A_{int,i}} \qquad (8.28)$$

The values calculated in this example are all provided in Table 8.2 so that the reader can check their calculations.

The flow Reynolds number, Re_D, and friction factor, f, are calculated from Eq. (8.5) and (8.7), respectively. The velocities are calculated from Eq. (8.28). The static pressure at location 1 is determined from the Bernoulli Equation, Eq. (8.29), neglecting the loss factor from the tank outlet for simplicity here. $P_{t,1}$ is therefore the tank pressure of 2.758 MPa (400 psia).

$$P_1 = P_{t,1} - \frac{1}{2}\rho V_1^2 \qquad (8.29)$$

The total (friction and minor) losses in each segment of line moving downstream from location 1 to 8 can now be calculated via Eq. (8.30), and the static pressure is determined from Eq. (8.31). To perform these calculations, start at location 2 and methodically work downstream. The pressures calculated using this method are provided in Table 8.2.

$$K_{total,i} = \frac{f L_i}{D_i} + K_i \qquad (8.30)$$

$$P_i = P_{i-1} - \frac{1}{2}\rho V_i^2 K_{total,i} + \frac{1}{2}\rho\left(V_{i-1}^2 - V_i^2\right) \qquad (8.31)$$

Once the conditions at location 8 have been calculated, the mass flow rate estimate can now be refined using Eq. (8.21) and the new, more accurate, estimate of total pressure at location 8. Note that the static pressure at location 8 must first be converted to total pressure via the Bernoulli equation (Eq. (8.29) applied at location 8). This calculation of mass flow rate produced a new value of 0.1531 kg/s, which is 5% below the initial estimate. All of the line conditions and eventually also the mass flow rate can now be re-calculated by re-calculating all of the previous steps with the new mass flow rate. The results

TABLE 8.3 Feed line conditions calculated after one iteration using a mass flow rate of 0.1531 kg/s. Note: * Friction factors are only calculated for segments of line where they are needed for friction calculations. Line segment 6 does not include a friction calculation as the minor loss factor for the 90-degree bend already accounts for friction losses in that segment.

Location	V	Re_D	f	P	
	[m/s]	[-]	[-]	[Pa]	[psia]
1	1.131	4.508E+06	0.0144	2.757E+06	399.87
2	1.131	4.508E+06	-	2.755E+06	399.64
3	2.055	6.076E+06	0.0152	2.753E+06	399.25
4	2.055	6.076E+06	-	2.747E+06	398.41
5	2.055	6.076E+06	0.0152	2.545E+06	369.08
6*	2.055	6.076E+06	-	2.536E+06	367.81
7	2.055	6.076E+06	0.0152	2.535E+06	367.69
8	2.055	6.076E+06	-	2.534E+06	367.48

from this iteration are provided in Table 8.3. The new mass flow rate calculated after the iteration is 0.1538 kg/s, within 0.5% of the estimate for this iteration, and so the calculations were not iterated upon any further. However, if greater accuracy is desired the designer can continue to iterate until the desired level of convergence is achieved.

8.3 Compressible feed line analysis

If a gaseous oxidizer or a pressurant gas is used, then the first step in designing the gas feed line is to evaluate where the gas flow choke point will be in the feed line. The gas choke point is generally used to control the gas mass flow rate in these systems. For oxidizer flow, the feed line should be designed to situate the choke point close to the combustion chamber for multiple reasons:

1. Flow velocities will be higher downstream of the choked orifice, which increases the risk of safety issues and pressure drops in the feed line.
2. Locating the choke point closer to the combustion chamber is also better for minimizing

the risk of combustion instabilities, see Section 8.4.

Typically, an orifice is used to control the mass flow rate. The choked orifice is sized per Eq. (8.32).

$$A_{orifice} = \frac{\dot{m}\sqrt{\gamma R T_t}}{C_{d,orifice}\gamma P_t}\left(\frac{\gamma+1}{2}\right)^{\frac{\gamma+1}{2(\gamma-1)}}$$
(8.32)

Here the subscript t refers to total conditions per the isentropic flow equations, Eqs. (8.33) and (8.34). The specific gas constant, R, is defined per Eq. (8.35); the ratio of specific heats, γ, is defined per Eq. (8.36); and the Mach number, M, is defined per Eq. (8.37). The discharge coefficient of the orifice, $C_{d,orifice}$, depends on the orifice geometry; it is best practice to experimentally calibrate the discharge coefficient for an orifice that will be used in a compressible gas feed line.

$$P_t = P\left(1+\frac{\gamma-1}{2}M^2\right)^{\frac{\gamma}{\gamma-1}}$$
(8.33)

$$T_t = T\left(1+\frac{\gamma-1}{2}M^2\right)$$
(8.34)

$$R = \frac{R_u}{M_w}$$
(8.35)

$$\gamma = \frac{c_p}{c_v} \qquad (8.36)$$

$$M = \frac{V}{a} = \frac{V}{\sqrt{\gamma R T}} \qquad (8.37)$$

As an initial first-pass estimate of the needed orifice size, it is generally reasonable to assume that the total temperature, T_t, is constant at the reservoir temperature and either ignore pressure losses or approximate pressure losses in the feed line as a percentage of the total pressure. Thus as a starting point, the designer may assume that the total pressure, P_t, at the orifice is the upstream reservoir pressure (if unregulated), or the regulated pressure. Keep in mind when doing this that Eq. (8.32) will give you an idealized orifice size. Either the orifice area will need to be increased slightly, or the regulator/reservoir set pressure will need to be increased to achieve the desired mass flow rate when accounting for pressure losses in the feed line. As the calculated pressure drops in a feed line are refined, the assumption about total pressure and temperature should be revisited. For short feed lines and high mass flow rates, T_t may be treated as constant along the length of the feed line, i.e., adiabatic flow. However, for long sections of feed lines and low mass flow rates, it may be more accurate to assume that there is sufficient heat transfer in the line to ensure that the gas temperature upstream of the orifice is the ambient temperature, i.e. it may be more appropriate to assume that the flow is isothermal. Isothermal flow with friction is not discussed here, instead the reader is directed to Ref. [295] for further reading on this topic.

For a given motor design chamber pressure and oxidizer mass flow rate, the minimum pressure upstream of the orifice to guarantee choking is approximately given by Eq. (8.38).

$$P_t|_{upstream} = P_c \left(\frac{\gamma + 1}{2} \right)^{\frac{\gamma}{(\gamma - 1)}} \qquad (8.38)$$

Note again that Eq. (8.38) ignores pressure drop between the orifice and the combustion

chamber. If a significant pressure drop, ΔP, is expected between the orifice and the chamber, then it should be accounted for, and P_c in Eq. (8.38) should instead become $P_c + \Delta P$. Eq. (8.38) also ignores the expanding area ratio of the orifice; if a smooth expansion is present, then it may lower the minimum upstream pressure required to guarantee choking.

To ensure that flow chokes at the orifice and not elsewhere in the feed line, care should be taken to make the orifice the smallest cross sectional area in the feed line. The propulsion engineer should pay special attention to the internals of the valves selected. When the internal geometry of a valve is not available (as is common), a simple flow test with pressure measurements can evaluate whether or not there is choked flow across the valve via Eq. (8.38).

Care should also be taken in the rest of the oxidizer feed line to ensure that flow velocities are not too high in order to reduce ignition hazards; for oxygen systems, a safety rule of thumb is to keep flow velocities below 30 m/s [301]. The feed line tubing can therefore be sized based on the design mass flow rate per Eq. (8.39). Here A_{tube} and D_{tube} refer to the internal area and diameter of the feed line, respectively. Eq. (8.39) also assumes that the oxidizer gas is ideal.

$$\dot{m} = \rho V A_{tube} \Rightarrow V = \frac{\dot{m}}{\rho A_{tube}} \leqslant V_{max}$$

$$\Rightarrow D_{tube} \geqslant \sqrt{\frac{4 \dot{m} R T}{V_{max} P \pi}} \qquad (8.39)$$

The Mach number immediately upstream of the choke point can be determined from continuity, i.e., by solving Eq. (8.43). Recall that:

$$\dot{m} = \rho V A = \frac{\gamma}{\left(\frac{\gamma + 1}{2} \right)^{\frac{\gamma + 1}{2(\gamma - 1)}}} \frac{P_t A}{\sqrt{\gamma R T_t}} f(M) \qquad (8.40)$$

Where:

$$f(M) = \left(\frac{\gamma+1}{2}\right)^{\frac{\gamma+1}{2(\gamma-1)}} \frac{M}{\left(1+\frac{\gamma-1}{2}M^2\right)^{\frac{\gamma+1}{2(\gamma-1)}}}$$

(8.41)

Continuity between the tube and orifice requires that:

$$\dot{m}_{tube} = \dot{m}_{orifice} \quad (8.42)$$

$$\Rightarrow f(M_{tube}) = \frac{C_{d,orifice} A_{orifice}}{A_{tube}} \quad (8.43)$$

Note that Eq. (8.43) produces two possible solutions, a subsonic and a supersonic solution. When looking at the upstream section of line the physical solution will always be subsonic, with a Mach number, M, less than one. Pressure drops upstream of the orifice can be evaluated using Fanno flow in straight sections of tubing, reported pressure drop curves across components, and an approach similar to that discussed in subsection 8.2 for incompressible oxidizers. This is discussed in more detail in the following subsections. Recall that when the Mach number is low, typically below approximately 0.3, the flow can often be approximated as incompressible.

The equations in this section typically assume ideal gas properties. A reasonable level of accuracy can generally be achieved by analyzing the feed line with these equations while using a real gas equation of state to determine upstream tank conditions. However, when greater accuracy is required, then modification of these equations using real gas properties can be worthwhile, as described in Ref. [129].

8.3.1 Fanno flow

Fanno flow describes adiabatic fluid flow through a tube of constant area in the presence of friction. This is often approximately the case for compressible oxidizer flow through a feed line. From the Navier-Stokes equations, it is possible to derive an expression for the change in

Mach number for such flow due to the presence of friction (see any good compressible flow textbook, such as Ref. [142], for the full derivation):

$$\frac{4C_f dx}{D} = \frac{1-M^2}{1+\frac{\gamma-1}{2}M^2}\frac{dM^2}{\gamma M^4} \quad (8.44)$$

Here the coefficient of friction, C_f, is defined per Eq. (8.45), where τ_w is the shear stress at the wall (equal to $\mu\frac{dV}{dy}$ for a Newtonian fluid). However, throughout the rest of this chapter, f shall be used instead of C_f to be consistent with the nomenclature used in the incompressible flow section. Recall that f is simply four times C_f, as shown again here in Eq. (8.45).

$$C_f = \frac{f}{4} = \frac{\tau_w}{\frac{1}{2}\rho V^2} \quad (8.45)$$

Integrating Eq. (8.44) between two points, 1 and 2, that are a distance l apart yields:

$$\frac{fl}{D} = \frac{fl^*}{D}\bigg|_1 - \frac{fl^*}{D}\bigg|_2 \quad (8.46)$$

Where:

$$\frac{fl^*}{D} = \frac{1-M^2}{\gamma M^2} + \frac{\gamma+1}{2\gamma}ln\left(\frac{(\gamma+1)M^2}{2(1+\frac{\gamma-1}{2}M^2)}\right)$$

(8.47)

Note that throughout this section, the superscript * is used to denote sonic conditions, which will be constant for a given section of the Fanno flow. Thus l^* refers to the length of tubing required to bring the flow to choked condition and should not be confused with L^*, the characteristic combustion chamber length.

Eqs. (8.46) and (8.47) can be used along with Eq. (8.9) or Eq. (8.6) to determine the change in Mach number in a feed line due to friction. Since the flow is assumed to be adiabatic in the derivation of these equations, the total temperature,

T_t, is assumed to be constant. The static temperature can be determined from Eq. (8.34). The change in static pressure and total pressure can then also be determined from Eqs. (8.48), (8.51), (8.52), or (8.49).

Static pressure:

$$\frac{P^*}{P} = \left(\frac{2}{\gamma+1}\right)^{1/2} M \left(1 + \frac{\gamma-1}{2} M^2\right)^{1/2} \quad (8.48)$$

Total pressure:

$$\frac{P_t^*}{P_t} = \left(\frac{\gamma+1}{2}\right)^{\frac{\gamma+1}{2(\gamma-1)}} \frac{M}{\left(1 + \frac{\gamma-1}{2} M^2\right)^{\frac{\gamma+1}{2(\gamma-1)}}} \quad (8.49)$$

Density and velocity:

$$\frac{\rho^*}{\rho} = \frac{V}{V^*} = \left(\frac{\frac{\gamma+1}{2} M^2}{1 + \frac{\gamma-1}{2} M^2}\right)^{1/2} \quad (8.50)$$

Pressure change as a function of temperature and Mach number:

$$\frac{P_2}{P_1} = \frac{M_1}{M_2} \sqrt{\frac{T_2}{T_1}} \quad (8.51)$$

Total pressure change as a function of temperature and Mach number:

$$\frac{P_{t,2}}{P_{t,1}} = \frac{M_1}{M_2} \left(\frac{T_2}{T_1}\right)^{\frac{\gamma+1}{2(1-\gamma)}} \quad (8.52)$$

The static pressure (or total pressure) value at some downstream location, 2, can be determined in terms of the conditions at the upstream location, 1, by remembering that $P_2 = P_1 \frac{P_2}{P^*} \frac{P^*}{P_1}$. Similarly, the change in velocity and density can be calculated with Eq. (8.50). Other compressible minor losses can also be calculated using the Fanno line equations, this will be discussed in Section 8.3.2.

8.3.2 Other minor compressible pressure losses

The conservation of mass equation and the Fanno flow equations can be used together to evaluate the pressure losses and flow conditions from the various minor losses discussed in Section 8.2. The approach used here is similar to the incompressible approach, but all losses are treated as equivalent Fanno friction losses and then solved with Eqs. (8.46)–(8.52).

Losses from flow through a given section of line with constant diameter can be calculated using loss factors, K_i, and directly adding them to the friction losses, per Eq. (8.53). These loss factors are similar to those in Section 8.2. If compressible loss factors can not be found in literature for the given flow geometry, then they may be initially approximated using the incompressible values provided in Section 8.2.2. However, test data shows that this can produce significant error. This is illustrated in Fig. 8.8 for sudden expansions and contractions, where it can also be seen that in the limit of flow velocity approaching zero, $M \rightarrow 0$, the two compressible and incompressible loss factors become equivalent, as is expected.

$$\left.\frac{fl}{D}\right|_{equivalent} = \left.\frac{fl}{D}\right|_{actual} + \sum_i K_i \quad (8.53)$$

Note that the conditions downstream of any area change will need to be calculated, as an additional step, since the Fanno flow equations only apply to constant area flow. Minor losses from valves can also be calculated with this method, but determination of the loss factor, $K_{valve,line}$, from a defined discharge coefficient and valve area, $C_{d,valve}$ and A_{valve}, needs to take into account the change in area/fluid density within the valve, per Eq. (8.54). Note that the subscript "line" is used in Eq. (8.54) to make it clear that the loss factor calculated here is in terms of the velocity in the line, not the velocity

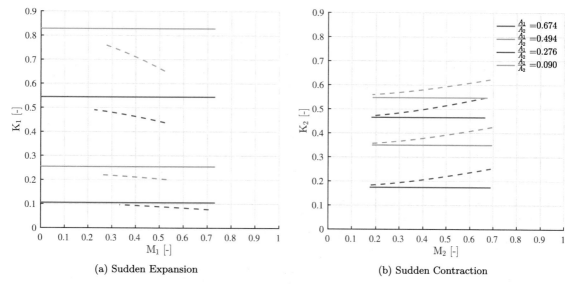

(a) Sudden Expansion

(b) Sudden Contraction

FIGURE 8.8 Compressible (dashed line) and incompressible (solid line) minor loss factor for flow through a sudden expansion (a) and sudden contraction (b). The legend included in (b) applies to both figures. The subscript on the reported minor loss factors refers to the location at which it applies, i.e. the pressure loss assumes the flow velocity at the subscript location, as defined in Figs. 8.2 and 8.3 for expansion and contraction, respectively. The compressible loss factors shown are for air with a ratio of specific heats, γ, of 1.4. The figure is generated using data from Refs. [298] and [302].

inside the valve.

$$K_{valve,line} = \frac{A_{line}^2 \rho_{line}}{C_{d,valve}^2 A_{valve}^2 \rho_{valve}} \qquad (8.54)$$

For Eq. (8.54), it is possible to estimate the approximate flow conditions in the valve using the ideal equations, Eqs. (8.55) and (8.56).

$$\rho_{valve} \simeq \rho_{line} \left[\frac{1 + \frac{\gamma - 1}{2} M_{line}^2}{1 + \frac{\gamma - 1}{2} M_{valve}^2} \right]^{\frac{1}{\gamma - 1}} \qquad (8.55)$$

$$f(M_{valve}) \simeq \frac{A_{line} f(M_{line})}{A_{valve}} \qquad (8.56)$$

The use of ideal equations to determine the density and Mach number of the gas in the valve can introduce significant error to the calculations. It will underpredict the pressure losses from flow through the valve. A more conser-

vative approach accounts for the pressure drop inside the valve when calculating density and Mach number, via Eqs. (8.57) and (8.58). Note that both of these equations can be derived from conservation of mass between the line and the valve with adiabatic flow.

$$\rho_{valve} \simeq \rho_{line} \frac{A_{line}}{C_{d,valve} A_{valve}} \frac{M_{line}}{M_{valve}}$$
$$\times \sqrt{\left(\frac{1 + \frac{\gamma - 1}{2} M_{valve}^2}{1 + \frac{\gamma - 1}{2} M_{line}^2} \right)} \qquad (8.57)$$

$$f(M_{valve}) \simeq \frac{A_{line} f(M_{line})}{C_{d,valve} A_{valve}} \qquad (8.58)$$

8.3.3 Compressible flow conditions from an area change

Total pressure losses from a change in cross-sectional area are already accounted for using the Fanno flow relations in the previous section. The change in flow Mach number is therefore calculated using the one-dimensional compressible flow equations that assume isentropic and adiabatic flow. Since total pressure, P_t, and temperature, T_t, are constant, the change in Mach number can be calculated from conservation of mass, per Eq. (8.59).

$$\dot{m}_1 = \dot{m}_2 \Rightarrow f(M_2) = \frac{A_1 f(M_1)}{A_2} \qquad (8.59)$$

The downstream Mach number, M_2 can be solved by recalling that $f(M_2)$ is defined per Eq. (8.41). Once the downstream Mach number, M_2, is determined, the new static pressure and temperature can then be calculated from the isentropic relations given in Eqs. (8.33) and (8.34).

8.3.4 Compressible flow calculation example

Using the tools presented in the preceding sections, the propulsion engineer can now calculate losses for compressible flow through a tube from a tank, with friction, an area change, a valve, and a choked orifice as shown in Fig. 8.9. As a reminder, it is assumed that the flow is adiabatic in order to apply the Fanno flow equations.

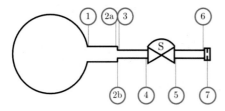

FIGURE 8.9 Example schematic for calculating pressure drop with compressible flow. The schematic is illustrative only and not to scale. The locations correspond to: 1. Tank outlet. 2a. Upstream of area change. 2b. At area change, fluid properties after expansion loss factor but without area change. 3. Downstream of area change. 4. Upstream of valve. 5. Downstream of valve. 6. Upstream of orifice. 7. At orifice sonic location.

For this example, assume gaseous oxygen (O_2) is in the tank at 2.068 MPa (300 psi). Oxygen has a ratio of specific heats, γ, of 1.4, which enables the use of Fig. 8.8 for K, if necessary. NIST [22] is an excellent source of information on gases. At room temperature ($T_t = 298.15$ K) and a pressure of 2.068 MPa, the viscosity of O_2 is $\mu = 2.090 \times 10^{-5}$ Pa s.

Flow is choked at the orifice (station 7). Assume the orifice diameter is 0.0015 m with a discharge coefficient of $C_{d,orifice} = 0.8$.

The large tubing between stations 1 and 2a has an internal diameter of 0.011 m, and the rest of the tubing has an internal diameter of 0.004 m. A surface roughness (ϵ) of 0.38 µm is assumed for the internal surface of all tubing, consistent with electropolished tubing. Take the lengths from station 1 to 2 to be 0.20 m, from station 3 to 4 to be 0.050 m, and from station 5 to 6 to be 0.055 m. The valve internal diameter is .003 m, and its discharge coefficient is $C_{d,valve} = 0.7$. These line lengths and diameters are summarized in Table 8.4.

Similar to the incompressible flow example, a full solution of flow through the compressible feed line requires iteration. First, determine the approximate mass flow rate. This can be done by initially ignoring the pressure losses in the feed line, and then using the equation for choked flow, Eq. (8.60). This produces an approximate mass flow rate of 0.00719 kg/s.

$$\dot{m}_0 \simeq \frac{\gamma C_{d,orifice} A_{orifice} P_{t,0}}{\sqrt{\gamma R T_{t,0}}} \left(\frac{\gamma+1}{2}\right)^{-\frac{\gamma+1}{2(\gamma-1)}}$$

$$(8.60)$$

Next, the conditions at the start of the feed line can be calculated by assuming that the Mach number there is sufficiently low to approximate the total conditions as static conditions (this assumption can be checked and corrected instantly).

$$T_1 = T_{t,0} \qquad (8.61)$$
$$P_1 = P_{t,0} \qquad (8.62)$$

TABLE 8.4 Internal line diameters and lengths for the schematic shown in Fig. 8.9.

Location	Internal Diameter		Downstream Line Length	
	[m]	[inch]	[m]	[inch]
1	0.0110	0.433	0.200	7.874
2a	0.0110	0.433	-	-
3	0.0040	0.157	0.050	1.969
4	0.0040	0.157	-	-
5	0.0040	0.157	0.055	2.165
6	0.0040	0.157	-	-
7	0.0015	0.059	-	-

TABLE 8.5 Mach number and static condition iterations at location 1 using the mass flow rate estimate of 0.00719 kg/s.

Iteration	Assumption	Calculated				
	M_1	P_1	T_1	M_1	M_1 Error	
[-]	[-]	[Pa]	[K]	[-]	[%]	
0	0	2068427.19	298.15	0.00860882	−100.000%	
1	0.00860882	2068319.88	298.145581	0.0086092	−0.004%	
2	0.0086092	2068319.87	298.145580	0.0086092	0.000%	

If the gas is also assumed to be ideal, then the density can also be easily calculated:

$$\rho_1 = P_1/(RT_1) \qquad (8.63)$$

The Mach number at location 1 is then calculated using Eq. (8.64).

$$M_1 = \frac{V_1}{\sqrt{\gamma RT_1}} \Rightarrow M_1 \simeq \frac{\dot{m}_0}{\rho_1 A_1 \sqrt{\gamma RT_1}} \qquad (8.64)$$

If improved accuracy is desired, the zero Mach number assumption of Eqs. (8.61) and (8.62) can now be avoided by iterating on the above calculations with the newly calculated Mach number and the isentropic flow equations, Eqs. (8.33) and (8.34). The results of this iteration are given in Table 8.5. We can see from this table that the results after the first iteration are very converged for this example (within 0.004%), so these values are used for the remainder of the calculations here. The results from the second iteration are also provided to show the rapid convergence achieved with this approach.

The Mach number, total pressure, and static pressure at location 2a, upstream of the area change, are calculated from the Fanno line equations between 1 and 2a. First, the Reynolds number is calculated to check if flow is turbulent or laminar using Eq. (8.5). In this example, flow will always be turbulent, but it is good practice to check this as the friction factor depends on it: Eq. (8.7) vs. Eq. (8.9).

Next, the friction in the lines can be determined using Fanno flow by solving for Eqs. (8.47) and (8.53) at station 1 and plugging them into Eq. (8.46) to get $\frac{fl^*}{D}$ at the next station (2a). The Mach number at 2a can now be solved for numerically from Eq. (8.47). There are multiple equivalent options to now calculate the conditions at location 2a via Eqs. (8.48) to (8.52). For this example, Eq. (8.48) is arbitrarily selected to start this process. It allows us to calculate $\left(\frac{P_{2a}}{P^*}\right)$ and $\left(\frac{P_1}{P^*}\right)$. P_{2a} is then calculated by plugging

TABLE 8.6 Feed line conditions calculated using an initial mass flow rate estimate of 0.00719 kg/s. The calculated mass flow rate using these results was found to be 0.00697 kg/s. Note: * The line Reynolds number and friction factor are only provided at those locations where their calculation is required in order to evaluate the downstream line conditions.

Location	M	Re_D*	f*	P		P_t	
	[-]	[-]	[-]	[Pa]	[psia]	[Pa]	[psia]
1	0.00861	3.98E+04	0.02228	2.0683E+06	299.98	2.0684E+06	300.00
2a	0.00861	-	-	2.0683E+06	299.98	2.0684E+06	299.99
2b	0.06527	-	-	2.0622E+06	299.10	2.0684E+06	299.99
3	0.06534	1.10E+05	0.01831	2.0600E+06	298.77	2.0661E+06	299.67
4	0.06539	-	-	2.0586E+06	298.57	2.0647E+06	299.46
5	0.06730	1.10E+05	0.01831	1.9999E+06	290.06	2.0063E+06	290.98
6	0.06736	-	-	1.9983E+06	289.83	2.0047E+06	290.75

these values into Eq. (8.65).

$$P_{2a} = \left(\frac{P^*}{P_1}\right)\left(\frac{P_{2a}}{P^*}\right)(P_1) \qquad (8.65)$$

The isentropic relations (Eqs. (8.33) and (8.34)) are then used to solve for the total pressure and static temperature. The results from all of these calculations are provided in Table 8.6 to allow the propulsion engineer to check their calculations.

The Mach number, total pressure, and static pressure at location 3 are calculated with two steps:

- First, the conditions at 2a are used to calculate the ideal change in flow conditions from the area change. For this first calculation, the total pressure at 2a is equal to the total pressure at station 2b. The Mach number at station 2b can be found by conservation of mass applying Eq. (8.59) between stations 2a and 2b and plugging in (8.41). The static pressure is then again found from (8.33).
- Second, the Fanno flow equations are used for the minor loss due to the expansion, giving the conditions for station 3. The Reynolds number is again checked, and a friction factor is determined. Fanno flow is used to determine the Mach number, though this time, K is from Eq. (8.12) for a contraction. Any line length from 2b to 23 is ignored, so L_{actual}

in Eq. (8.53) is zero. The Mach number at 3 is again calculated using Eq. (8.47), and the static and total pressures can also be found as above.

The pressure losses and Mach number change between point 3 and 4 are again calculated with the same approach and Fanno equations. To calculate flow conditions through the valve, the density at location 4 is needed. This can be calculated from continuity, Eq. (8.66), where T is found using the isentropic relation (Eq. (8.34)) and assuming adiabatic flow (T_t is constant).

$$\rho_4 = \frac{\dot{m}_4}{A_4 V_4} = \frac{\dot{m}_0}{M_4 A_4 \sqrt{\gamma R T_4}} \qquad (8.66)$$

Once the conditions at location 4 are known, this becomes the "line" conditions in Eqs. (8.54), (8.57), and (8.58). These equations are used with Eq. (8.53) and Eqs. (8.46)–(8.52) to determine the total pressure and Mach number at location 5.

The pressure losses and Mach number change between point 5 and 6 are then directly calculated with the Fanno equations. Once the total pressure is known at point 6, the mass flow rate at this location can be determined from Eq. (8.60).

Re-calculating the mass flow rate through the feed line via Eq. (8.60) now produces a mass flow rate of 0.00697 kg/s, 3.1% lower

TABLE 8.7 Iterated feed line conditions calculated using an initial mass flow rate estimate of 0.00697 kg/s. The calculated mass flow rate using these results was found to be 0.00699 kg/s. Note: * The line Reynolds number and friction factor are only provided at those locations where their calculation is required in order to evaluate the downstream line conditions.

Location	M	Re_D*	f*	P		P_t	
	[-]	[-]	[-]	[Pa]	[psia]	[Pa]	[psia]
1	0.00834	3.86E+04	0.02244	2.0683E+06	299.99	2.0684E+06	300.00
2a	0.00834	-	-	2.0683E+06	299.98	2.0684E+06	299.99
2b	0.06325	-	-	2.0626E+06	299.16	2.0684E+06	299.99
3	0.06331	1.06E+05	0.01841	2.0605E+06	298.85	2.0663E+06	299.69
4	0.06336	-	-	2.0591E+06	298.65	2.0649E+06	299.49
5	0.06509	1.06E+05	0.01841	2.0042E+06	290.68	2.0101E+06	291.55
6	0.06514	-	-	2.0027E+06	290.46	2.0086E+06	291.33

than the initial estimate. To improve the accuracy of the predicted conditions, the calculations can now be iterated through until convergence is achieved. The results of an iteration using 0.00697 kg/s as the initial mass flow rate estimate are provided in Table 8.7. Following this iteration, the mass flow rate was found to be 0.00699 kg/s. This is within 0.2% of the mass flow rate estimate for this iteration, which is sufficiently converged for this example, so no further iteration is included here.

Another check for accuracy/convergence is to evaluate the Mach number in the line at point 6 via Eq. (8.67), it should be approximately equal to that given by Eq. (8.43). If not, then this entire calculation approach may need to be applied in reverse, moving from the choke point, 6, back upstream towards the tank, point 0. The forward calculation approach was laid out here since it can be used even in the absence of a choke point in the line. After the second iteration, the Mach number at location 6 calculated via Eq. (8.43) was within 0.2% of that calculated using the forward calculation approach, equivalent to the error in the calculated mass flow rate.

$$f(M_6) = \frac{A_7 C_{d,orifice}}{A_6} \qquad (8.67)$$

8.4 Combustion instabilities

8.4.1 Introduction to hybrid rocket instabilities

Combustion instabilities are the undesired and often violent pressure and thrust oscillations that can occur during the operation of chemical rockets [47]. They are typically defined as chamber pressure oscillations with an amplitude greater than 5% of the mean value [303]. Designing to minimize combustion instabilities is a key but challenging part of the hybrid motor design (and indeed of any new rocket propulsion system). It is important to understand and limit combustion instabilities in order to reduce the risk of failure (e.g., from a large chamber pressure spike, large vibrational loads, unintended resonance, or a localized burn-through) as well as to limit the vibrational loads that will be experienced by the payload [40]. Here a brief overview of common combustion instabilities observed in hybrid rocket motors is provided. An in-depth discussion of instabilities typical in hybrid rocket motors is provided by Karabeyoglu in Ref. [40]. The categorization presented by Karabeyoglu is adopted here such that observed instabilities are broadly classified according to their frequency of oscillation.

Low frequency instabilities generally have an oscillation frequency less than 200 Hz. Common

TABLE 8.8 Summary of hybrid rocket instabilities categorized by their typical frequency of oscillation.

Instability Class	Name	Cause
Low Freq. (2 - 200 Hz)	Feed System Coupled Instability	Acoustic coupling between the feed system and combustion processes as well as a form of combustion lag (such as from oxidizer vaporization).
	Intrinsic Low Frequency Instability	Thermal transients in the solid fuel coupled to boundary layer combustion and chamber gas dynamics.
	Bulk Mode/Helmholtz Instability	Bulk acoustic resonance within the combustion chamber.
	Chugging	Flame-holding instability.
	Chuffing	Char layer formation/break-off when operating at low oxidizer mass flux.
Medium Freq. (200 - 2000 Hz)	Longitudinal Acoustic Modes	Acoustic resonance along the longitudinal length of the hybrid motor combustion chamber.
	Pressure Coupling Instability	Coupling between the fuel regression rate and the combustion chamber pressure.
High Freq. (>2000 Hz)	Transverse Acoustic Modes	Acoustic resonance along the transverse direction of the hybrid motor combustion chamber. These transverse modes can be tangential or radial.
	Higher Order Acoustic Modes	Higher order transverse and longitudinal acoustic resonance modes.

low-frequency instabilities are intrinsic low frequency instabilities, feed coupled instabilities, chuffing, chugging, and the chamber bulk mode (acoustic) instability. The intrinsic low frequency instability is unique to hybrid rocket motors. It is observed to some degree in almost every hybrid test program and is believed to be a function of thermal lag in the solid fuel grain, boundary layer combustion dynamics, and transient gas dynamics in the combustion chamber [40]. Feed coupled instabilities are observed in both hybrid rocket motors and liquid rocket engines and result from a coupling of the feed system with the combustion processes within the chamber. Chuffing instabilities are observed in hybrid rocket motors operating at very low oxidizer mass flux and are attributed to the char/melt layer formation and break-off processes that take place only in this flux regime. The chamber pressure oscillations produced by chuffing typically occur at very low frequency on the order of a few Hz. Chugging instabilities in hybrid rocket motors are attributed to a lack of flame holding at the fore end of the fuel grain. The bulk mode, also known as the Helmholtz mode, is an acoustic instability. Low-frequency instabilities are often the dominant mode observed in hy-

brid rocket motors; however, they are frequently accompanied by medium frequency modes at lower amplitude [40].

Medium frequency instabilities within hybrid rocket motors typically range from 200 Hz to 2000 Hz. They are most commonly associated with the excitation of longitudinal acoustic modes of the chamber [40]. However, the magnitude of these instabilities is typically small compared to their low-frequency counterparts. A pressure coupled instability is also in this moderate frequency range but is only present in hybrid motors operating in a regime where fuel surface regression rate is a function of chamber pressure. This could be caused by operation at mass flux extremes or where radiation cannont be ignored, for example, with metallic additives.

High frequency instabilities, which are typically those above 2000 Hz, are due to higher longitudinal and transverse modes. High frequency instabilities are generally considered to be inactive in hybrid rocket motors. All of these instabilities are summarized in Table 8.8 with the most important instabilities also discussed in more detail in the following subsections.

Issues with the feed system design, combustion chamber design, or operating conditions can lead to instabilities. Large volumes between the feed line isolating element and the combustion chamber have been associated with an adverse influence on motor stability for systems using a gaseous oxidizer [40]. In typical hybrid rocket motors with a gaseous oxidizer, the choked orifice is generally situated close to the combustion chamber injector in an effort to minimize the volume between the feed line isolating element and the combustion chamber. When doing this, however, the propulsion engineer should avoid injecting supersonic flow directly into the combustion chamber. The stability level of a hybrid rocket motor is often controlled by conditions in the precombustion chamber [40].

If present, successfully mitigating these combustion instabilities can be a lesson in determination and patience for the propulsion engineer. When a strong combustion instability is present in a hybrid motor, it will generally prevent the establishment of other instabilities. Thus when a first instability is resolved through changes to the system design or operation, it is typical for other instabilities to then become apparent. The propulsion engineer must diligently work through resolving each one in turn until the motor is stable. Once a motor has been demonstrated to be stable at nominal operating conditions, the stability of the motor should then be demonstrated across the potential range of operating pressures and mass fluxes. Additionally, all transient conditions should be studied for potential excitations. The same approach is used for liquid rocket engines where a dynamic stability rating is established per NASA SP-194 [304].

8.4.2 Feed system coupled instability

As the name implies, feed system coupled instabilities are the result of acoustic coupling between the feed system and the combustion processes in the combustion chamber. In order to be present, there must be a form of combustion lag, such as an oxidizer vaporization delay,

and acoustic coupling between the combustion chamber and feed system. Feed coupled instabilities are most commonly observed in hybrid motors that use a liquid oxidizer since a simple solution exists in hybrids using a gaseous oxidizer via a sonic choke point to isolate the combustion chamber dynamics from the feed line. Feed system coupled instabilities have posed particular challenges for propulsion engineers working with cryogenic oxidizers, and those with high vapor pressure, like nitrous oxide. It has been demonstrated that this instability can be mitigated in motors using liquid oxygen [45] or nitrous oxide [47] via careful design of a cavitating venturi or an integrated venturi and injector as an isolating element, see Ref. [47] for details.

Feed system coupled instabilities typically have a very regular pressure oscillation, with a narrow bandwidth peak in the Fourier domain [40]. Higher-frequency modes at multiples of the fundamental frequency will typically also be seen in the Fourier spectra. The regular pressure oscillations at the fundamental frequency are seen both in the combustion chamber and in the feed system.

An example of a chamber pressure trace, and the corresponding spectrogram (also referred to as a waterfall plot), for a hybrid motor with a feed system coupled instability is shown in Fig. 8.10. This test data is taken from a developmental hot-fire test of the Peregrine motor, a nitrous oxide and paraffin hybrid described in detail in Refs. [47], [177], [305]. The feed system coupled instability is most notable in the first half of the test shown, where a regular sinusoidal signal at 15 Hz can be clearly seen. At around 55 seconds, the first longitudinal mode is excited, making the feed system coupled instability less coherent but still present for the duration of the test. In this motor, the feed system instability was overcome by increasing oxidizer supercharge/saturation pressure and injector pressure drop in order to choke the flow

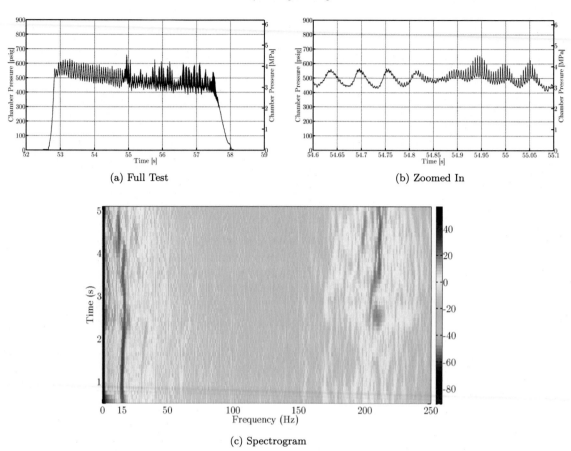

(a) Full Test

(b) Zoomed In

(c) Spectrogram

FIGURE 8.10   Chamber pressure data showing a feed system coupled instability from a developmental hot-fire test of the Peregrine motor, a nitrous oxide and paraffin hybrid. The feed system coupled instability can be seen around 15 Hz. The figure is used with permission from Ref. [47].

at injector holes [177], but a cavitating venturi is suggested to minimize system mass.

For further information on the feed system coupled instability and approaches to mitigate it, see Ref. [47].

8.4.3 Intrinsic low frequency instability

The intrinsic low frequency instability is present in nearly all hybrid rocket motors. It is the result of thermal transients in the solid fuel coupling with the boundary layer combustion and

chamber gas dynamics. Ref. [40] provides a thorough description of the underlying physics driving this instability and develops a linearized model for it. This linearized model is used to derive Eq. (8.68) for the primary frequency of the hybrid motor intrinsic low frequency instability. The temperature, T, and gas constant, R, used in this equation are the average gas values in the port, G_o is the oxidizer mass flux, L_F is the fuel grain length, and O/F is the average oxidizer to fuel mass ratio. Suggested values for the average temperature gas product, RT, of 6.38×10^5

m^2/s^2 and 4.47×10^5 m^2/s^2 for systems using high-energy oxidizers (e.g., GOx/LOx) or low-energy oxidizers (e.g., N_2O), respectively, are recommended by Ref. [40]. Eq. (8.68) holds for both liquefying and classical fuels as well as gaseous and liquid oxidizers. Unfortunately, this linear model can only predict the frequency of oscillation. It cannot determine the amplitude of the instability. However, it is frequently used to identify the mode if it has been observed in testing.

$$f_{ILFI} = 0.234 \frac{G_o RT}{LP_c} \left(2 + \frac{1}{O/F} \right) \quad (8.68)$$

The intrinsic low frequency instability is typically low frequency like the feed coupled instability but will generally have a broad bandwidth peak in the Fourier domain. The instability has been known to vary in intensity during a single test, even appearing, or disappearing mid-test [40]. The presence of this instability will typically excite other acoustic modes within the combustion chamber.

For further information on this instability and the analytical approach to modeling it, see Refs. [40] and [53].

8.4.4 Acoustic instabilities and bulk mode

The subject of acoustic instabilities and the bulk mode is covered in detail in liquid and solid rocket engine literature, such as Refs. [306], [125], [307] and [308]. These instabilities are often present in hybrid rocket motors but are typically excited by one of the other low frequency instabilities and rarely have an amplitude large enough to cause damage to the motor (unlike liquid systems where transverse acoustic instabilities are known to be a particular challenge). However, they are still undesirable for motor performance. A brief description of acoustic instabilities is provided such that the propulsion engineer becomes familiar with the concept of

these instabilities. However, resolution of issues associated with excitation of these instabilities within a motor will require consultation of relevant literature, several references are provided here.

The bulk mode corresponds to instability with a uniform spacial pressure distribution but oscillating chamber pressure with time. This instability is also referred to as Helmholtz instability, as it corresponds to the vibration of a Helmholtz resonator [309]. The lack of a spatial pressure gradient results in almost no velocity fluctuations within the chamber as a result of this instability [309].

Acoustic instabilities form when there is a resonance between energy release from the combustion process and the natural acoustic modes of the combustion chamber internal geometry [47]. The transverse, radial, and longitudinal acoustic modes all cause spacial pressure variation within the combustion chamber, see Fig. 8.11. It is important for the propulsion engineer to have an idea of an approximate value of these resonant chamber frequencies before conducting hot-fire tests. Eq. (8.69) gives a rough estimation of the longitudinal, transverse and radial acoustic mode frequencies, f_{ijk}. For clarity, the equation for only longitudinal modes, Eq. (8.70), is also provided here, as are the expressions for the first acoustic modes, see Fig. 8.11. In these equations, D can be treated as the average internal chamber cross-section diameter, a is the speed of sound, L is the acoustic length which is often approximated as the distance from the motor fore end to the nozzle throat, λ_{ij} is the i-j mode eigenvalue and i, j, k are the mode number for the tangential, radial and longitudinal mode, respectively. Note that the acoustic length is generally a little less than the chamber length due to gradual contraction in the combustion chamber diameter towards the nozzle throat [2]. The speed of sound in the chamber is generally approximated by that predicted from combustion equilibrium

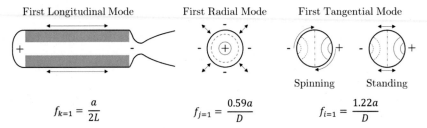

$$f_{k=1} = \frac{a}{2L} \qquad\qquad f_{j=1} = \frac{0.59a}{D} \qquad\qquad f_{i=1} = \frac{1.22a}{D}$$

FIGURE 8.11   Schematic of the first primary mode of each of the acoustic instabilities. The plus and minus refer to the location of maximum and minimum pressure, respectively, in the chamber at a given snapshot in time. The location of the plus and minus would be reversed a half cycle later. Note that the arrows show the notional direction of (typically cyclic) loading on the motor as a result of the instability. The equations given can be used to calculate the frequency of each of these first modes per Ref. [2].

calculations, though it is not uncommon for this approximation to produce significant error [2].

$$f_{ijk} = a\sqrt{\left(\frac{\lambda_{ij}}{D}\right)^2 + \left(\frac{k}{2L}\right)^2} \; [2] \qquad (8.69)$$

$$f_k \simeq \frac{ka}{2L} \; [308] \qquad (8.70)$$

More detailed analysis approaches that account for the variation in combustion chamber geometry are described in the literature, see, for example, Ref. [308] or [306]. Internal combustion chamber geometry changes can be used to mitigate acoustic excitation by changing the mode frequencies. As an example of this, it has been shown that a diaphragm can be used to alter the first longitudinal mode excitation frequency [310,311].

An example of the excited first longitudinal mode in a hybrid motor is shown in Fig. 8.12. Again, this test data is taken from a developmental hot-fire test of the Peregrine motor, the nitrous oxide and paraffin hybrid described in detail in Refs. [47,177,305]. Some internal geometry changes were made to resolve this acoustic instability, and a subsequent stable test of the motor was completed, as shown in Fig. 8.13.

8.4.5 Chugging instability

Chugging instabilities are most often associated with the excitation of the low frequency bulk mode instability in liquid rocket engines. They can be present in hybrid rocket motors in the form of a flame-holding instability at the leading edge of the fuel grain. They have been visualized in a slab burner configuration, [85], [312], and observed in more standard (at a moderate scale of 11 inch and 24 inch outer diameter) hybrid rocket tests [45], [313]. If the flame is not anchored at the front of the fuel grain then unburned fuel can accumulate. This is followed by subsequent higher-rate combustion. The description of this instability as applied to liquid rocket engines provided by Huzel and Huang [2] reads as nearly perfect description of the observed processes in Ref. [312] for a hybrid slab burner. The flame recedes, allowing fuel to accumulate then detonate, or rapidly combust, at which point "the resulting chamber pressure spikes cause a reduction, or even reversal, of the propellant flows and a consequent rapid collapse of chamber pressure, allowing propellants to rush in and repeat the cycle [2]." The schlieren images of the free stream recorded for the pressurized slab burner visualization tests of Ref. [312] show the regression of the flame along with the explosive combus-

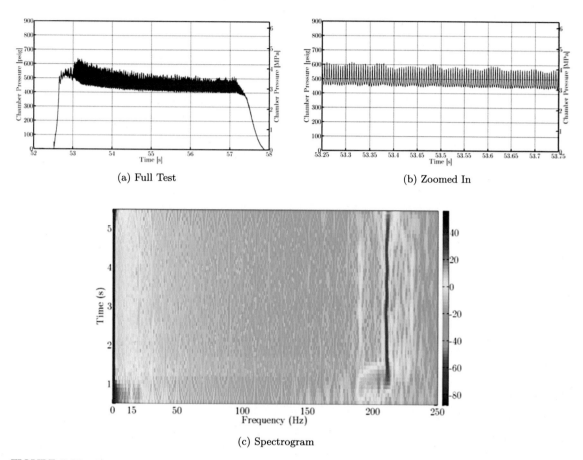

(a) Full Test

(b) Zoomed In

(c) Spectrogram

FIGURE 8.12 Chamber pressure data showing an excited first acoustic mode from a developmental hot-fire test of the Peregrine motor, a nitrous oxide and paraffin hybrid. The first mode can be seen around 200 Hz. The figure is used with permission from Ref. [47].

tion events, a rapid forward motion of the flame and a reversal in the oxidizer flow.

There are multiple approaches to avoiding or mitigating this instability in hybrid motors. It is important for the propulsion engineer to ensure that they achieve complete vaporization of the oxidizer in the motor fore end, this can be achieved passively such as through a cavitating injector, or more actively via heat addition in the motor fore-end [45], [48]. The propulsion engineer should also ensure that the oxidizer injection geometry promotes stabilization of the

flame at the motor fore end (e.g., the slab burner used in Ref. [312] was inherently unstable due to the uniform oxidizer flow encountering a forward facing step).

8.4.6 Chuffing instability

The chuffing instability is a very low-frequency hybrid rocket motor instability that only occurs at very low fuel regression rates (i.e., at low oxidizer mass flux regimes). It is an instability that is due to fuel char/melt layer formation and

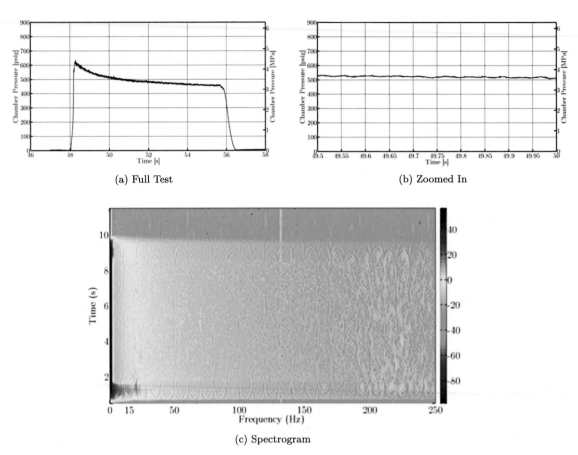

(a) Full Test   (b) Zoomed In

(c) Spectrogram

FIGURE 8.13   Chamber pressure data showing a stable hot-fire from a developmental hot-fire test of the Peregrine motor, a nitrous oxide and paraffin hybrid. The figure is used with permission from Ref. [47].

break off when the thermal layer in the fuel becomes thick [40]. The fuel surface regression rate is sufficiently fast to keep the surface thermal layer thin under normal operating conditions. When the oxidizer mass flux is extremely low, the thermal layer extends down into the fuel surface, a phenomenon is known as "cooking". The softened layer will increase in thickness before being mechanically removed from the fuel surface, where it combusts and causes a pressure spike. The thermal lag times for heat penetration into the fuel surface dictate that this instability is very low frequency, typically around 1–5 Hz [40]. While this instability can typically

be avoided by the selection of operating conditions, it is especially important to consider during transient events, e.g., motor shut down and throttling.

8.5 Propellant budgeting in flight

The authors recommend that the approach for oxidizer propellant budgeting in flight follow that used in liquid systems. There are multiple methods used to predict propellant remaining in the tank throughout an extended duration flight mission:

1. Flight telemetry data: remaining propellant budget is calculated from flight telemetry data and thermodynamics.
2. Book keeping: the amount of oxidizer remaining in the tank is determined based on the original loaded mass prior to launch and the calculated mass transferred out during each burn.
3. Thermal capacitance: the amount of oxidizer remaining in the tank is determined based on the change in temperature in the tank when a known amount of heat is supplied via heaters. This method relies upon a good thermal model of the spacecraft and reliabile insturmentation to determine the thermal mass and therefore the mass of oxidizer remaining.
4. Direct flow meter measurements: this is similar to book keeping but is more accurate as it uses direct flow measurements via a flow meter (in flight).

The flight telemetry method requires accurate pressure and temperature readings. Unfortunately, this can be a challenge for long-duration flight missions where sensor accuracy can drift or in some spacecraft where analog to digital conversion may not be very precise. Thus the book keeping method is often also used concurrently to monitor propellant usage. It is recommended to do both methods, as described in depth in Ref. [314]. The thermal capacitance method can be a preferred approach when the thermal model of the spacecraft is known to be accurate. The direct flow measurement is an improvement over the book-keeping method but requires that a flowmeter be flown. This approach is planned for the Europa Clipper propulsion system [315].

C H A P T E R

9

Hardware design

9.1 Introduction

Hardware is broken into flight hardware and ground test/ground support equipment. Mass, reliability, and performance play critical roles in flight hardware, which will be discussed in the following sections. The specifications for a ground test or ground support equipment (GSE) can often be relaxed from these more stringent flight hardware requirements to reduce cost. However, the test as you fly, fly like you test principle should be applied whenever high reliability is necessary. This ensures that failures do not arise from differences between what was tested on the ground and what actually flies.

The following sections will discuss hardware relevant to hybrid rocket propulsion systems. This includes descriptions of valves, tanks, filters, venturis, regulators, pumps, standard instrumentation (pressure transducers, thermocouples, etc.), injectors, combustion chambers, and nozzles. Parameters that are critical to component selection will be discussed, along with considerations for reliability, qualification, and manufacturability.

9.2 Valves

Valves are necessary to move propellant through-feed systems. The most common types used for hybrid rockets are discussed here. In flight systems, common valves include pyrotechnic and latching valves for isolation, solenoid valves for thruster operation, and service valves (aka fill and drain valves) for propellant loading. Check valves can be used to limit flow in one direction. Throttle and needle (or pintle) valves can be used to control the amount of propellant flow. Additional flow control valves, including butterfly valves and ball valves, are also discussed along with relief valves, which are only used for safety in ground systems.

Pyrotechnic and solenoid valves come in either a normally open or normally closed configuration. This designation describes the initial state of the valve (when unpowered). Most schematics will indicate whether a pyrotechnic valve is normally open or normally closed, whereas solenoid valves are usually just assumed to be normally closed unless otherwise noted.

9.2.1 Pyrotechnic valves

Pyrotechnic valves (aka pyrovalves) provide excellent long-term isolation and rapid actuation. They are classified as either normally closed or normally open based on their initial state. When employed, a ram shears through a single piece of the parent metal, either opening a flow path that did not exist before (normally

Hybrid Rocket Propulsion Design Handbook
https://doi.org/10.1016/B978-0-12-816199-9.00017-2

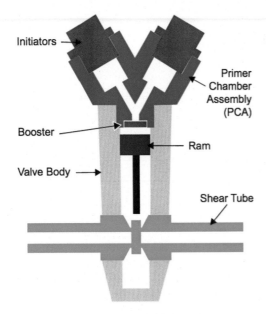

A normally closed pyrovalve

FIGURE 9.1 Cartoon depicting the internals of a pyrovalve [316]. Once activated, the ram is propelled downwards and shears the green material blocking the tube. The ram is then retained in place. A hole through the ram allows flow through the shear tube.

closed) or closing off a previously open path (normally open). The actuation time is on the order of 10 ms. These isolation valves are single use and are therefore, generally, only used in flight applications. Pyrotechnic valves are activated by an electronic signal, which ignites a small explosive change in the initiator. Some pyrotechnic valves, such as that shown in Fig. 9.1, use dual initiators to provide redundancy. A note of caution: pyrovalves with dual initiators should not be fired simultaneously as the charges can reflect off each other and cause a misfire [316]. Not all propellants are suitable for use with pyrovalves. For example, firing a pyrovalve in a high-pressure oxygen environment has a significant risk of fire. Additionally, shock caused by the pyrotechnic actuation can drive

loads in the system and damage other components.

9.2.2 Poppet valves

Poppet valves are valves which use a valve element (called the poppet) that travels perpendicular to the valve seat [272]. There are various types of poppet designs, as shown in Fig. 9.2. Poppet valves are used in both flight and ground test applications. They are often used as a form of on/off pressure isolation valve and are well suited to applications where rapid changes from closed to full flow are required [272]. Poppet valves are generally not well suited to throttling unless the poppet is carefully contoured due to the significant flow area gain when the poppet is close to the valve seat [2]. Poppet valves can generally maintain low leak rates when closed but have large pressure drops across them when open due to the complicated internal flow geometry [2]. They can use hard or soft seat materials. However, poppet valves using hard seats are particularly sensitive to leakage from particulate contamination.

Solenoid valves

Solenoid valves are a form of a poppet valve that use a solenoid to drive the poppet. A schematic of a solenoid valve is shown in Fig. 9.3. Power is provided to the solenoid to generate a magnetic field, which in turn moves the poppet to either open (normally closed) or close (normally open) the valve.

Solenoid valves must be powered constantly to maintain a change of state from their normal position. The most common form of solenoid valve is a normally closed valve, but normally open solenoid valves also exist and are used. Solenoid valves can be designed to provide a fast response (on the order of a few ms), and as such, they are regularly used for situations requiring fast-acting open/close response, including pulsed operation.

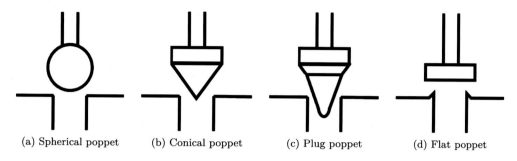

(a) Spherical poppet   (b) Conical poppet   (c) Plug poppet   (d) Flat poppet

FIGURE 9.2   Cartoon depicting common poppet shapes used in poppet valves. The figure is based on Ref. [272].

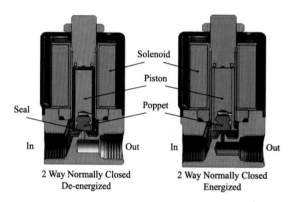

FIGURE 9.3   The internals of a normally closed solenoid valve in both the de-energized and energized states. Adapted from Ref. [317].

The power required to operate a solenoid valve depends on the valve's design. Higher pressure valves that are capable of fast response generally require more power. The power through the solenoid is sufficiently high to cause substantial heating in the valve coil. Therefore, solenoid valves will generally be designed with maximum duty cycles or with a specified drop-down voltage for scenarios where the valve is to be held open (for a normally closed valve). Solenoid valves generate back ElectroMotive Force (EMF) when they change state. System electronics should either be designed to tolerate this or preferably to mitigate it (e.g., using a Zener diode to significantly reduce the back EMF).

Latch valves

Latch or latching valves are a type of poppet isolation valve that only need to be powered to change position (from open to closed or closed to open). Latch valves do not require power to remain in a given state. They can use solenoids to change the valve state but are designed such that they can remain in either the open or closed state indefinitely without current. Latching in a given state is typically achieved via mechanical means for large valves or magnetic means for small valves [272]. An example of the internals of a magnetic latching valve is shown in Fig. 9.4. Latch valves are typically used as a form of isolation for scenarios where the valves may be required to change state multiple times, but generally not in a high frequency or pulsed operation mode.

Pilot valve

A pilot valve is a small valve used to control the flow of fluid, which in turn is used to actuate larger valves. An example of an integrated piloted solenoid valve is discussed below as its own class of poppet valve. Stand-alone pilot valves are also regularly used to actuate valves with high flow capacity, such as ball valves and butterfly valves. There are many different pilot valve designs, and they can be actuated by solenoids (and therefore just a form of solenoid valve), pneumatically actuated, or actuated via a mechanical connection with other components

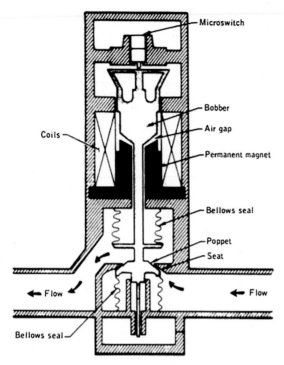

FIGURE 9.4 The internals of a magnetic latching valve from Ref. [318].

port. The pilot port is designed to be larger than the bleed port, so fluid from behind the main poppet is dumped through the pilot port. This causes a pressure differential which drives the main poppet open [272].

9.2.3 Butterfly valves

Butterfly valves consist of a disk valve element that is supported on a shaft such that it can be rotated in a housing between open, partially open, or closed positions [272]. The simplest form of butterfly valve uses a clamshell design where a circular sealing disk is rotated on a shaft. The internal circle must be smaller than the external housing to allow rotation. Unfortunately, this means that even in the closed position, there is an internal leakage path, making this form of butterfly valve ill-suited as a shutoff valve. Two alternative designs, one using a canted elliptical sealing disk and the other using an offset section of a sphere in the place of the disk, overcome this issue and allow the valve to be used for isolation [272]. In all three configurations, an actuator is required to rotate the disk between the open and closed positions. Butterfly valves generally use a pneumatic line controlled by a pilot valve to actuate a piston-type actuator for the main valve.

Butterfly valves are used in large systems, e.g., they are typically used in feed lines with nominal diameters from 5 cm to 43 cm (2 inches to 17 inches) [2]. Butterfly valves have heritage in both cryogenic and storable oxidizer feed lines. They are favored for their low hydraulic resistance (low pressure drop), relatively compact design, light weight, and their ease of service [2,272]. One downside of butterfly valves is that they require at least one external shaft seal which can provide a possible path for external leakage [272].

[2]. However, regardless of the actuation mechanism pilot valves are generally required to have a fast response and low leakage rates [2].

Piloted solenoid valves

Piloted solenoid valves are used when the actuation force required to open the valve exceeds the capabilities of the solenoid. Thus, piloted solenoid valves are typically used for high pressure or higher flow (larger poppet) applications [272]. An example cut-away of a piloted solenoid valve is shown in Fig. 9.5. In this valve, and in general for pilot valves, upstream pressure is supplied behind the main poppet via a bleed line. The solenoid valve opens the small pilot valve, which allows flow through the pilot

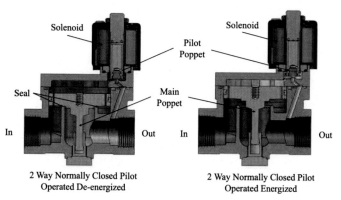

FIGURE 9.5 Internals of a piloted normally closed solenoid valve, shown both in a de-energized and energized state. Adapted from Ref. [317].

9.2.4 Ball valves

Ball valves are valves that use a rotating ball with a central channel for fluid flow, see Fig. 9.6. These valves use a spherical sealing surface that can be precisely manufactured to provide low leakage rates when closed. Ball valves typically allow for high fluid flow rates with a minimal pressure drop. In fact, ball valves using an internal channel area equal to that of the surrounding lines can provide essentially unrestricted fluid flow when in the full open orientation [2]. Ball valves have a heritage in cryogenic and storable propellant feed lines [2]. Similar to butterfly valves, ball valves require an actuator of

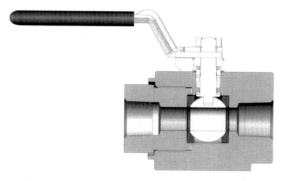

FIGURE 9.6 The internals of a ball valve. Source: Swagelok.

some sort to rotate the ball between the open and closed positions.

9.2.5 Needle valves

Needle valves are a form of control valve used to meter the fluid flow rate. They typically rely on manual adjustment of the needle position to achieve a given flow throttle and therefore are usually only used in ground systems. Needle valves generally have a small flow capacity (due to a small control orifice) and have a long, slender, valve element that facilitates tight throttling [272], see Fig. 9.7. As such, needle valves are typically used to control low flow rates. Needle valves are useful when a range of low flow rates is required during a test campaign and when precise replication of flow conditions is not required between tests as it is difficult to obtain the same settings between multiple tests. If exact conditions must be replicated across a number of tests, then orifices or cavitating venturis are recommended as an alternative to needle valves.

9.2.6 Check valves

Check valves are intended to allow fluid flow in a single direction. These valves operate via internal springs that open to flow above the de-

FIGURE 9.7 The internals of a needle valve. Source: Swagelok.

sign cracking pressure and close upon backflow (negative pressure). There are multiple internal geometries possible for check valves which are classified according to the seat design to include ball, poppet, swing, flapper and cone types [272]. A cross section of a typical poppet check valve is shown in Fig. 9.8.

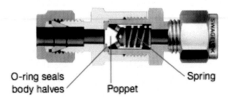

O-ring seals   body halves   Poppet   Spring

FIGURE 9.8 The internals of a check valve. Source: Swagelok.

The design cracking pressure is the key parameter for check valves. It is defined as the difference between the inlet and outlet pressures needed to achieve a specified minimum flow through the check valve. If the cracking pressure is too high, then the pressure drop across the check valve needs to be large to maintain flow. If the pressure differential across the check valve is not always higher than the cracking pressure when fluid is flowing, then the check valve can "chatter". Chattering refers to rapid opening and closing of the check valve seat. It is highly undesirable as it introduces unsteady flow, is known to generate particulate, and can

be an ignition hazard. Conversely, if the cracking pressure is too low, then it is more likely that backflow through the check valve may occur. In general, check valves are very susceptible to contamination and often leak. They have been known to stick in either the open or closed position in flight and can have unpredictable pressure drops across them.

9.2.7 Relief valves

Relief valves are used as a safety feature in ground testing. They relieve pressure from a system in order to avoid damaging other downstream hardware if there is an overpressure event or failure. The set pressure of the relief valve is the pressure at which the valve will open and divert flow. It is always chosen to be lower than the lowest-rated component or line can tolerate, this pressure limit is referred to as the Maximum Allowable Working Pressure (MAWP). For example, if the downstream components and line are designed for maximum operating pressures of 6900 kPa (1000 psia), 10,300 kPa (1500 psia), and 5500 kPa (800 psia), then the MAWP is 5500 kPa (800 psia), and the relief valve set point should be below this value. In general, for ground systems, the planned Maximum Expected Operating Pressure (MEOP) for the system will be below the MAWP. In this case, it is recommended to set the relief valve at a pressure at least 20% above the MEOP as long as it is still below the MAWP; this avoids inadvertent activation of the relief valve during pressure fluctuations due to normal operation.

Relief valves, such as the example shown in Fig. 9.9, can often be purchased with an adjustable (via spring) set point. This allows the user to select a specific relief pressure within the rated range of the spring. Relief valves like this are not single-use; they can be re-used after opening to relieve pressure. Therefore the relief valve set points can and should be verified via test. Once opened, the relief valve will typically re-seat (close) once the pressure has dropped be-

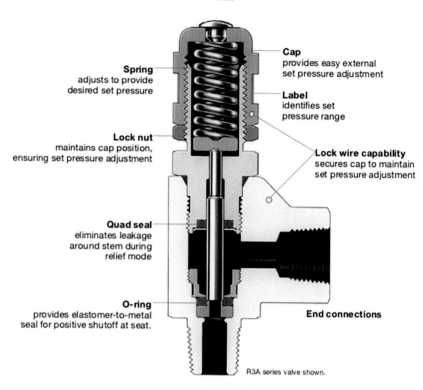

Cap
provides easy external
set pressure adjustment

Spring
adjusts to provide
desired set pressure

Label
identifies set
pressure range

Lock nut
maintains cap position,
ensuring set pressure adjustment

Lock wire capability
secures cap to maintain
set pressure adjustment

Quad seal
eliminates leakage
around stem during
relief mode

O-ring
provides elastomer-to-metal
seal for positive shutoff at seat.

End connections

R3A series valve shown.

FIGURE 9.9   The internals of a relief valve from Ref. [319]. Source: Swagelok.

low approximately 85% of the set pressure. The relief valve set pressure is generally repeatable to within ±5%, but can vary with environmental conditions, such as temperature. For example, Swagelok relief valves are specified to be repeatable within ±5% between 15–27 °C (60–80 °F), but ±20% outside that range [319]. For this reason, it is often recommended to use a relief valve with at least 20% pressure margin above the desired set pressure. In order to properly protect downstream hardware, relief valves need to be designed with sufficient flow area/flow capacity to relieve the entire flow rate in a line, including, for example, in a scenario where the regulator fails open.

When working with toxic or hazardous propellants, it is critical to ensure that the relief valves direct the flow safely by either scrubbing and/or venting it to an environmentally responsible and safe location.

Rupture disks are a high-flow, single-use alternative for pressure relief. These disks burst when the design pressure has been reached and can typically relieve pressure very quickly. However, they are unforgiving of transient spikes in pressure and some types of fragment, which can contaminate the feed system for future use.

9.3 Tanks

Pressure vessels in hybrid rockets include the oxidizer tank(s), pressurant tank(s) and the combustion chamber. Tanks will be discussed here, and combustion chamber hardware will be discussed in Section 9.7. In hybrid rocket propulsion systems, tanks are typically used to store oxidizer and pressurant. They must be designed to provide the required storage volume and design pressure. Tanks storing liquids must also provide a form of propellant management to minimize any gas transfer into the feed line, maximize the usable mass of propellant, and damp out propellant motion, aka slosh. A list of possible propellant management options is provided in Section 9.3.1.

Tanks consist of a main storage volume (typically cylindrical or spherical for metal tanks or cylindrical or near sphere for composites), boss(es), and possibly a skirt, mounting brackets and/or a propellant management device. A boss or bosses are needed to hold the tank during manufacture and operation. This is also usually where the opening into the tank resides. Often, a skirt or mounting supports need to be added to the pressure vessel design to integrate it with a flight system. Note that the way a tank is held can drive mechanical loads and therefore the design. Tanks are often primary structure in rocket applications, meaning that they carry the main thrust loads, and it is not sufficient to design them to solely withstand pressure loads. The mounting design can also affect the thermal design of the system since it dictates the amount of thermal conduction to the surrounding vehicle and should therefore be examined relatively early in the design phase.

The recommended approach to estimate tank mass, to first order, is discussed in Chapter 7. The tank mass approximately scales with a design pressure and volume in this method. As such, the efficiency of a given tank design is measured in terms of $P \Psi / m_t$, where P is the rated pressure, Ψ is the volume, and m_t is the mass of a tank. Tank design should first be evaluated based on the available volume envelope, the maximum design pressure, and the required storage volume. More advanced design options, including structural and thermal loads, can then be considered.

Many tank suppliers list their heritage tanks online, providing information on the tank mass, storage volume, and pressure rating. Refs. [268], [270], and [320] provide examples of this for three large United States flight tank manufacturers. When existing tank specifications line up with the requirements of a new system, it circumvents the need to estimate mass in the early design phase, and generally reduces the component risk and cost. However, since tanks make up the large portion of any rocket system (in terms of both volume and mass), there is often a convincing rationale to customize them for a given application. To minimize mass, it is generally desirable to configure tanks to be as close to spherical as the system envelope allows. However, aerodynamics or packaging considerations may prevent this. The minimum required thickness of the tank wall is dependent on the size of the tank and the design pressure. For tanks using high-performance materials and with a low design pressure or small diameter, the minimum wall thickness for manufacturability will dictate the tank wall thickness.

Brittle failure is the most common failure mode for thin-walled pressure vessels like tanks. Thus, fracture mechanics must be considered in the tank design. The simplified equations for generating a first-order tank mass presented throughout this and other general rocket design books do not begin to delve into this rich subject area. The recommended approach for testing and analyzing the potential for crack growth and failure of a tank are discussed in more detail in Refs. [7] and [8] for metallic and composite overwrapped pressure vessels, respectively.

A note on toroidal tanks: toroidal tanks are tanks in the shape of a toroid (donut). These tanks always seem like a good idea on paper for packaging around a hybrid rocket motor. However, a single toroidal tank typically weighs more than multiple cylindrical tanks but is dependent on how the tanks are mounted. There may also be more residual oxidizer as draining a toroidal tank is more complicated. Also,

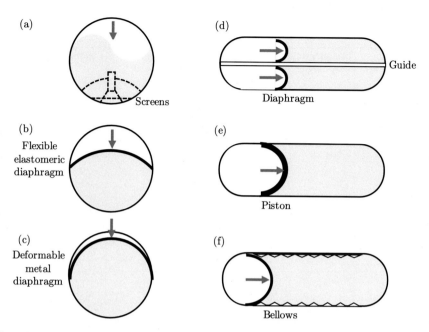

FIGURE 9.10 Schematic of common propellant management devices: (a) Capillary Retention PMD, (b) Elastomeric Diaphragm, (c) Metal Diaphragm, (d) Rolling Cylindrical Metal Diaphragm, (e) Piston, and (f) Metal Bellows. Note that in all images the direction of pressurization force is indicated with an arrow, pressurant gas is white, and propellant is gray. The figure is derived from Refs. [1], [2] and [3].

it is typically far less expensive to use multiple identical tanks. One of the often forgotten costs for hardware is that of qualification. New tank designs generally require at least one destructive burst test to verify that the tank meets or exceeds the burst pressure requirement. Additional flight tanks will therefore be required to prove that the new design works as expected, leading to a large increase in cost for the harder to manufacture, toroidal tank.

9.3.1 Propellant management

Propellant Management Devices (PMDs) are required for liquid oxidizer tanks in order to minimize the likelihood of gas transfer into the oxidizer feed line. A summary of commonly used propellant management approaches is provided in Table 9.1. These approaches are also shown schematically in Fig. 9.10. The expulsion

mechanism is classified as either positive or passive. Positive expulsion uses a physical barrier between the pressurant gas and liquid oxidizer. Passive devices do not have a physical separation between the oxidizer and pressurant gas but instead use surface tension to keep fluid in contact with the tank.

Key design properties of the various propellant management approaches are summarized in Table 9.1 and explained briefly here:

- Expulsion efficiency: propellant management approaches are designed to maximize the usable liquid oxidizer. Even in very efficient tanks, there will be some residual propellant in the tank when it is emptied, and this residual is calculated as a percentage of the liquid storage capability of the tank.
- Center of gravity control: liquid moving around inside the tank can transmit forces

TABLE 9.1 Summary of propellant management approaches for liquid tanks. Table is derived from information in Refs. [1], [2], [3], and [4].

Name:	Capillary Retention PMD	Elastomeric Diaphragm	Metal Diaphragm	Rolling Cylindrical Metal Diaphragm	Piston	Metal Bellows
Expulsion Mechanism	Surface Tension (Passive)	Positive expulsion	Positive expulsion	Positive expulsion	Positive expulsion	Positive expulsion
Mass	Low	Low	High	Low	Moderate	High
Cost	High	Moderate	High	Low	High	High
Heritage	Spacecraft	Launch vehicles, spacecraft, upper stages	Launch vehicles, spacecraft, upper stages, missiles	Missiles	High acceleration missiles	Launch vehicles, spacecraft, missiles
Expulsion Efficiency	High (up to 99.5%)	High (99.5%)	Moderate to High (95-99.5% for well designed tanks)	High (99.5%)	High (99.5%)	Moderate (97–99%)
Volume Efficiency	High	High	High	Moderate	Low	Low
Center of Gravity Control	Low	Moderate	Good	Moderate	Moderate	Good
Expulsion ΔP	Low	Low	High	Low	Low	Moderate
Sliding Seals	None	None	None	None	Uses a sliding seal, possible leakage risk	None
Ullage Needs	Some needed	Some needed	None	Some needed	Some needed	Some needed
Packaging Limitations	Can be used with most tank configurations	Can be used with most tank configurations	Design optimizes only for special envelopes (typically close to spherical)	Cylindrical tanks. Design adapts easily to growth	Cylindrical tanks. Design adapts easily to growth	Cylindrical tanks
Pressurant Ingestion	Dissolved pressurant	Permeation of pressurant	None	None	None if seal holds	None
Propellant Off-Load Limitations	None	Off-load not desirable.	Off-load not desirable	Off-load not desirable	None	None
Other	Mass can grow if large damping is needed	Compatability constraints with propellants and diaphragm	-	Internal weld inspection difficult	Critical tolerance on tank shell to reduce leakage risk	-

to the system structure. This phenomenon is known as propellant slosh. Without mitigation, slosh forces can be large enough to overwhelm the attitude control of the spacecraft/rocket. The ability of a given propellant management approach to minimize slosh is captured in Table 9.1 as the center of gravity control.

- Expulsion ΔP: There is always some pressure drop across a tank when the propellant is flowing out of it. Some devices are more efficient at reducing this pressure drop.
- Ullage needs: the ullage of a tank is generally defined as the volume of gas in the tank as a percentage of the total internal usable volume of the tank. Many propellant management approaches require at least some initial ullage in the tank.
- Packaging limitations: some propellant management approaches can only be used for specific tank shapes, this is captured under packaging limitations.
- Pressurant ingestion: refers to the possibility of gas transfer into the feed line even if the propellant management approach is well designed to avoid direct gas transfer.
- Propellant off-load limitations: the ability to off-load propellant can be critical for launch preparation in the event of a scrubbed launch. However, not all propellant management approaches allow for this.

9.3.2 Insulation

Tanks that are used to store cryogenic propellants, or that are using propellants that may freeze in space, must be insulated. Two insulation approaches are typically used: the first is foam insulation that is covered by metallic foil, and the second is nonmetallic honeycombs [1].

9.4 Other components

Common feed system components in addition to valves and tanks include filters, regulators, pumps, and tubing/fittings.

9.4.1 Filters

Filters are widely used in both ground and flight feed systems to remove particulate or other contaminants from feed systems and protect delicate downstream components such as regulators and flow control valves. (Particulate can lodge in valve seats creating a leak path.) Hybrid propulsion systems commonly use mechanical filters designed to remove particulate that could otherwise damage the feed system (e.g., by getting caught in valve seats and leading to leakage). These filters are generally classified by their absolute micron rating, which is defined as the size in microns (10^{-6} m) above which all hard spherical particles will be removed under static blowdown conditions [272]. The other key aspects to selecting a filter are 1) its pressure drop at a given flow rate, both at the beginning of life when it is clean and at the end of life when it is contaminated, and 2) the filter capacity. Filter capacity is the maximum weight of contaminant with a specified particle distribution that can be accumulated on the inlet side of the filter without exceeding a specified pressure drop [272]. Care must be taken with all filters to ensure material compatibility and verify that the filter is initially clean and will not introduce undue particulate into the feed line by virtue of its design/manufacture.

There are various types of mechanical filters, categorized by their filtration approach. Square mesh filters use woven metal fibers (in a similar manner to woven cloth) to create a square opening pattern. These filters are commonly used when only large particulates need to be filtered from the system, they typically have an absolute rating greater than 40 microns. Sintered mesh filters use similar woven metal fibers, but the fibers are sintered together where they cross. This sintering process fixes the pore size and reduces the possibility of changes in pore size under mechanical stress and vibration [272]. Unfortunately, the sintering process does not allow for a reduction in absolute rating size, and it also introduces additional contamination into the filter, making it challenging to achieve high cleanliness levels with these filters. Etched disk filters have historically been the filter of choice for applications that require high cleanliness levels and/or low absolute micron ratings. Etched disk filters are all metal filters. They rely on flow through a stack of etched disks designed to introduce a convoluted flow path. Some etched disk filters can be used with strong oxidizers, including fluorine [272].

9.4.2 Venturis

Venturis are used in flight and ground systems. They can be used for flow rate control or for measurement of the fluid flow rate. The venturi profile for both purposes is designed to minimize the total pressure loss of the fluid.

Cavitating venturis are used to control the flow rate of fluids. The flow rate equations for these venturis are provided in Chapter 8. The design of a cavitating venturi starts by first calculating the venturi throat area needed for a given flow rate using these equations. Minimization of total pressure loss is achieved by ensuring that there are well-rounded transitions at the upstream and downstream sides of the throat. A maximum divergence angle of 5-6 degrees is recommended for the downstream diverging section [321]. The convergence angle can be more abrupt, an angle around 37 degrees is common [321]. If the feed line length is limited, then shortening of the venturi can be achieved by making the converging section more abrupt. It is not recommended to make the diverging segment a wider cone. If it must be altered at all, then it is better to instead cut it off at the downstream end [321]. The design approach for a critical flow control venturi for gas flow (choked flow at the venturi throat) is similar, further details on this type of venturi design can be found in Ref. [322].

Similar rules of thumb apply to the profiles of noncritical venturis intended for flow rate measurement but with different angle limita-tions and a cylindrical throat. Ref. [323] recommends a converging angle of 21±1 degrees and a diverging angle between 7 and 15 degrees. A straight cylindrical section of throat with a length equal to the throat diameter is also recommended [323]. Also, for these venturi flow meters, the throat diameter should be sized such that the pressure differential between the throat and line is within the range of the differential pressure transducer that will be used. More details are provided on testing with venturis in Chapter 11.

9.4.3 Regulators

Gas systems require some sort of pressure regulation. This can be achieved by a conventional regulator or by active pressure control, sometimes called "bang-bang pressure control," where valves are opened and closed to control gas flow and maintain a desired pressure. Here we discuss conventional regulators, which are widely used in both ground and flight systems. Conventional regulators are generally used to modulate the supply of gas to achieve a given set pressure immediately downstream of the regulator. Regulators use a reference force (loading) which is applied the metering element to oppose the regulated pressure acting on the sensing element [272]. Different loading approaches can be used; a regulator can be weight loaded, spring-loaded, or pressure loaded. Fig. 9.11, shows an example of a mechanical spring-loaded regulator.

The key parameters used to size/select a regulator for a given application are the regulator set pressure, upstream pressure range, lockup pressure, leakage rates (internal and external), operating range/flow rate range, droop, and step response time. The regulator set pressure is the nominal regulated outlet pressure. In most ground regulators, this can be adjusted within some specified range. For flight systems, the set pressure must be selected early in the design process, as it cannot be changed in flight.

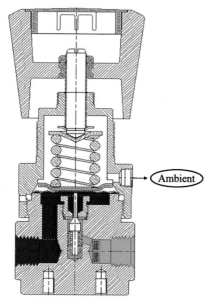

FIGURE 9.11 Illustrative internals of a spring-loaded vented regulator. Flow moves from right (green inlet) to left (blue outlet). Source: Tescom.

All regulators will have an allowable dead band around the set pressure. The upstream pressure range is the minimum and maximum pressure at the inlet of the regulator at which the regulator will still regulate to the set pressure. Regulator lockup occurs when the pressure downstream of the regulator increases to the point where the regulator closes completely [272]. As such, the lockup pressure (which is always higher than the set point) is the regulator outlet pressure when the regulator is closed, and the only flow is due to internal leakage [272]. Some ground regulators are venting regulators, meaning that the regulator will vent gas if the downstream pressure is higher than the lockup pressure. Regulator leakage rates are defined in the same manner as for other valves. Regulators can have significant internal leakage and should not be relied upon as a pressure isolation device. The regulator operating range, or mass flow rate range, refers to the minimum and maximum gas flow rate that the regulator can deliver, while ensuring that the regulated outlet pressure is within the allowed dead band. Many US suppliers will list the allowable range in standard cubic feet per minute instead of SI units. This is still a measurement of mass flow rate and can be converted using the density of the specified gas at standard conditions (typically 20 °C and 101.3 kPa). Regulator droop is defined as the decrease in regulator outlet pressure when the regulator is regulating with full gas flow, at the upper limit of its operating range. In an ideal regulator, there would be no droop. The regulator step response time is the time it takes for a regulator to respond to a sudden demand in regulation from the lockup condition. The step response time will be specified as the maximum time it will take for the regulator to change from zero to maximum flow for a given maximum liquid flow rate [272].

9.4.4 Quick disconnect

Quick disconnects are essentially a combination of a mechanical tube joint and a valve. When connected, they join two sections of tubing, ideally with unrestricted internal fluid flow, and zero external flow. When disconnected, they typically close off the previous internal flow path, essentially sealing each side of the tubing to prevent external flow. In reality, there will be a pressure drop across the connected quick disconnect and an external leakage rate, specified for each of the connected and disconnected configurations. The quick disconnect coupling will have a plug and a socket for rapid (less than one second) coupling or decoupling. Quick disconnects have been used in both liquid and gas feed lines and have been used with hazardous propellants and in cryogenic systems. They are used in both ground and flight feed lines. A range of quick disconnect configurations are used, see Ref. [272] for details.

TABLE 9.2   Standard Fractional and Imperial Sizes in Stainless Steel Tubing [5].

Tube OD [in]	Tube Wall [in]	Working Pressure [psig]	Minimum Bend Radius (to center of tube) [in]
1/16	0.014	8100	
	0.020	1200	
1/8	0.028	8500	
	0.035	10,900	
1/4	0.035	5100	
	0.049	7500	7/8
	0.065	10,200	
3/8	0.035	3300	
	0.049	4800	1-3/16
	0.065	6500	
1/2	0.035	2600	
	0.049	3700	
	0.065	5100	1-3/4
	0.083	6700	
5/8	0.049	2900	1-3/4
	0.065	4000	
3/4	0.049	2400	2-1/2
	0.065	3300	
1	0.083	3100	3
1-1/4	0.095	2800	3-1/2
	0.12	3600	
1-1/2	0.12	3000	4
	0.134	3400	
2	0.134	2500	6
	0.188	3600	

9.4.5 Tubing/fittings

Tubing is sized to withstand the nominal and transient pressures in the feed system, keep the flow velocity within an acceptable range, see Chapter 11, and match the size of components whenever possible. Tubing sizing equations for incompressible fluids are provided in Section 7.13.2 and for compressible fluids in Section 8.3. Standard size seamless tubing is available in both English and metric sizes with varying wall thicknesses/pressure ratings, see sizing options from Swagelok in Table 9.2 for imperial sizes and Table 9.3 for metric sizes. In the United States, tube stubs/connectors on standard propulsion components are most commonly in fractional inches, making tubing of the same size desirable. Pressure drop calculations for tubing require knowledge of the surface roughness, see Chapter 8. This parameter can be estimated by the manufacturer or discerned via testing.

Flange fittings or welds are the preferred connection method for tubing. Bends are typically required in a feed system. The minimum bend radius should not decrease the outer diameter of the tube by greater than 6% [2].

Fittings are used to transition between tube sizes, and change the flow direction (elbows) or junctions to join multiple components (tee's or crosses). These also play an important role in pressure drops across the feed system, see Chapter 8.

TABLE 9.3 Standard Metric Sizes in Stainless Steel Tubing [5].

Tube OD [mm]	Tube Wall [mm]	Working Pressure [bar]
3	0.5	330
	0.7	560
6	1	420
	1.5	710
8	1	310
	1.5	520
10	1	240
	1.5	400
12	1	200
	1.5	330
	2	470
16	1	140
	1.5	230
	2	330
18	1	120
	1.6	200
	2	290
20	2	260
22	2	230

Threaded fittings include straight threads (e.g., MS) or tapered threads (e.g., NPT). Straight threads rely on a secondary sealing mechanism, such as an o-ring or gasket. A tapered thread is expected to seal on the threads. A nonmetallic compound, such as Teflon tape, is typically used on the threads of tapered fittings to facilitate a tight seal.

Fittings come in either plug (aka male) or socket (aka female) options. Plugs have threads on the outside diameter of the fitting, while sockets have them on the inside diameter. The pitch of a fitting refers to the number of threads per unit of length (in or mm). The geometry of the crests (peaks), roots (valleys) and flanks (flat area between the crest and root) of the threads and the angle between each can vary by the standard used to define the fitting, see Fig. 9.12.

Note that material compatibility with the oxidizer and other feed line components is again critical. Most spacecraft have welded joints, meaning that the tubing would need to be made of a material that can be welded with that of the component. When that is not possible, a transition or braze joint would be necessary to move between materials. Alternative connections may be possible for ground testing, such as AN, NPT, Swagelok, etc. However, the associated leak rates of these joints make them less attractive for space applications. The propulsion engineer also needs to understand the relative strength of the materials being joined and the possibility of galling one or the other in this case.

9.4.6 Pumps

The purpose of a pump is to take in fluid at low pressure and expel it at high pressure. Key pump performance characteristics are their flow capacity, efficiency, and the pressure head increase they can impart to the fluid. Pumps require an additional system to drive the turbine. This has typically been done pneumatically by

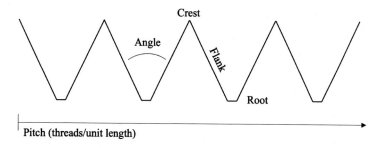

FIGURE 9.12 Geometry of a fitting thread shown here on a straight thread.

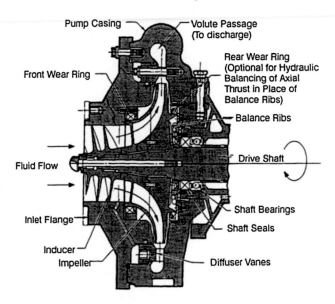

Pump Casing

Volute Passage
(To discharge)

Rear Wear Ring
(Optional for Hydraulic
Balancing of Axial
Thrust in Place of
Balance Ribs)

Front Wear Ring

Balance Ribs

Fluid Flow

Drive Shaft

Shaft Bearings

Inlet Flange

Shaft Seals

Inducer

Impeller

Diffuser Vanes

FIGURE 9.13   Internals of a single-stage centrifugal pump. Figure is from Ref. [2] with permission.

hot gas, though an increasing number of electrically driven pumps are becoming available. Pumps can often reduce mass in larger rockets but at increased cost and complexity. Modern electric pumps are making this option trade well for smaller systems.

Chapter 7 describes an approach to estimating the mass of a centrifugal turbopump. Chapter 10 provides an example of trading a pumped design against a pressure-fed system. This section focuses on a high-level description of turbopump hardware.

Positive displacement pumps may be used for small systems with low oxidizer mass flow rates (and therefore low thrust). This class of pumps are efficient at low speeds and include diaphragm pumps, piston pumps, and rotary displacement pumps. One drawback of this class of pump is that their discharge pressure varies with each stroke [6].

Centrifugal pumps are the most common pump used in medium and large rockets as at this scale they can be efficient in mass and volume whilst delivering large propellant flow

rates at high pressure [6]. Centrifugal pumps are comprised of both stationary hardware, the stator, and rotating hardware, the rotor. Centrifugal pump systems have the following main elements [6,2,1]:

- Turbine - the turbine is used to drive the pump by providing power to the shaft
- Power source - a power source is needed for the turbine
- Inducer - an axial flow rotor that is used to increase the total pressure of the fluid as it enters the pump to reduce the potential for cavitation across the impeller
- Impeller - a spinning wheel within a casing that accelerates the fluid to high velocity in the radial direction
- Volute - a collection passage to take the high-velocity flow from the impeller and direct it to the diffuser
- Diffuser - converts the kinetic energy to potential energy, taking the high-velocity fluid and converting it to high pressure

Single-stage pumps only have a single impeller and are therefore limited in the pressure head they can impart to the fluid [6]. Most centrifugal pumps are only single stage, but multistage centrifugal pumps do exist, such as the Space Shuttle Main Engine high-pressure fuel turbopump [2]. A cut-away of an example single-stage centrifugal pump is shown in Fig. 9.13. The design equations of Chapter 7 provide a means of calculating whether a single stage might be sufficient for a given application.

The performance of centrifugal pumps depends upon the pump speed and the design of the impeller, vanes, and casing [6]. The impeller tip speed determines the maximum pressure head that can be added to the flow across a single stage of a centrifugal pump.

This speed is limited by the material strength of the impeller. Centrifugal pump performance is also limited by cavitation, which will occur whenever the fluid static pressure decreases below its vapor pressure. Cavitation in centrifugal pumps is most likely to occur behind the inlet pump impeller vane. Cavitation in pumps is problematic as it causes variations in fluid flow rate (from unsteady vapor bubble formation and collapse), which can make combustion and thrust erratic and dangerous [6].

The main requirements for any oxidizer pump system in addition to performance are high reliability, low mass, stable fluid flow rates in the design operating range, long life and ideally low cost [2]. Unreliable pumps can be potentially dangerous in an oxidizer feed system. Particular care has to be taken to prevent any leakage through shaft seals in the pump. Leakage in this area can be very hazardous, and thus, multiple seals and drainage provisions for leaks are often used in the pump design [6].

9.5 Instrumentation

Instrumentation is one of the areas where the fundamental architecture is quite different between flight and ground. On the ground, propulsion engineers typically want to collect as much data as possible, measuring pressure (pressure transducers), temperature (thermocouples and thermistors), thrust (load cells), and flow rate. In flight, the instrumentation is typically stripped down to the minimum needed to monitor faults or reconstruct the performance. This section discusses common sensors used. Chapter 11 provides more information on how they are used in ground testing.

9.5.1 Pressure measurements

Pressure transducers

Pressure transducers convert pressure into an electrical signal. This is often done by measuring the deformation of a diaphragm caused by exposure to pressure. Specifications include the pressure range, accuracy (as a percentage of the maximum operating pressure), operating temperature, pressure port, electrical input/output, and frequency/response time.

Pressure transducers measure either absolute pressure - pressure relative to vacuum, gauge pressure - pressure relative to atmospheric pressure, or differential pressure - pressure between the two sides of the sensor. Note that absolute pressure and gauge pressure are related to each other by the atmospheric pressure: $P_{absolute} = P_{Gauge} + P_a$. The aptly named differential pressure transducer, is used to measure the difference in pressure between two points. A common application of differential pressure transducers within hybrid rockets is measuring the change in pressure across a venturi in order to calculate the mass flow rate, see Chapter 11 for details. This could also be achieved with two absolute or gauge pressure transducers; however, taking the difference in measurement of two transducers is generally less accurate than a direct differential measurement.

Static pressure is the most common pressure measurement; however, high-frequency response (>1 kHz) pressure transducers exist in order to capture instabilities. This latter type of measurement is more difficult to achieve, typically requiring precise calibration or even a companion static measurement.

Pressure transducers are used to measure feed line and chamber pressure, with the latter being more difficult. Pressure signals can be affected by temperature or attenuated by stand-off distances or other means of protecting the instrumentation from combustion gases. Much like all flight hardware, each pressure transducer design will also have specified shock and vibration levels that they are able to tolerate. Pressure transducers are known to take a set point shift, or peg high/low under large shock loads, as is believed to have happened on the Mars Science Laboratory Sky Crane Descent Stage [21].

Strain gauges

Electrical resistance strain gauges are used on the surface of pressure vessels as an alternative method for measuring pressure. Strain gauges are designed to deform along with the part to which they are attached. They are essentially just resistors with a variable resistance proportional to the strain they experience. Strain gauges are often configured as Wheatstone bridge circuits in order to accurately measure small changes in resistance. The change in resistance is measured as a voltage drop across the gauge. The measured strain must then be converted to the load experienced by the pressure vessel. In the case of a thin-walled cylindrical pressure vessel, the internal pressure, P, is related to the measured strain, ϵ, by Eq. (9.1) and (9.2) for the hoop and longitudinal directions, respectively [324]. In order to do this, the gauges must be accurately aligned with the principal stress directions. See Fig. 9.14 for how these strain gauges must be aligned in order to apply these equations. In these equations, ϵ_{long} and ϵ_{hoop} refer to the longitudinal and hoop strain, respectively, E is the elastic modulus of the tank, v is Poisson's ratio for the tank, t is the tank wall thickness, and D is the tank diameter.

$$P = \frac{4t E \epsilon_{hoop}}{D(2 - v)} \quad\quad (9.1)$$

$$P = \frac{4t E \epsilon_{long}}{D(1 - 2v)} \quad\quad (9.2)$$

FIGURE 9.14 Strain gauge alignment on a tank to measure hoop and longitudinal strain.

9.5.2 Temperature measurements

Thermocouples

Thermocouples work by making a circuit by joining two dissimilar metals (metals with different Seebeck coefficients), see Fig. 9.15. When one side of the circuit is heated, it produces a voltage, which is correlated to the temperature at the junction (T_{meas}). This voltage is measured at the point denoted with a "V" in the image. Any contribution from the coppers is canceled out. However, T_{ref} must be known. No excitation voltage is required.

Thermocouples are named by the combination of dissimilar metals used, with K-type (chromel/alumel) being the most common because it is rugged and can be used over a very wide range of temperatures, typically from about −270 to 1260 °C. They can also be used

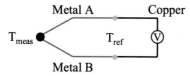

FIGURE 9.15 Thermocouple.

in oxidizing environments. However, there are a number of other options that can also be used. Thermocouples are an inexpensive way to measure temperature; however, they usually require an amplifier to measure their signal and can have a slow response time.

Thermistors

Thermistors use a highly temperature-dependent resistor to give a temperature measurement. They are one of the cheapest and simplest ways to measure temperature but generally have a narrow temperature sensing range, can be highly nonlinear, and are often less accurate than other options without calibration. A benefit of thermistors is that they do not need a sophisticated sensing circuit, their signal can be read without amplification.

PRTs

Platinum Resistance Thermometers (PRTs) use platinum wires that have a resistance that is dependent on temperature. A current is applied to the wire, and the voltage is measured, then converted to temperature with a calibration equation. PRTs are the surface temperature sensor of choice for many spacecraft because of their simplicity, accuracy, and robustness. In a hybrid rocket application, a PRT might be used to monitor tank temperature, for example.

9.5.3 Thrust measurements

Thrust measurements are incredibly useful, but notoriously difficult to achieve with a high degree of accuracy. A thrust stand must measure the force imparted by the rocket while also re-

straining it. The combustion chamber or rocket is placed on a floating stand, which is then connected to a rigidly mounted structure. Stiffness in the system due to the feed system, restraints, and instrumentation can affect the measurements. For example, the feed system typically bridges the gap between the floating and fixed parts of the system, which adds resistance to the motor as it pushes on the load cell. If the resistance is not repeatable, the thrust measurements will not be useful.

The design of a thrust stand depends on the physical size of the motor, the thrust level, whether thrust vectoring or off-axis thrust measurements are required, the desired accuracy, stiffness, and environmental conditions, including temperature, pressure, and spin rate. Large thrust stands have small natural frequencies, resulting in larger thrust deflections. These deflections are absorbed by the supports and are therefore not measured by the load cell [325]. Therefore the smallest stand that can accommodate a given motor is desirable. However, since thrust stands are expensive to develop and calibrate, they are often made to accommodate a variety of motors, and it is uncommon to optimize a stand for particular motor, unless it is lab scale.

Thrust stands can be designed to measure thrust in single or multiple axes and in either a vertical or horizontal configuration. The authors have also witnessed stands on an angle, though they are less common. Horizontal thrust stands are often used for single-axis measurements of hybrid rockets because the load measurement avoids capturing the changing mass of the propellant within the motor. Thrust vectoring can be difficult to measure in this configuration, making a vertical test stand, with its typically undesirable complication of having to accurately estimate the expended fuel over time, more appealing. The instantaneous change in mass of a motor can be measured with a load cell during the burn. However, it should be noted that any off-axis thrust component can introduce error in this measurement. Further information on design choices to minimize thrust stand error is provided in Chapter 11.

9.5.4 Flow measurements

There are various methods available to measure or control fluid flow at a known rate, such as venturis, turbine flow meters, Coriolis flow meters, ultrasonic flow meters, magnetic flow meters, vortex flow meters, etc. Here, a brief overview of the most common options is provided. For any hybrid rocket application, the propulsion engineer should follow the approach used for other component selection to ensure that the flow meter is compatible with the fluid it will be measuring; material compatibility can be a particular concern for some of these options.

For liquid flow, a cavitating venturi can be used to give an accurate measurement of the flow rate, knowing only the upstream pressure and the fluid temperature. For gaseous flow, a sonic venturi can also be used in a similar manner, see Ref. [322]. A noncritical venturi can also be used for subsonic flow; this relies on a differential pressure measurement between the throat and venturi inlet, as well as knowledge of the upstream temperature and pressure. Chapter 11 provides more details on this.

Turbine meters have been widely used for decades by the natural gas industry and are very reliable. They can be used with gases, liquids, or with cryogenic fluids. Turbine flow meters use a rotor parallel to the direction of flow and measure the volumetric flow of fluid via the speed of rotation of the rotor. Further detail on using these flowmeters with gases is provided in Ref. [326].

Coriolis flow meters are used to measure the mass flow rate of liquids or gases. They do this by measuring the Coriolis force generated by the fluid. They are widely used in a range of industrial applications such as oil and gas, water, chemical, and food and beverage. Coriolis flow meters are extremely accurate when measuring

steady-state flow, but they are typically expensive. Ref. [327] provides more details on using these types of flow meters.

Ultrasonic flow meters measure the volumetric flow rate of gases or liquids and introduce no pressure drop to the fluid flow. Further details on using ultrasonic flow meters with liquids is provided in Refs. [328] and [329].

Magnetic flow meters can only measure the flow rate of conductive liquids and therefore are not used in hybrid rocket motor applications.

There are also small electronic flow meters for low gas mass flow rate applications that use differential pressure across a laminar flow field to measure the mass flow of the gas. These devices are typically limited to low-pressure applications, less than approx. 1 MPa (150 psia).

9.5.5 Regression rate measurements

The most common way to determine the regression rate of a hybrid fuel (in a rocket configuration) is to measure the mass of the fuel grain before and after a test. This gives a space and time-averaged regression rate. Additional detail can be determined by measuring the dimensions of the fuel grain before and after each test. This can be done using a micrometer at the fore and aft ends of the grain or by measuring the three-dimensional topology with a camera/laser.

If more precise temporal knowledge of the regression rate is required, several direct measurement techniques have been employed to determine the instantaneous regression rate during a test. The Penn State University has used x-ray measurements to monitor hybrid fuel regression in real-time [191]. Ultrasonic pulse echo measurements have been used to measure regression rate in solid fuel ramjets [330] and hybrids [331] and can allow measurement of a fuel surface over a short distance, about 12.7 cm (5 in) [332]. See Fig. 9.16. Finally, sensors can be embedded into the fuel grain to measure the instantaneous regression rate, the electrical resistance of these embedded wires changes as they burn

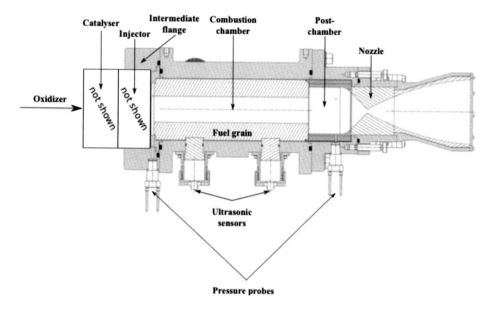

FIGURE 9.16   Example of Ultrasonic Regression Rate Measurement Sensors [331].

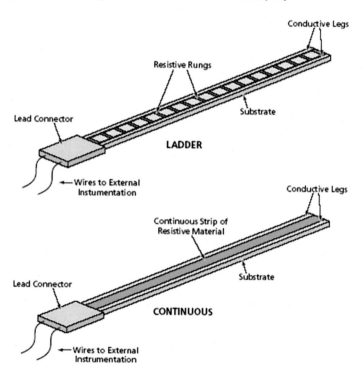

FIGURE 9.17   Regression Rate Measurement Sensors [333].

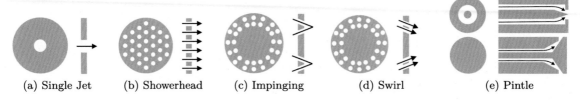

(a) Single Jet   (b) Showerhead   (c) Impinging   (d) Swirl   (e) Pintle

FIGURE 9.18   Common hybrid injectors.

with the fuel. The change in resistance with the remaining length can be translated to give the regression of the fuel [333]. Two types of sensors have been used: one is continuous, and the other uses a ladder-like design, giving discrete measurements as each rung is burned through, see Fig. 9.17.

9.5.6 Inertial measurement unit

An Inertial Measurement Unit (IMU) is used to determine a rocket's attitude, angular rates and velocity in space. It achieves this information using accelerometers (three axis linear acceleration) and gyroscopes (rotational velocity). IMUs constantly integrate the acceleration/velocities over time; therefore small errors accumulate, leading to drift in measurements. These measurements can be coupled with other sensors (e.g., GPS or magnetometers) to reduce this error.

9.5.7 Shock measurements

Shock sensors are often used for transporting sensitive equipment, such as flight hardware. Shock sensors are a form of accelerometer. They are single-use devices that will trip when exposed to a shock in excess of the rated amount. They can be attached inside and outside of equipment packaging to ensure hardware is not exposed to any potentially damaging accelerations (e.g., by being dropped or mishandled). These sensors are directional; therefore multiple sensors will be used on a single compo-

nent to ensure that all directions are being monitored.

9.6 Injection

There is no more important part of hybrid rocket design than the injector. They are used to deliver oxidizer from the feed system and distribute it in the combustion chamber. Depending on the oxidizer being used, the injector may also be required to set the mass flow rate, help atomize or vaporize the oxidizer, or to reduce the likelihood of a feed-coupled instability by isolating the feed system from the combustion chamber. There are two main types of injectors: axial (oxidizer is introduced along the same axis as the fuel port) and radial (flow is introduced perpendicular to the fuel port). Axial injectors are the most commonly used, as radial injectors (for both liquid and gaseous oxidizer) more frequently suffer from instabilities (see, for example, [17], [334], [335], and [225]). Fig. 9.18 shows common axial hybrid injectors.

Many propulsion engineers have made the mistake of overlooking the criticality of injector design for hybrid rockets, mistakenly thinking they play less of a role in mixing and stability than they do in liquid rocket engines. However, they have been shown to be critical to downstream mixing and stability [336]. Frequency characteristics within the feed system or manifold of an injector can couple with chamber acoustic modes. The rule of thumb for injector pressure drop in liquid engines is $15 - 20\%$ of

TABLE 9.4 Discharge coefficients for a variety of injector options. Adapted from [6].

Type	Cartoon	Hole Diameter [mm]	C_d
Sharp Edged Orifice		< 2.5	0.65
		> 2.5	0.61
Sharp Edged Cone		1.00	0.69-0.70
		1.57	0.72
Short Tube with Rounded Entrance (L/D~1)		1.00	0.70
Short Tube with Rounded Entrance (L/D > 3.0)		1.00	0.88
		1.57	0.90
Short tube with a Conical Entrance		0.50	0.70
		1.00	0.82
		1.57	0.76
		2.54	0.80-0.84
		3.18	0.78-0.84
Short Tube with Swirl		1.0-6.4	0.2-0.55

the chamber pressure for stability [2]. A pressure drop sufficient to choke the injector elements [177], usually at least 20%, has be recommended for self-pressurizing propellants (e.g., N_2O) [47]. A hybrid motor is typically considered stable if it has pressure oscillations of less than ±5%. Therefore pressure drops approaching the chamber pressure ±5% could have catastrophic results.

The design of the injector affects the discharge coefficient (see Chapter 8 for a definition of C_d).

Table 9.4 shows common injector designs and their corresponding discharge coefficients.

The types of injectors include the following:

Showerhead: The oxidizer is injected axially through a pattern of holes, as depicted in Fig. 9.18b. This method is not particularly good at atomization [2]. However, it is frequently used with gaseous oxidizers.

Nozzles/Fan Formers: Oxidizer is injected in a cone or fan shape. Commercially available nozzles, typically employed in air-breathing systems, can be used for this purpose. These injec-

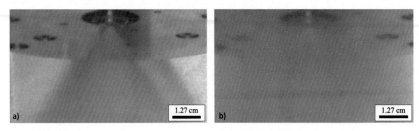

FIGURE 9.19 Nozzle spray injector at a) ambient and b) low pressure (about 3 kPa) using EnSolv (MON simulant) from [146]. The flow spreads out much more in the low pressure environment.

tors atomize the incoming flow and can result in cone-shaped (with or without swirling flow) or a flat fan. Fig. 9.19 shows an example of a nozzle creating a conical flow. It should not be a surprise that the performance of the injector is dependent on the downstream pressure condition, as depicted visually in Fig. 9.19.

Self Impinging/Like Doublets: Two streams of oxidizer impinge at a single point, creating a fan. Atomization is accomplished due to the energy dissipation from the impingement. If the incoming mass flow rate ($\dot{m}_{inj}$), velocity (V), and angle (α) are the same, the resulting angle of the stream will be zero degrees (it will continue in the axial direction). If any of these are different, by design or manufacturing tolerances, the resultant angle, also known as the β angle, is given by Eq. (9.3).

$$\tan(\beta_{inj}) = \frac{\dot{m}_{inj,1} V_1 \sin(\alpha_1) - \dot{m}_{inj,2} V_2 \sin(\alpha_2)}{\dot{m}_{inj,1} V_1 \cos(\alpha_1) + \dot{m}_{inj,2} V_2 \cos(\alpha_2)}$$
(9.3)

Here, β_{inj} is the angle between axial and the direction of the resulting stream, $\dot{m}_{inj,i}$ is the mass flow rate of the oxidizer through hole "i", V_i is the velocity of the oxidizer through hole "i", and α_i is the angle between the axial direction and the stream injected through hole "i", as can also be seen in Fig. 9.20. Remember that the velocity is a function of the pressure drop across the injector.

Like Triplets: Similar to the previous case, three streams of oxidizer impinge at a single

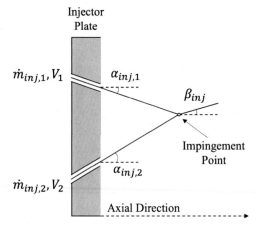

FIGURE 9.20 Injector beta angle for impinging doublets. Adapted from Ref. [2].

point. Like triplets are easier to package than doublets since more triplets can be fit into a given area than doublets. However, they generally have worse performance as they create larger droplets and a more narrow fan.

Cavitating Venturis: Cavitation within the injector changes the discharge coefficients as it creates a vena contracta within the port. If the local static pressure in the venturi drops below the vapor pressure of the oxidizer, it will vaporize (aka cavitate). The oxidizer exiting the injector element will be in a very different state depending on the pressure relative to the vapor pressure of the oxidizer, see Fig. 9.21.

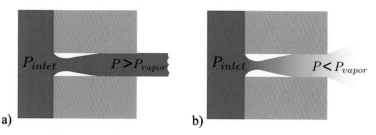

a)   b)

FIGURE 9.21 Cavitation through an Injector for a a) low vapor pressure propellant and b) high vapor pressure propellant. Adapted from Ref. [337].

If the entrance to the injector port is rounded as little as $r > 0.14D$, the formation of a vena contracta can be prevented [338]. To further complicate the design, Ref. [47] found that the discharge coefficient varied for different pressure drops across the same injector configuration (using N_2O) due to two-phase flow effects. Nuclear power engineers have created useful models for two-phase flow through orifices like injectors. Ref. [47] gives a good overview of this work.

Cavitation, specifically in its application for flow control, was discussed in detail in Chapter 8, Section 8.2.3.

The desire to throttle a hybrid rocket motor may drive the injector design. Throttling of a hybrid rocket can be accomplished by reducing the oxidizer flow. Since the regression rate, and therefore mass flow of fuel, depends on the mass flux, reducing the oxidizer mass flow will also reduce the fuel mass flow. There is no need for complex momentum matching of the throttled propellants as in liquid bi-propellant rockets. Pintle injectors have variable $C_d A$, enabling throttling of the oxidizer mass flow rate even if the oxidizer tank pressure is fixed. Throttling can also be achieved with other injectors by changing the tank pressure (e.g., via an active pressure control system, see Chapter 7).

9.7 Combustion chamber design

9.7.1 Pressure vessel/chamber case

Combustion chamber design is similar to tank design with the quantity of fuel and MEOP driving the design. However, as discussed in Chapter 7, the regression rate of the fuel, oxidizer-to-fuel ratio, and mixing requirements drive the geometry. There are four common materials used for combustion chamber structure: aluminum, steel, titanium, and carbon-fiber composite, though others are also possible. Steel is often used for ground-based testing because it is inexpensive and strong. Aluminum is similarly inexpensive, but has improved strength per mass capability. Titanium and carbon fiber are much more expensive, but also have even higher strength per mass and are commonly used in flight applications. Table 9.5 provides a summary of material properties for common hybrid rocket combustion chamber materials.

The material used for pressure vessels like combustion chambers must be well known. The propulsion engineer needs to be familiar with material certifications and obtain one for all the materials used for a build. Basis tolerance limits are used for aerospace grade tank materials, with A-basis (99% of the test data meets or exceeds specification with 95% confidence) accepted as the standard and B-basis (90% of the test data meets or exceeds specification with

TABLE 9.5 Common combustion chamber materials. General material properties are provided for reference only, material certifications should be used for fabrication.

	Density (ρ) [kg/m^3]	Ultimate Tensile Strength (F_{tu}) [MPa]	Yield Strength [MPa]	CTE [μm/m-C]	Efficiency $\left(\dfrac{F_{tu}}{\rho g_0}\right)$ [km]	Cost
Aluminum (6061-T6)	2700	310	276	24	12	Low
Titanium (Ti-6Al-4V)	4480	1170	1100	8 to 10	27	High
Steel 304	8000	640	235	17	8	Low
Graphite (Carbon Fiber) [1]	1550	895		−0.6 [266]	59	High

95% confidence) also used on occasion. Materials data can be found in individual specs or a handbook such as the *Metallic Materials Properties Development and Standardization* [339].

9.7.2 Insulators

Insulation in the combustion chamber is utilized to protect the chamber case in the fore end, aft end, in the event of asymmetric fuel burning, or to prevent damage due to excess oxidizer flowing through the hot chamber after the fuel has been expended. Tubes in standard sizes of various low regression rate fuels and ablatives are readily available and can be used to make cartridge loaded fuel grains for ground testing. These include phenolics, fiberglass (e.g., G10), plastics, etc. High-performance ablative materials, such as carbon cloth phenolic (CCP) and silica cloth phenolic (SCP) are commonly used in flight applications, adopted from solid rocket motors. Ablation is the degradation of a material under the application of heat. A char layer, rich in carbon and therefore low in thermal conductivity, forms on the surface of the material, protecting the layers below from further degradation for some time. Erosion is then the mechanical process in which material is removed under the high-velocity gases. The fore end of the combustion chamber can often be maintained at temperatures that do not require an ablative insulator, but ablative materials are often required in the aft end of the combustion chamber.

9.7.3 Seals

Any time two external materials are joined, a seal of some form is required, such as a weld, a braze joint, or an o-ring/elastomeric seal. Welds are generally desired over elastomeric seals as they tend to be more robust and once shown to be leak free, generally remain so. Welding is a rich topic area that is beyond the scope of this handbook. For information on the various types of welds, and design guidelines for each, a dedicated text such as Ref. [340] should be consulted. Welding is not always possible or desirable (as in the case of a joint that needs to be repeatedly opened and re-sealed) and as such, the propulsion engineer should be familiar with the design approach for elastomeric seals.

Designing for an elastomeric seal must account for the environmental operating conditions of the seal, the gland geometry into which the seal will be installed and the seal material and geometry [341]. Seals on the combustion chamber are generally static seals, meaning that there is no relative motion between the seals and the parts they are sealing. O-rings are most commonly used to ensure a leak-tight elastomeric seal, and as such, are described in more detail in the following sub-section, though gaskets can also be used following similar design principles.

O-ring design

O-rings seal together two surfaces by deforming under pressure. Sealing occurs when a pressure difference develops across the ring that

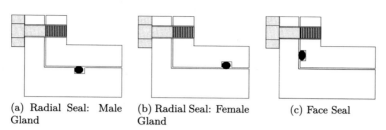

(a) Radial Seal: Male Gland   (b) Radial Seal: Female Gland   (c) Face Seal

FIGURE 9.22 Common o-ring seal definitions: radial seals (male and female glands) versus face seals. The o-ring cross section is shown as a black oval in each image. The center line (axis of rotation) is assumed to be horizontal at some distance below each image.

does not exceed the contact stress of the o-ring [342]. Most seals can generally be defined as face (aka axial) seals or radial seals, see Fig. 9.22.

There are several companies that make o-rings and the authors are not endorsing any particular one. However, Parker-Hannifin Corp. has a very well-written o-ring handbook that, at the time of writing, is readily available for free, see Ref. [343]. This reference is an excellent source of information on o-rings, including materials (their compatibility with various fluids, temperatures, radiation, etc.), applications, and design. Gas permeability rates for o-ring materials with common gases are included in the handbook as well.

In hybrid rocket design, o-ring material selection is typically driven by material compatibility with the oxidizer. A compatible material should be selected along with one that can handle the thermal environment. It is challenging to find an o-ring material that can seal against high temperature for an extended period of time. To mitigate against a catastrophic seal failure in the combustion chamber, it is recommended to consider protecting the seal by use of an insulator, a labyrinth seal, and/or using an additional/redundant o-ring seal.

O-rings are specified by their inner diameter, cross-section diameter, and their material hardness/durometer [342]. Standard o-ring dimensions exist. Many Computer Aided Design (CAD) programs have databases with standard o-ring grooves pre-programmed. All o-ring grooves are sized based on three main parameters:

- Stretch: O-ring stretch is defined as the percentage increase in the inner diameter of the o-ring when it is installed in the o-ring groove. Typically a stretch of 0-5% is recommended. Occasionally a larger stretch is possible, but only in short lifetime applications [343]. For radial seals, the recommended range is 1–5% with a preferred value is 2% [344]. Face seals can use stretch values down to zero, allowing for materials that are brittle under tensile stress to still be used. Note that face seal grooves for sealing internal pressure (a pressure vessel) will be sized differently than those used to seal a vacuum (external pressure) [343]. Pressure vessels dimension the groove using the mean o-ring outer diameter, whilst vacuum applications dimension the groove using the inner mean o-ring inner diameter [343].
- Squeeze: O-ring squeeze is defined as the percentage decrease in o-ring cross section when it is compressed in the o-ring groove. The maximum squeeze for dynamic applications is approximately 16%–25% depending on the size of the o-ring. For static seals, this can increase to approximately 30% for most elastomers, especially for face seals. The minimum squeeze for all seals is given as about 0.2 mm (0.007 inches) [343].
- Fill: Most designs call for a gland fill of 60–85% [343]. The accounts for tolerance

stacking and swell due to fluid exposure and/or temperature.

There are numerous failure causes for o-rings. As previously mentioned, they can fail as a result of exposure to excessive heat or a chemically incompatible fluid, they can also fail if the o-ring gland is poorly designed or the gap is too large between the two sealing surfaces, and they can fail during installation. The use of backup o-rings, and following the design guidelines laid out here and discussed in more detail in Refs. [343] and [341], help to mitigate such failures.

9.8 Nozzles and thrust vector control

Nozzles for hybrid rocket motors can be a challenging and potentially expensive element of the system. The design of hybrid motor nozzles is somewhat unique as compared to liquid bi-propellant engine nozzles or solid rocket motor nozzles. Low-cost solutions utilizing readily available materials work for some hybrid rocket motors but not for others.

Compared to solid rockets, hybrids typically have lower thrust (therefore have a longer burn time), burn at a higher temperature, and have more oxidizing species in the exhaust gas. As a result, it is believed that the nozzle erosion is probably more severe in hybrid rockets than in solid rockets. However, since hybrids also typically optimize at lower pressures, and pressure plays the dominant role in nozzle erosion, the effect can be minimized through design. Solid rocket motors generally utilize multiple different materials within the nozzle. Solid rocket motor nozzles are not generally designed to withstand multiple burns, where thermal soak back into the nozzle can cause de-bonding and thermal warping between dissimilar materials.

Liquid engine nozzles are often designed for multiple re-starts and can also be used in systems with high-temperature combustion prod-

ucts. These engines can typically use heavier materials for their combustion chamber and nozzle since the chamber is relatively small compared to that of an equivalent solid or hybrid rocket motor (liquid engines do not store propellant in their combustion chamber). Further, these engines often utilize film cooling inside the combustion chamber to reduce the chamber and nozzle wall temperatures. The film cooling approach may be an option for hybrid rocket motors but has not yet been studied in detail with optimized and documented design approaches. As such, the thermal and oxidizing environment in a hybrid rocket motor is often seen as more extreme than that of liquid engines.

Hot-fire testing is recommended to verify any nozzle design. If the hybrid rocket is to be used in space or as an upper-stage motor, then the hot fire testing should be conducted in a vacuum chamber to correctly capture the nozzle thermal environment (radiative cooling rather than convective).

9.8.1 Nozzle profile

Nozzle design was introduced in Chapter 2 and further discussed in Chapter 7. Fig. 9.23 shows common nozzle profiles. The typical conical nozzle, and a Rao or bell nozzle, are in various lengths. A simple conical nozzle is often used for low-cost applications or as a starting point for design. A bell nozzle, found by applying the method of characteristics, increases performance over a conical design. Bell nozzles can also be truncated for applications with limited length at the cost of a nozzle correction factor. For example, an 80% bell nozzle has a nozzle correction factor in the 98.5-99% range, depending on the area ratio [6]. The nozzle efficiency approaches one as the exit angle approaches zero. An aerospike nozzle maintains high efficiency over a range of altitudes because it allows the exhaust flow to expand with ambient pressure conditions. With such a benefit, the question becomes why these nozzles are not more common.

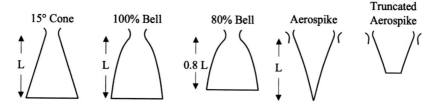

FIGURE 9.23 Examples of common nozzle contours.

The answer lies in finding suitable materials that can provide both strength and thermal protection throughout the burn. Also, it is nearly impossible to cool the tip of the aerospike, making a truncated version desirable for practical applications.

Rocket nozzles can be external or submerged. In a submerged design, the nozzle throat is inside the combustion chamber. This reduces the overall length but exposes both sides of the nozzle to hot gases and leads to a recirculation region in which slag may accumulate.

9.8.2 Nozzle materials

Hybrid rocket nozzles typically utilize multiple types of materials. The first is structural, maintaining the shape and rigidity of the nozzle under the combustion loads. The list of structural materials is the same as that for the combustion chamber, tanks, etc. The second type of material transfers heat out of the nozzle throat. Nozzle throat materials are prized for their ability to withstand and remove heat either through conduction (e.g., in a regeneratively cooled nozzle) or ablation. The former can be enhanced in ground testing, where thicker materials and cooling using a separate liquid are possible. Nozzle erosion depends on the chamber temperature and pressure as well as the oxidizing species in the exhaust gases. Mechanical ablation also occurs.

The most common lab-scale nozzle throat material is graphite (either ATJ or pyrolytic) for cost, availability, and machinability. Other options include 3D or 4D carbon-carbon, refractory metals (e.g., Molybdenum, Tungsten), and ceramics. Treatments such as Polymer Infiltration and Pyrolysis (PIP) have been used to enhance the performance of graphite and carbon-carbon. PIP has been found to postpone notable erosion for tens of seconds.

An insulator is used to isolate the heat generated in the nozzle throat from damaging the rest of the nozzle. A flame barrier (e.g., insulator or ablative) is used to protect the nozzle (other than the throat) from high-temperature exhaust gases.

Heavy-weight ground test motors often enjoy less nozzle erosion than flight motors, since the motor structure can conduct a substantial amount of the heat away from the nozzle. However, once the design progresses to a flight weight, no additional material is available to provide cooling, which will correspond to an increased erosion rate. In a lab-scale design, the authors observed negligible erosion with a heavy weight design, even after multiple minutes of testing. The same motor was lightweighted to be more representative of flight, and the nozzle erosion became so substantial that the mission objectives could not be met without a substantial redesign of the nozzle for 60-s tests. It is therefore recommended that the nozzle area be as flight-like as possible for ground testing if long-duration burns are required. This difference between flight and ground nozzle erosion can be further exacerbated in upper-stage and in-space motors where nozzle cooling only occurs via radiation in flight, rather than convec-

TABLE 9.6 Nozzle and Insulating Materials. Note: Fiber is the direction of the fiber, and fill is the direction orthogonal to it. The erosion rate varies with operating conditions and propellant combination. Values are given here for comparison only.

	Density [kg/m^3]	Thermal Expansion [μm/m C]	Thermal Conductivity [W/m K] at room temperature	Melt Temperature [°C]	Erosion Rate if available [mm/s]	Note	Source
ATJ Graphite	1540-1780	25	27.8 - 95	3650	0.05-0.3	Brittle	[6], [345], [265], [346], [347], [264]
Pyrolytic graphite	2187	20 (fiber), 0.5 (fill)	3.5 (fiber), 300 (fill)				[348]
3D Carbon-Carbon	1750	1.8 (fiber) - 16.2	0.46 - 4.85 (fiber)		0.24		[6], [349]
Copper	8930	16.4	385	1083			[350]
Titanium	4500	8.9	17	1660			[350]
Tungsten	19,350	4.4	163	3370		Oxidizes	[350]
Molybdenum	10,280	5.35	138	2617	0.11		[350], [264]
Rhenium	21,030	6.2	39.6	3180		Oxidizes	[350]
Iridium	22,650	6.8	147	2443			[350]
Paper Phenolic	1820	9.9	0.3	N/A		Chars	[351]
Carbon Cloth Phenolic	1420	14.4	50.8 (fiber)	N/A	0.65		[6], [349]
Silica Cloth Phenolic	1750	13.7	25.4 (fiber)	N/A	1.1		[6], [349], [352]

tion on the ground. Mission profile testing in a vacuum chamber is recommended to reveal any such potential issues.

Note that the regression of the nozzle material depends on the operating conditions (pressure, O/F) and propellant combination. For example, Ref. [264] found the addition of aluminum (to paraffin fuel) to increase the onset time of nozzle erosion and decrease the overall rate. Modeling of nozzle erosion of HTPB with N_2O, H_2O_2, N_2O_4, and O_2 (in order of increasing regression) was shown to be about 1.5–3 times that of solid rocket propellant [265] based on the increased oxidizing species in the exhaust gas. The choice of oxidizer drives this difference in nozzle erosion, with little impact due to the fuel selection.

Ref. [265] found a peak nozzle erosion with stoichiometric combustion of HTPB/O_2 to be about 2.9 times that for a solid rocket (69% AP, 12% HTPB and 19% Al). From the same study, HTPB with N_2O_4, H_2O_2, and N_2O, were approximately, 2.2, 2.0 and 1.6 times that of the solid rocket motor, respectively.

The numbers included in Table 9.6 should be used for comparison and verified for the specific application.

9.8.3 Thrust vector control

There are several types of thrust vector control that can be applied to a hybrid motor. Solid motors use flex seal or trapped ball nozzles to deflect thrust. These basically move the nozzle into the desired position. Flex seals are not good for low-temperature applications. Several of the larger Northrop Grumman Star motors (e.g., Star 37 and 48) use flex seals [353]. Trapped balls can have the joint at the subsonic or supersonic portion of the nozzle. The subsonic option is more common, but can have trouble with metallic slag if it exists in the design. Both of these options could be applied to a hybrid design.

Liquid Injection Thrust Vector Control (LITVC) is practical in hybrids since they already have a liquid (typically the oxidizer) on board. Unlike the methods described previously, LITVC offers only discrete thrust angles in predefined directions. It must be controlled like a "bang-bang" type system. Liquid is injected in multiple ports in the supersonic section of the nozzle. This creates shocks inside the nozzle, which translate into known deflections that can be used to control the rocket. An example of a system using LITVC is the Minuteman II.

Most liquid engines gimbal the entire motor. While this may be possible for a hybrid motor, it becomes more difficult as the size of the rocket increases. Since the motor case houses all of the fuel, it will quickly not make sense from a mass standpoint to move the whole motor.

9.9 Ignition

Ignition is a transient process that generates heat, pressure, and free radicals (atoms or molecules with unpaired electrons) to initiate combustion. Ignition depends on the initial temperatures of the propellants, heat transfer, pressure, and mixing of the fuel and oxidizer. There are multiple ways to add heat to the system, which will be introduced in the following sections.

There are several design considerations relevant to selecting the type of igniter. Total mass and the environments have a first-order impact on the igniter design. Mission objectives may require single or multiple starts, which also affects the choice of ignition mechanism. Finally, igniter performance (e.g., ignition delay time and repeatability) is used for motor ballistics modeling and is especially critical if multiple, short burns are required, since the ignition time could represent a large fraction of the overall burn and directly influence the impulse imparted. The ignition delay depends on the oxidizer run valve opening time, the time to fill the lines and combustion chamber with oxidizer, power to the ignition source, as well as the time to vaporize and

heat a small amount of fuel and initiate a reaction. Ignition systems with mass flow rates of 0.1–0.3% the main flow have shown to be sufficient for large motors [2]. Some smaller hybrid designs have used closer to 0.5% of the mass flow rate.

Ignition timing is system dependent. Hybrid propulsion systems using liquid oxidizers (e.g., LOx) typically vaporize some of the fuel before the oxidizer is introduced into the combustion chamber to discourage pooling. Many rockets using gaseous oxidizer open the oxidizer valve first. Timing of the igniter versus oxidizer flow is not an issue for systems that use catalytic decomposition of the main oxidizer flow (e.g., H_2O_2) since the catalyzed oxidizer is used to vaporize the fuel.

There are three potential issues that can arise during the ignition process:

- Blowout is when the main oxidizer flow lifts the diffusion flame away from the fuel surface. Heat transfer to the fuel grain is then inhibited.
- Quenching occurs when the flame is extinguished upon entering a small passageway.
- A hard start, or overshoot, is a large pressure spike that occurs during the ignition process. This can occur when a large amount of fuel is heated/vaporized prior to the introduction of the oxidizer.

While the first two issues keep the motor from achieving successful combustion, the third can damage the fuel grain and lead to a Rapid, Unplanned Disassembly (RUD) of the combustion chamber.

The most common ignition options are discussed in the following sections.

9.9.1 Pyrotechnic and pyrogen devices

A NASA Standard Initiator (NSI) is an example of a pyrotechnic initiator, see Fig. 9.24 (a). It translates an electrical signal into pyrotechnic actuation of various devices or ignition of mo-

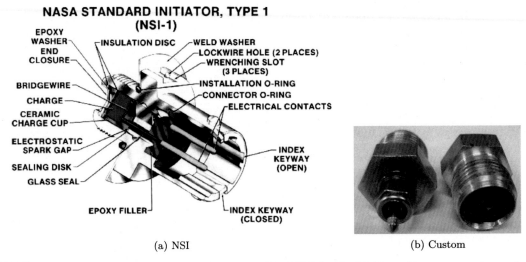

(a) NSI   (b) Custom

FIGURE 9.24   Example pyrotechnic initiators for ignition: (a) a NASA Standard Initiator, (b) a custom made pyro igniter using a model aircraft glow plug.

tors. It was originally developed for the Apollo missions and uses a bridgewire to ignite a solid propellant mixture (zirconium, potassium perchlorate, and graphite in a Viton B binder) [354]. These are highly reliable and used in many space flight applications. However, the rigorous testing that led to their reliability and repeatability also means that they are typically quite expensive. Therefore many institutions have developed custom igniter formulations and use commercially available glow plugs, like those for model aircraft, to provide the needed electrical impulse, see Fig. 9.24 (b).

It is typical for these igniters to be staged, meaning a progressive series of chemical reactions is used to evaporate enough fuel and oxidizer to initiate combustion. They can burn for a long time, on the order of a second. Drawbacks include that they are only capable of a single start and require the same safety precautions as solid rocket propellant (e.g., they are ElectroStatic Discharge sensitive). They can also be sensitive to environments. For example, black powder has not been shown to ignite at low pressures. Also, pyrotechnic igniters have suf-

fered problems (as severe as not starting) when used with cryogenic propellants [355].

9.9.2 Through bulkhead initiator

Through Bulkhead Initiators (TBIs) achieve ignition by detonating a donor charge, which creates a shock wave that propagates through an integral metal diaphragm. An acceptor charge on the other side of the diaphragm is detonated by the shock wave. This, in turn, sets off the ignition charge (deflagration) that ignites the rocket. A major benefit of this method is enhanced safety since there is no direct electrical path into the motor. See Fig. 9.25.

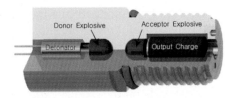

FIGURE 9.25   Notional image of a Through Bulkhead Initiator (TBI) showing the solid parent metal through which the initiation is transferred [356].

9.9.3 Monopropellants

Catalytic decomposition of monopropellants/oxidizer can be used to ignite hybrid rockets. Hydrazine is the most commonly used monopropellant in aerospace applications. However, several alternative propellants also decompose catalytically and produce hot, oxidizing biproducts that can ignite and combust with a hybrid fuel. The monopropellant is introduced into a chamber containing a high surface area structure, such as a 3D-printed web or small spheres/pellets, coated with a material that increases the rate of decomposition. The catalyst bed is typically heated to facilitate the reaction, though "cold starts" are possible with some monopropellant thrusters. A monopropellant ignited hybrid can operate in one of two ways: 1) a small amount of oxidizer is catalyzed to initiate combustion, or 2) the entire oxidizer flow can undergo decomposition, and the hot products are then reacted with the fuel grain. See, for example, Ref. [49].

The two options presently identified for hybrid rocket applications are hydrogen peroxide (H_2O_2) and nitrous oxide (N_2O). These propellants were discussed in greater detail in Chapter 5 and the focus of the following several paragraphs is their application to hybrid rocket ignition. Through this decomposition process, peroxide becomes H_2O vapor and gaseous O_2, while nitrous oxide produces about a two-to-one ratio of gaseous N_2 and O_2.

Hydrogen peroxide has been used as a monopropellant in space applications dating back to the 1960s [184], while nitrous oxide is a newer addition, without flight pedigree. The monopropellant ignition transient of hydrogen peroxide has been well characterized, with repeatable ignition transients of 50–150 ms depending on the peroxide grade and temperature [184]. This time is just the ignition transient of the catalytic decomposition; there would be an additional ignition delay to ignite the hybrid motor. Nammo's Unitary Motor demonstrated catalytic decomposition of liquid H_2O_2 at ambient temperature

to gaseous O_2 and water vapor at temperatures high enough (about 940 K) to ignite and sustain the HTPB/C motor [357]. Ignition delays of less than 100 ms were reported for subscale tests using H_2O_2 and paraffin or polyethylene fuels [358].

An effective catalyst is required to both strip oxygen from the monopropellant in question and then release it. Common catalysts used with High-Test Peroxide (HTP) are aqueous solutions of calcium or potassium permanganate (injected as a liquid) or silver [184] or platinum (solid on a metal substrate). Group 9 elements, most specifically, rhodium and iridium, have been shown to be the most reactive with N_2O [359]. Unfortunately, N_2O has also been shown to burn hot enough to destroy iridium-based catalyst (such as the commonly used hydrazine catalyst, S-405).

This highlights one of the greatest challenges with monopropellants: the temperatures reached during decomposition can actually damage the catalyst material. The adiabatic decomposition temperatures are 1022 K for 90% H_2O_2 and 1573 K for N_2O [358] (for comparison, hydrazine's is about 1366 K). Therefore some catalyst beds are designed to be consumable (e.g., PE/hydrogen peroxide lab scale testing of Ref. [360]) or to use this ignition method for only one or two starts [153].

Another disadvantage of monopropellant ignition is the energy required to preheat the catalyst bed. This is not an issue for a ground launch application; however, the batteries required for spaced-based ignition (or restarts) would be appreciable.

9.9.4 Augmented spark igniters

Augmented spark igniters are relatively common. It is essentially a tiny bipropellant engine that uses a spark plug to ignite the gas mixture. Ignition is carried out in a small chamber as opposed to the main combustion chamber and the hot exhaust products are used to ignite the main motor. The hybrid oxidizer flows axially into the small chamber where it is combined with an additional gaseous fuel (that usually has some swirl when injected), e.g., ethylene/N_2O [361] or methane/O_2 [10]. Multiple ignitions are achievable on both the ground and at altitude [282]. Fig. 9.26 shows the axial oxidizer and radial fuel injection. In this image, redundant spark igniters are included for reliability.

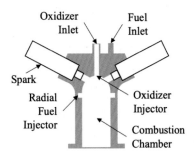

FIGURE 9.26 Cartoon of an Augmented Spark Igniter, adapted from [2].

Highly reliable spark plugs have been developed for rocket applications. However, most lower-budget/ground-based systems rely on automotive or marine varieties. The choice of spark plug is critical here as sooting (the collection of combustion biproducts between the tip of the electrode and the ground, impeding the arc) can be a major issue. If the spark plug is too cold, it will soot. If it is too hot for the application, it will overheat. A cold spark plug conducts heat well and has less insulation. Cold spark plugs also have a smaller surface area exposed to combustion gases. A hot spark plug is better insulated and retains more heat. Some spark plugs come with a resister to reduce Radio Frequency (RF) noise from the sparking.

9.9.5 Pyrophoric and hypergolic systems

Pyrophoric and hypergolic ignition is highly dependent on the propellant choice as was discussed in Chapter 5. In pyrophoric ingition, the substance lights on contact with oxygen (e.g., triethylaluminium aka TEA). Hypergolic ignition is achieved when two substances ignite on contact with each other. Hypergolic ignition is commonly used in liquid rockets, e.g., MMH/N_2O_4. Recent research into hypergolic solid fuel additives has made oxidizers such as NTO or MON's interesting for this application. See Table 5.4 in Chapter 5 for the most promising options and their ignition delays. It is important to note that hypergolic reactions occur in the liquid (or solid) phase, which is counterintuitive as standard ignition occurs in the gas phase. Note: a hypergolic reaction is often poor if the liquid has vaporized.

9.9.6 Laser ignition

Laser ignition has been demonstrated for hybrid rocket applications in both ambient and vacuum conditions [362]. One of the major advantages of such a system is that multiple ignitions are possible. Slab burner experiments suggest that, while it also vaporizes fuel, the laser actually heats embedded soot which provides the ignition mechanism for the propellants. Since the soot is relied upon for ignition, not every hybrid fuel can be ignited in this manner. However, a small disk of compatible fuel (e.g., Acrylonitrile Butadiene Styrene, ABS) has been successfully used to ignite fuel without a blackener (e.g., PMMA). A number of other fuels have been tested for laser ablation propulsion that could potentially be utilized as hybrid rocket igniters with some focused work, e.g., [363].

There are several challenges to laser ignition. An example of the hardware required is shown in Fig. 9.27. Optical access is necessary; therefore window sooting due to combustion is a concern if multiple ignitions are planned. The laser requires a decent amount of power (10–12 W in the tests by Ref. [362]). Therefore battery mass must be traded against competing ignition systems. The laser system also generates heat that the system must be designed to survive. Finally, the ignition delay can be quite long, on the order of a second, and variable over the number of starts. This makes it difficult to predict the impulse from a burn of a given length of time. Marked differences in ignition delay were evident from the different hardware configurations used by [362].

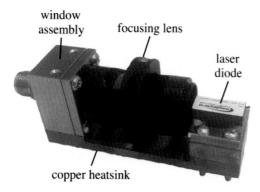

FIGURE 9.27   Photo of a Diode Laser Igniter from Ref. [364].

9.9.7 Arc ignition

Arc ignition relies upon electrical energy applied to a layered, 3D-printed fuel grain. The concentrated charge produces localized arcing, which pyrolyzes some of the fuel and can provide sufficient energy to ignite the hybrid rocket [152]. This method has so far been constrained to additively manufactured ABS fuel grains [153]. Note that the print geometry of the fuel grain is important to ensure multiple ignitions can be achieved [153].

9.9.8 Steel wool

Steel wool heats up when current is applied to it and will ignite readily in an oxygen-rich environment. For a hybrid ignition application, the steel wool is fed through the aft end of the motor. Cabling runs from the external power source, through the nozzle, and to the steel wool.

Gaseous oxygen is introduced into the combustion chamber through flexible hose in the same manner as the cabling (unless it is the main oxidizer), see Fig. 9.28. The steel wool is either consumed or expelled through the nozzle. The advantages of this system include that it is inexpensive, simple, and can be less massive since the hardware, including the secondary oxidizer and battery to provide the electrical impulse, is left on the ground. The disadvantages are that the ignition system can only be used for a single start, it is only effective for ground launches, and it is possible that the igniter could damage the rocket as it exits the through the nozzle.

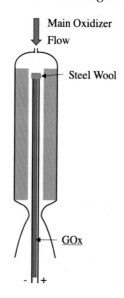

FIGURE 9.28   Cartoon of a steel wool ignition system.

9.10 Component selection

Component selection is critical for propulsion engineers. Practical considerations, such as cost and schedule, drive the desire to use Commercial Off the Shelf (COTS) hardware whenever possible. These typically compete with the desire to minimize mass. Other performance requirements will also dictate the selection of

TABLE 9.7 Recommended proof and burst factors of safety for metallic pressure components. Table is based on Ref. [7]. For proof and burst factors of safety for tanks see Refs. [7] and [8] for metallic and COPV tanks, respectively.

Component	Proof Pressure / MEOP	Burst Pressure / MEOP
Lines: D < 38 mm (1.5 inch)	1.5	4
Lines: D ⩾ 38 mm (1.5 inch)	1.5	2.5
Fittings: D < 38 mm (1.5 inch)	1.5	4
Fittings: D ⩾ 38 mm (1.5 inch)	1.5	2.5
Fluid Return Sections	1.5	3
Fluid Return Hose	1.5	5
Other Pressure Components	1.5	2.5

components. The key performance parameters for component selection are mass, power requirements, pressure rating, flow rating/capacity, leakage (without contamination), temperature range, actuation timing/response time, and material compatibility. Finally, environmental loads, including temperature, shock, vibration, acceleration and radiation, are often driving. It should be noted that vibration typically tests workmanship as opposed to the component design itself.

Performance parameters are discussed in more detail in the following sub-sections.

9.10.1 Mass and power

The main driver for any flight application is mass. Power needs typically translate into battery requirements and therefore also affect the mass of the system. Both mass and power are typically not drivers for component selection in-ground systems.

9.10.2 Pressure rating

Any components to be used in a feed line should be selected to be compatible with the Maximum Expected Operating Pressure (MEOP) in that feed line. As described in Section 9.2.7,

this is equivalent to saying that the system Maximum Allowable Working Pressure (MAWP) - the lowest rated operating pressure of any component, fitting or tubing in the system - should be equal to or larger than the system MEOP. The system MEOP is generally defined as the highest operating pressure that will be experienced under normal operation. Transient single events, such as priming water hammer, are not generally used to define the MEOP. Instead, care is usually taken to ensure that system-proof tests expose the system to higher pressure than any of these events. In such cases, the burst pressure rating of all parts of the system should also still be significantly higher than any surge pressure it will see. A discussion of proof and burst testing of components is provided in Chapter 11. When reviewing hardware pressure ratings for component and tubing selection, recall that the rated burst pressure is required to be significantly larger than the operating pressure, see Table 9.7 for details for metallic components.

9.10.3 Flow rating/capacity

All components in a feed system will introduce a potential source of pressure drop into the feed line and will have a maximum flow rate through them for defined inlet conditions. The flow capacity gives a measure of how unrestricted the flow is through a given component. Flow capacity for valves is typically described by the flow coefficient, C_V or discharge coefficient C_d. The definitions of flow capacity and discussion/examples of calculating pressure drops through components, including valves, are provided in Chapter 8.

When sizing components for a feed system using a gaseous oxidizer, recall that if the pressure drop across the component is sufficiently high, the flow will choke. If a liquid oxidizer is used, then the flow can also be effectively choked if cavitation is present. The equations used to define the flow capacity of a valve may not apply during choked operation, so care should be taken to check whether the pressure drop across the component is high enough for choking to occur. More details on this for both liquids and gases are provided in Chapter 8.

9.10.4 Leakage

Component leakage is defined as either internal leakage or external leakage. Internal leakage refers to leakage through the feed line. In the case of a valve, it is the leakage across the closed valve seat, allowing fluid to move from upstream of the valve to downstream of the valve under a positive pressure differential. External leakage refers to leakage from inside the component to the external environment.

Tight tolerances around the valve seat are critical for minimizing leakage across the seat. Valve seat materials are also key to operation at temperature since large differences in the coefficient of thermal expansion can open leakage paths with changes in temperature.

9.10.5 Temperature range

The selection of components to meet temperature requirements is fairly straightforward. The propulsion engineer should remember to add some margin to the expected operating temperature range and to consider the temperature of

the component throughout its full life. For a long hybrid rocket motor burn using a liquid oxidizer, the oxidizer temperature is likely to be fairly stable, but for the gaseous oxidizer, the gas temperature will likely be significantly colder at the end of the burn. Thermal soak back will need to be considered to evaluate the maximum component temperature for components located close to the combustion chamber.

9.10.6 Actuation timing/response

The actuation timing or response of a valve refers to how quickly it opens or closes. For a normally closed solenoid valve, the opening response is the time it takes for the valve to open when given full power and the closing response is the time it takes to close when power is removed. Depending on the valve design, the response times will often have some variation with fluid pressure and temperature. If response times are critical, then it is important to measure the response at representative conditions.

9.10.7 Material compatibility

Material selection deserves special attention to ensure compatibility with the propellants and sufficient strength. The propulsion engineer should independently evaluate the full list of wetted valve materials and evaluate the previous use of any lubricants inside the valve before deciding that it is suitable for use with a specific oxidizer.

An initial guide to material compatibility is included in Appendix B. This is provided as a starting point for the propulsion engineer. All materials selected for use with a given oxidizer should be thoroughly reviewed using additional reputable literature sources prior to use.

9.11 Reliability

Reliability is critical for flight hardware as there are rarely opportunities to fix issues during/after launch. Redundancy is recommended whenever it can be afforded. However, more commonly, a policy of selective redundancy shores up the areas most likely to see failures. An example of this would be dual coils in a valve. If one coil fails, the other can be used to open the valve. Mission success or safety requirements will drive the redundancy policy, potentially requiring the system to be single or dual-fault tolerant and determining if selective redundancy is acceptable.

The propulsion engineer is urged to think through failure modes as a system is designed. A Failure Mode Effects Analysis (FMEA), or if adding a criticality analysis, a Failure Mode, Effects, and Criticality Analysis (FEMCA) is a useful tool for identifying potential issues. A FMEA analyzes all potential failure modes, identifying their effects and severity. A criticality analysis ranks potential failure modes based on a combination of their probability of occurrence and severity. Steps in conducting an FME(C)A's are as follows [365]:

- Define the system
- Construct a block diagram
- Identify the failure modes and effects on the component, system and mission
- Evaluate worst potential consequences and assign severity
- Identify failure detection methods
- Identify corrective design/ways to minimize risk
- Identify effects of corrective actions
- Document the analysis and identify issues/controls to minimize risk

9.12 Range safety

Range safety requirements are imposed in the aerospace industry to protect people from hazards inherent in spacecraft being launched. It is recommended that the propulsion engineer be familiar with these safety precautions and that they be met or exceeded in ground test situations. According to [366], the seminal reference for range safety in the United States, systems with the potential to fail in a catastrophic manner must be dual fault tolerant (three independent inhibits). Those that may fail in a critical manner must be single fault tolerant (two independent inhibits). System failures that lead to marginal hazards do not require fault tolerance (single inhibit). The definition of the hazard class is based on risk to both human safety and cost. Other safety measures, such as remote operation with termination and fail-safe capabilities, are recommended for ground testing. It is easy to become complacent and downplay the risks associated with ground testing of a hybrid rocket. However, the lessons learned from accidents, such as the scaled composites nitrous tank explosion in 2007, which killed three people and injured three more, should not be forgotten.

The hardware necessary for safety considerations could be as simple as redundant valves. However, specific additions such as a mechanical safe and arm device or a Flight Termination System may be necessary.

9.12.1 Safe and arm

A Safe and Arm (S&A) device is placed within the pyrotechnic firing train and disables the ignition circuit. The initiator is physically rotated out of line with the circuit. S&A devices have a visual indicator of their state. Electromechanical or purely mechanical safe and arm devices are common in missile and rocket applications. They are less common for space because they tend to be quite massive, on the order of one to several kg. Fig. 9.29 shows an example of an electromechanical safe and arm device (max mass of 1.7 kg), but there are many other options.

FIGURE 9.29 Electromechanical safe and arm device with a manual safety key [367].

9.12.2 Flight termination system

A Flight Termination System (FTS) is typically an explosive that will stop the rocket from potentially destructive performance. A FTS is needed for experimental/sounding rockets with thrust vector control (and potentially others depending on the range and the potential for the rocket to harm people if it doesn't follow its predicted flight path/performance/etc.). In a solid rocket application, the FTS would blow up the rocket. For a hybrid rocket, it may be possible to separate the combustion chamber from or simply "unzip" (aka blowup) the oxidizer tank, since thrust cannot continue without the oxidizer. A safe and arm device can also be used on an FTS.

9.13 Qualification

Qualification is required for flight hardware to ensure that the hardware can survive launch vibration, any potential pyroshock events, and the expected thermal environments without a significant degradation in performance. Qualification can be achieved via test, analysis, or by similarity to previously qualified hardware (heritage). Where qualification of hardware is achieved by test, then two approaches can be taken. The first is lower risk but higher cost. It involves the use of a dedicated qualification component/model, which is exposed to qualification environments that are generally more extreme than the expected flight environments. The qualification component should be as identical to the flight hardware as possible, ideally taken from the same flight lot of hardware. The higher risk, but lower cost, the approach is to qualify the hardware using protoflight testing. With this approach, the flight hardware is exposed to the qualification levels in ground testing prior to then being used as flight hardware. The protoflight test durations are generally not as long as full qualification tests would be, but the levels are still higher than the expected flight levels.

9.14 Design for manufacturability

The propulsion engineer should be aware that the best design on paper is not always easily built. For example, calculations may result in non-achievable wall thicknesses for tanks and tubing. The minimum wall thickness for machinability, which varies by material, must be taken into account. Also, it is possible to spend an inordinate amount of money making custom thickness tubing, when the use of standard tubing would work and have a minimal impact on mass. Another common mistake that drives up cost is to design a hole size other than that made by a standard drill bit. This would require a technician to spend a considerable amount of time trying to machine the hole within tolerance when a single mill operation may have sufficed. The use of other commonly available components (fittings, o-rings, etc.) instead of custom designs will also minimize the cost of the system.

Parallel to design for manufacturability is the ease of assembly. There are many things one can design in a CAD program that cannot physically be manufactured and assembled. The propulsion engineer should think through each step of the integration and test procedures as the design is finalized. For example, consider:

- If a torque wrench is required to install a bolt, it must fit around the bolt and be turned without obstruction.
- If a system is welded, then the number of welds typically drives the assembly schedule.
- If the system must be welded, then the propulsion engineer must ensure that the weld head can fit within the available space.
- Can the assembly be divided into sub-assemblies so that parallel assembly can occur?
- Many flight propulsion components are long lead items, which can be delayed at the vendor. It can be helpful to have contingency assembly plans in place (e.g., the ability to switch the order of operations) to ensure that assembly can continue even in the event of some component delays.
- Issues with valve leakage or component failure can be uncovered late after the full assembly is complete. Wherever possible, design the feed system such that individual high-risk components can be replaced late if required.

- If the feed system must be clean, then consideration of the assembly process is critical to ensuring the overall cleanliness of the system. Every mechanical connection will generate some particulate, as will every weld. Ensuring that positive pressure is maintained, that tubing sub-assemblies are cleaned after welding where possible, that assembly is conducted in a clean environment whenever possible, and that dead-end welds have minimized all help to maintain system cleanliness.
- Final verification of an assembly process will typically require proof and leak testing. X-rays of welds may also be required. The propulsion engineer should ensure that such processes are accounted for in the assembly, integration, and test plan. This is especially critical when x-ray is required as some x-rays may be hard to complete at the conclusion of the assembly process and instead will need to be conducted during assembly.

In all of the above cases, asking the opinion of the machinists, welders, or technicians actually doing the work is often invaluable. Their input is key to successfully turning the paper design into hardware.

10

Design examples

This chapter applies the materials covered in the rest of this text to a selection of real-world examples. Each section in this chapter is a different worked example.

10.1 Mission requirements

Mission requirements, such as the required ΔV, can be estimated via the approach and examples given in Chapter 2. In practice, a model with six degrees of freedom is required to accurately understand the impacts of the trajectory, including aerodynamic drag, etc.

10.2 Propellant selection

Hybrid rocket motors have often been considered for upper-stage motors because they can be shut off after the desired impulse has been delivered, thereby minimizing orbital dispersions. Consider the following four propellants: paraffin, HTPB, LOx, and H_2O_2. Which is the best propellant combination for an upper stage of an Earth launch vehicle launched from NASA Kennedy Space Center in Florida, USA?

This example is intended to help the propulsion engineer free him or herself from the confines of a specific propellant combination. It is common to gravitate towards the familiar option, the one with which she or he has the most experience. However, it is not always the best solution. Additionally, the best solution is rarely straightforward. It is simple to grab the highest I_{sp} propellant combination from a Table, such as Table 5.5 of Chapter 5, but that misses many of the nuances that this book has been presenting.

While it is important not to rely on performance alone, it is a great place to start. Table 10.1 presents the vacuum I_{sp}'s for the four potential propellant combinations from Table 5.5. It shows that the ideal performance is very similar for all of the options, with a slight preference for LOx.

TABLE 10.1 I_{sp} comparison for the propellant options.

	HTPB	Paraffin
LOx	353 s	355 s
H_2O_2	329 s	328 s

For the case where the upper-stage is not volumetrically constrained, Table 10.1 shows that as in existing upper stages (e.g., Centaur) LOx would be the preferred oxidizer.

The next step is to look at the oxidizer to fuel ratios, compiled here in Table 10.2. There is a clear difference in the O/F ratios between the oxygen and hydrogen peroxide.

The relative benefit of a high or low O/F depends on the propellant combination. If the

TABLE 10.2 O/F comparison for the propellant options.

	HTPB	Paraffin
LOx	2.4	2.7
H_2O_2	6.5	7.2

oxidizer density is higher than the fuel density, as is the case here, a higher O/F allows for more of the propellant to be stored in a smaller volume. The densities of each propellant can also be found in Chapter 5 and are: $\rho_{HTPB} = 919$ kg/m^3, $\rho_{paraffin} = 924.5$ kg/m^3, $\rho_{LOx} = 1142$ kg/m^3, and $\rho_{H_2O_2} = 1443$ kg/m^3. Using the O/F ratios given in Table 10.2, the average density of the propellant can be found using Eq. (2.23). The density impulse is then easily calculated, see Table 10.3.

TABLE 10.3 Density Impulse comparison for the propellant options.

	HTPB	Paraffin
LOx	3.76×10^5 kg s/m^3	3.81×10^5 kg s/m^3
H_2O_2	4.41×10^5 kg s/m^3	4.43×10^5 kg s/m^3

The impulse density shows a clear benefit to using H_2O_2 over LOx if the propulsion system is volume constrained. The high O/F ratio associated with H_2O_2 for both fuels allows for more of the much more dense oxidizer to be used for optimal performance. This may lead to a reduced propulsion system mass as well (all of the propellant can be stored in a smaller volume).

Before moving on to the fuel selection, now is a good time to take a step back and evaluate any other considerations that may impact the oxidizer portion of this trade. Hydrogen peroxide is storable, which is convenient for an upper stage, though not required. However, it is usually catalyzed, with the byproducts (water vapor and O_2) becoming the actual oxidizers. Note that the calculations in Table 5.5 do not take catalytic decomposition into account. There is a mass impact associated with the physical system. LOx is a cryogen, so it will require boil off losses to be

accounted for until its use. It is assumed that the impacts of these considerations will be comparable for this example. Methods for achieving low to no boil-off to enable cryogenic propellants to be used for in-space applications are an area of active research but have not yet been demonstrated. Thus the fact that an Earth-based system is being evaluated is the only reason LOx can be considered.

The theoretical performance for paraffin and HTPB is quite similar; therefore other considerations are used to make a determination between the fuels. The propulsion engineer was not given any information about the launch-induced loads; however, this needs to be considered. Paraffin is brittle and requires strength additives, while HTPB, a form of rubber, is much more stress tolerant. Average summer temperatures in Florida exceed 30 °C. Both fuels should be required to survive that temperature, including the margin for heating on the launch pad (think about how much hotter it gets inside a car than outside in the summer). However, paraffin starts to soften at elevated temperatures and can flow in the direction of the gravity vector (slump), see Ref. [156].

The main difference between paraffin and HTPB is the combustion process. Paraffin is a high regression rate, liquefying fuel. The thrust would be expected to be higher for paraffin unless the surface area of the HTPB fuel grain is increased accordingly, forcing the propulsion engineer to deal with the challenges (including mass penalty) associated with multiple fuel ports.

As stated at the beginning of the example, there is rarely a straightforward solution to a propellant selection problem. Table 10.4 shows the case in which each combination makes the most sense.

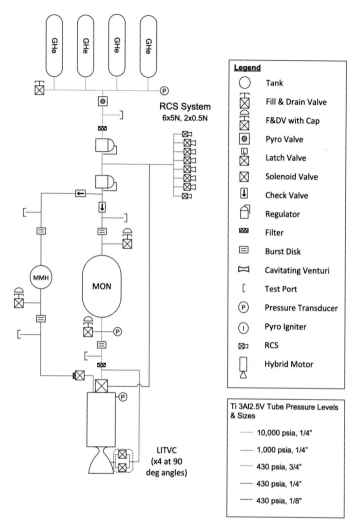

FIGURE 10.1   Schematic for a potential Mars Ascent Vehicle [48].

TABLE 10.4   Propellant choices.

	HTPB	Paraffin
LOx	Not volume constrained, low thrust, high stress loads or temps	Not volume constrained, high thrust
H_2O_2	Volume constrained, low thrust, high stress loads or temps	Volume constrained, high thrust

10.3 Staging - Mars Ascent Motor

A hybrid propulsion option has been studied for a Mars Ascent Vehicle (MAV) as part of a potential robotic Mars Sample Return [368], see Fig. 10.1. A hybrid design showed promise for this application because of the availability of low-temperature compatible propellants. The major challenges are 1) Entry Descent and

Landing loads and 2) Coefficient of Thermal Expansion (CTE) of the fuel grain over large temperature ranges 3) mass and geometry constraints imposed by the lander. Previous work has focused on a paraffin-based MAV. This example instead considers an HTPB/H_2O_4 hybrid propulsion system for a Mars Ascent Vehicle. What differences are there between an HTPB motor from a paraffin-based motor?

The MAV needs to launch a 10 kg payload to Mars orbit. On top of the actual samples and their container, assume 30 kg of additional mass for avionics, telecom, thermal control, etc. These two values together give $m_L = 40$ kg. Assume the ΔV required is 4,400 m/s. The $I_{sp,ideal}$ for HTPB/H_2O_4 can be found from Table 5.5 to be 329 s. For a single stage or first stage motor, assume a dry mass fraction of $\epsilon_1 = 0.2$, C^* efficiency of $\eta_{C^*} = 0.95$ and a nozzle efficiency of $\eta_n = 0.98$. If a second stage is required, assume a dry mass fraction of $\epsilon_2 = 0.23$, C^* efficiency of $\eta_{C^*} = 0.93$ and $\eta_n = 0.98$.

Note that the second stage has been penalized in this example in terms of both dry mass fraction and performance efficiency. Typically it is easier to make larger motors slightly more mass-efficient, but considerations such as thrust vector control might reduce the real value. The η_{C^*} and η_n are dependent on the motor design and testing would be required to validate these assumptions. The values used in this example are not conservative.

A minimum of two burns are required to achieve insertion into low Mars orbit. Luckily, this does not preclude a single-stage to-orbit design, since it is possible to restart a hybrid rocket. Therefore both single-stage and two-stage options will be considered to minimize initial mass (m_i).

First, Eq. (7.7) can be used to find the real specific impulse for each stage $I_{sp,real1} = 306.3$ s and $I_{sp,real2} = 299.9$ s. Then Eq. (2.44) can be expanded for one or two stages, see Eq. (10.1) for how this is done for two stages.

$$m_{i,1} = \left(\frac{1}{1-\epsilon_1}\right) m_{p,1} + \left(\frac{1}{1-\epsilon_2}\right) m_{p,2} + m_L$$

(10.1)

The resulting two-stage equation can be plugged into the rocket equation to solve for ΔV. Either an optimization routine or brute force plotting of multiple variables, e.g., propellant masses of each stage, see Fig. 10.2, leads the propulsion engineer to a total initial mass of approximately 355 kg. Note that Fig. 10.2 shows a range of optimum stage 1 and 2 propellant masses (the range between the two stars), this gives the propulsion engineer some design flexibility.

The mass of a single stage to orbit MAV would be just over 1,000 kg using the stage 1 values from before. Driven by their curiosity, the propulsion engineer would naturally start trying reduced dry mass fractions and evaluating other potential performance enhancements to make this a more competitive option. She/he will find that at an $\epsilon = 0.16$, the total mass is about 469 kg, which is much more paletable from a Mars lander accomodation perspective. However, squeezing all the required dry mass into less than 70 kg (payload not included) will be a challenge. A similar thought experiment could be undertaken looking at increasing performance (eta_{C^*}) and the effect it would have on the mass.

From this example, the two-stage to-orbit design looks much better than a single-stage one. This is because the same assumptions were applied to both cases (the first stage of the two-stage design would be comparable to the single-stage design). However, a two-stage design needs many things that would not be necessary for a single-stage design, such as a separation system, harnessing to both stages, and potentially extra batteries or avionics to fire/control the second motor. Therefore a decreased dry mass fraction, as compared to the first-stage motor design, is reasonable. Also, the dry mass

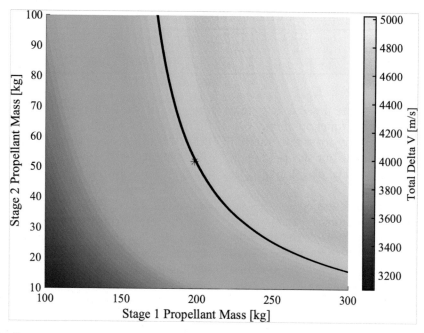

FIGURE 10.2 Two stage ΔV optimization. The black line shows where $\Delta V = 4,400$ m/s. The black star shows the location of minimum total mass.

fraction is introduced as a percentage of the total mass of the stage, adding a somewhat unfair disadvantage to the single-stage design since it has a higher mass. Propulsion components such as valves usually have a fixed mass over a range of motor sizes. In general, dry mass fraction decreases with motor size. Finally, the development of a single motor enables the team to focus resources on additional testing. This makes it feasible to assume that the C^* efficiency in the single-stage case could be improved, making the discussion of the last paragraph relevant. However, it should be noted that the mechanisms applied to increase efficiency (such as mixing devices, precombustion or post combustion chambers) come at the cost of added mass, making it very difficult to achieve both high performance and low mass.

Assume now that the single-stage design could achieve $\epsilon = 0.15$ and $\eta_{C^*} = 0.98$. The new total mass becomes 369 kg, making the single and two-stage options comparable. These are very aggressive values, bordering unrealistic. However, this shows that the devil is in the details. The better information the propulsion engineer has about motor performance and propulsion system mass (see Chapter 7), the more useful the trade results will be.

10.4 Test/performance prediction

All propulsion engineers need to be able to predict the performance of a rocket they have designed either for ground test or for flight. This example of a CubeSat motor test is intended to show the reader how to apply the equations of Section 3.4.2 and 3.6.1 in Chapter 3 to predict the expected chamber pressure over a predefined burn time. For this example Test 67 of Refs. [9] and [10] is considered. This test was selected as the chamber pressure history versus time and key initial conditions were previously

TABLE 10.5 Test prediction example inputs from Ref. [9] and [10].

$\dot{m}_O$ [kg/s]	r_i [m]	L_f [m]	t_b [s]	Oxidizer [-]	Fuel [-]	ρ_F [kg/m^3]	$\bar{h}_f$ [kJ/mol]	a_o [S.I.]	n [S.I.]
0.0068	0.00511	0.1778	65.9	$O_2(g)$	$C_5H_8O_2$	1.1834×10^3	-166.9	1.44×10^{-4}	0.24

published. It is assumed here for simplicity that the fuel regression rate is a function of only oxidizer mass flux. The oxidizer mass flow rate for this test was kept constant but the prediction method adopted here could be used for variable oxidizer mass flow rate.

The key initial variables needed for this assessment can be taken from Refs. [9] and [10], see Table 10.5. The internal volume of the combustion chamber is taken to be 105 cubic centimeters, and the nozzle throat diameter is assumed to be 4.2 mm (0.166 inch). The motor is ignited from ambient pressure, assumed to be 101.3 kPa [14.7 psi]. The oxidizer for this test is gaseous oxygen. It is assumed that the initial temperature of the oxygen and fuel is 20 °C.

Two approaches are used here to predict the combustion chamber pressure versus time. The first approach uses the steady-state design equations of Section 3.4.2 in Chapter 3.

The burn is modeled as a series of j time steps. The oxidizer mass flow rate, $\dot{m}_O$, is constant at 6.8 g/s per Table 10.5. At any given time, t_j, the fuel port radius, $r(t_j)$, is given by Eq. (3.59). Eq. (3.59) assumes that the oxidizer mass flow rate is constant throughout the burn. If the oxidizer mass flow rate is not constant, then the fuel port radius can be calculated using a simple forward Euler method, see Eq. (3.80) with the fuel regression rate calculated from Eq. (3.48) using the oxidizer mass flux at the previous time step. For constant $\dot{m}_O$, the fuel regression rate, $\dot{r}$, is also calculated from Eq. (3.48). The fuel mass flow rate can be determined using this regression rate and the fuel density (in Table 10.5) via Eq. (3.62). The instantaneous O/F can be calculated using the oxidizer mass flow rate and the newly calculated fuel mass flow rate. A chemical equilib-

rium solver is used to determine the equilibrium fluid conditions in the combustion chamber at this O/F ratio, specifically, the average molecular weight, M_w, ratio of specific heats, γ, and the fluid temperature $T_{t,c}$ (which is equal to the adiabatic flame temperature). For this example, the solver described in Ref. [74] is used. Note that the equilibrium solver will need the fuel chemical composition and enthalpy of formation, h_f, (both of which are in Table 10.5) as well as the combustion chamber pressure. The chamber pressure from the previous time step, $P_{t,c}(t_{j-1})$ should be used. At the first time step, it is reasonable to use the initial starting ambient pressure. Thereafter, the chamber pressure can be calculated using Eq. (3.64), or equivalently Eq. (10.2). For simplicity, it is assumed that the discharge coefficient, C_d for flow from the combustion chamber to the nozzle throat, is 1. It is also assumed that the velocities in the combustion chamber are low, such that the static temperature and pressure are equal to the total, or stagnation, values. The combustion efficiency, η_{C*} was assumed to be constant at 0.85 (85%) throughout the burn, based on previous history with this motor. If the test was being conducted for the first time with this motor, then a value of 1 would have been assumed for combustion efficiency to ensure that an upper bound on expected chamber pressures was determined. This approach was used to predict the chamber pressure for the full duration of the burn. The results of the approach can be seen in Fig. 10.3 overlaid with chamber pressure test data for Test 67.

$$P_{t,c} = \eta_{C*} \frac{(\dot{m}_O + \dot{m}_F)}{\gamma C_d A_{th}} \sqrt{\gamma R T_c} \left(\frac{\gamma + 1}{2} \right)^{\frac{\gamma + 1}{2(\gamma - 1)}}$$

(10.2)

The second approach that can be used to predict chamber pressure over time accounts for the acoustic fill/emptying times of the combustion chamber per Section 3.6.1 in Chapter 3. To utilize the equations in that section to predict chamber pressure over time, the propulsion engineer again starts with knowledge of the planned oxidizer mass flow rate versus time, $\dot{m}_O$, initial chamber pressure, $P_{t,c}$, fuel grain properties (density, ρ_F, length, L_f, initial port radius, r_i, enthalpy of formation, h_f), the burn time, t_b, and for the quasi-steady analysis also the initial chamber volume, V_C.

The approach to determining the fuel radius over time and fuel regression rate is the same as for the steady-state approach since the fuel regression rate is not a function of chamber pressure. Given that the fuel port radius is known from these equations, the instantaneous chamber volume, V_C, can be calculated using Eq. (3.78). It is assumed for this example that the precombustion and post combustion chamber volumes remain constant throughout the burn (this is equivalent to saying that there is minimal ablation of the materials in the precombustion and post combustion chamber volumes).

The change in chamber pressure at each time step can be calculated via Eq. (3.79), and the chamber pressure at the following time step then can be calculated via Eq. (3.81). Note that at each time step, the chamber temperature, $T_{t,c}$, average gas molecular weight, M_w, and ratio of specific heats, γ, are again calculated using the chemical equilibrium combustion solver CEA [74] for the O/F in the chamber at that time. For simplicity, the O/F of the propellant being added to the combustion chamber at each time step is used. During ignition and shutdown, the combustion chamber pressure changes much more rapidly than during the steady-state phase of the burn. In order for this method to capture these transients, and for the solution to remain stable, small time steps must be used during these times. For this example, time steps of 0.001 s were used during ignition and shut

down, and time steps of 0.01 s were used during steady-state combustion. The results of this quasi-steady approach are shown in Fig. 10.3 overlaid with chamber pressure test data and the steady-state predictions for Test 67. It can be seen that, as expected, the quasi-steady and steady-state chamber pressure predictions are essentially the same during steady-state combustion, but differ during ignition and shutdown. It can also be seen in Fig. 10.3 that the quasi-steady predictions still underpredict the length of time of these transients. This is likely due to a number of factors. Firstly, Test 67 used a nitrogen purge gas at the end of the burn. The mass flow rate of nitrogen was not recorded in Refs. [9] and [10], and so it was not included in this analysis. The use of this purge gas would lengthen the shutdown transient compared to the quasi-steady prediction that did not include that gas mass being injected in the combustion chamber. Secondly, as discussed in Chapter 3, the quasi-steady equations used here do not capture changes to the motor ballistics from changes in the assumed fuel regression rate behavior, such as from establishment of the thermal layer in the fuel or boundary layer formation during ignition, or thermal soaking into the fuel during shut down. Again, overlooking such effects likely serves to underpredict the length of time that the transients persist. The final factor that should be considered is that the combustion chamber volume and the nozzle diameter used in this example were rough approximations. Despite these shortcomings, the quasi-steady approach can still be seen to accurately capture the steady-state combustion trends and can be used to estimate acoustic fill/emptying times and capture these acoustic effects on chamber pressure.

Note that the two approaches outlined above involved running the equilibrium solver at every time step. If speed is desired in the computations, and the change in O/F is not expected to be large across the burn, then an assumption of constant combustion products and temperatures

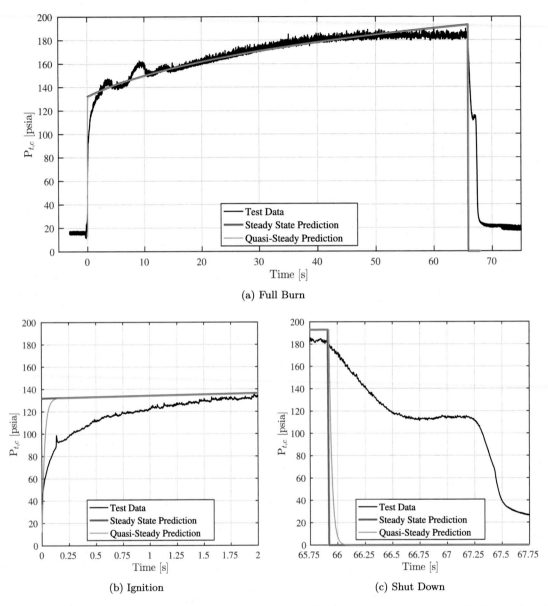

FIGURE 10.3 Predicted and measured chamber pressure versus time for Test 67 of Refs. [9] and [10]. Note that the steady-state and quasi-steady predictions are essentially perfectly overlaid in image (a). Images (b) and (c) are included to show where they deviate during the ignition and shutdown transients.

can be assumed, since the equilibrium results are usually only a weak function of chamber pressure. Fig. 10.4 is included here to show how little

some key equilibrium code outputs (C^*, $T_{t,c}$, and $I_{sp,vac}$) varied over the course of the fairly long (~66 s) burn being considered.

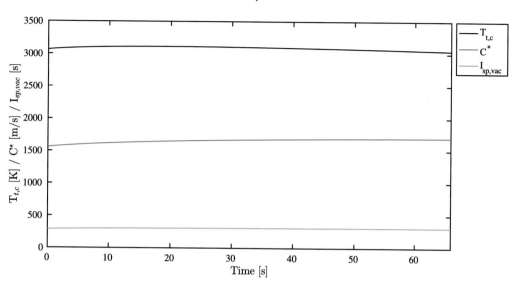

FIGURE 10.4 Predicted ideal equilibrium C^*, $T_{t,c}$, and $I_{sp,vac}$ versus time for Test 67 of Refs. [9] and [10]. $I_{sp,vac}$ is calculated assuming an area ratio of 10. They are determined using CEA [74].

10.5 In-space motors

This example focuses on applying the design approach of Chapter 7 to calculate primary propulsion system masses and key motor/tank dimensions. The goal of this example is to design an orbit insertion motor for a 500 kg (wet) spacecraft. Orbit insertion requires $\Delta V = 1000$ m/s at a thrust of $F = 6kN$. The payload can tolerate a maximum acceleration of 3g's. The sponsor specifically requested a space storable propellant combination: HTPB/MON3. Use a helium-stored gas pressurization system to maintain a constant oxidizer mass flow rate. The AFT range for the motor is 5–50 °C, with nominal operation at 20 °C. The chamber pressure is 2.07 MPa. The sponsor is willing to accept a η_{C^*} of 95%. Assume an 80% bell nozzle with an expansion ratio of $\mathcal{R} = 40$, inlet radius of $1.5r_{th}$ and η_n of 97.5%. See Appendix C for a summary of design equations that may come in handy here.

A chemical equilibrium solver can be run at this point. Or, these just happen to be the conditions (P_c and $\mathcal{R}$) presented in Chapter 5, so performance data can also be found there. For simplicity, the performance values for N_2O_4 are used instead of MON3. The results will be similar, if slightly conservative, with this approximation. For HTPB/MON3: $I_{sp,vac} = 327$ s at an $O/F = 3.9$ and $\gamma = 1.13$. The density is $\rho_F = 919$ kg/m^3 from Table 5.2, and $\rho_O = 1433$ kg/m^3 from Table 5.3. Published equations for the variation of density with temperature are available in the literature and will be used for the pressurization system sizing later [369]. As an aside for nozzle development later on, the adiabatic flame temperature is 3391 K, and water vapor makes up 21% of the exhaust gases (by moles).

The mass of the propellant can be found by rearranging Eq. (2.27) and recalling that $m_i = m_f + m_p$, where m_p is the usable propellant mass, and m_i is the total initial spacecraft mass given above. This equation is also given as Eq. (7.10) in Chapter 7.

$$m_p = m_i \left(1 - e^{\frac{-\Delta V}{I_{sp,real}\, g_0}} \right)$$

$$m_p = 500 \text{ kg} \Bigg(1$$

$$- e^{\dfrac{-1000 \text{m/s}}{327 \text{ s} \times 0.95 \times 0.975 \times 9.81 \text{ m/s}^2}} \Bigg)$$

$$m_p = 142.9 \text{ kg}$$
$$(10.3)$$

The next step is to solve for the nozzle exit pressure, P_e, using Eq. (7.8). The nozzle exit pressure is also an output of a chemical equilibrium solver, such as CEA. The simplifying assumptions (one-dimensional, frozen equilibrium, isentropic flow, etc.) made here typically overpredict P_e and C_F as compared to an equilibrium solver.

$$\left(\frac{\gamma+1}{2}\right)^{\frac{1}{\gamma-1}} \left(\frac{P_e}{P_c}\right)^{\frac{1}{\gamma}} \sqrt{\frac{\gamma+1}{\gamma-1}\left[1-\left(\frac{P_e}{P_c}\right)^{\frac{\gamma-1}{\gamma}}\right]}$$

$$-\frac{1}{\mathcal{R}} = 0$$

$$P_e = 5.7 \text{ kPa}$$
$$(10.4)$$

The thrust coefficient can be found from Eq. (7.9).

$$C_F = \left\{ \left(\frac{2\gamma^2}{\gamma-1}\right)\left(\frac{2}{\gamma+1}\right)^{\frac{\gamma+1}{\gamma-1}}\left[1-\left(\frac{P_e}{P_c}\right)^{\frac{\gamma-1}{\gamma}}\right]\right\}^{\frac{1}{2}}$$

$$+ \left(\frac{P_e}{P_c} - \frac{P_a}{P_c}^{0}\right)\frac{A_e}{A_{th}}$$

$$C_F = 1.97$$
$$(10.5)$$

Alright. There is now enough information to size the nozzle throat using Eq. (7.14), given here as Eq. (10.6).

$$A_{th} = \frac{F}{\eta_n C_F C_d P_c}$$

$$A_{th} = \frac{6000 N}{0.975 \times 1.97 \times 1 \times 2.07 \times 10^6 \text{ Pa}}$$

$$A_{th} = 0.0015 \text{ m}^2 \quad (10.6)$$

Great! Now on to the propellant mass flow rate, using Eq. (7.16)

$$\dot{m}_p = \frac{F}{I_{sp} g_0 \eta_{C^*} \eta_n}$$

$$\dot{m}_p = \frac{6000 N}{327 \text{ s} \times 9.81 \text{ m/s}^2 \times 0.95 \times 0.975}$$

$$\dot{m}_p = 2.02 \text{ kg/s} \quad (10.7)$$

At this point, it is possible to calculate the C^*. It will be needed later, so may as well go for it. (It is also an output of most chemical equilibrium solvers, so that is also an option.)

$$C^* = \frac{P_c A_{th}}{\dot{m}_p}$$

$$C^* = \frac{2.07 \times 10^6 \text{ Pa} \times 0.0015 \text{ m}^2}{2.02 \text{ kg/s}}$$

$$C^* = 1549 \text{ m/s} \quad (10.8)$$

Note that this value is slightly less than the peak C^* for HTPB/MON3 of 1660 m/s. This is because the operating conditions here have been maximized for peak I_{sp}, and the motor is therefore running at a higher O/F ratio than it would for peak C^*.

Since the total mass flow rate and the O/F ratio are known, the oxidizer mass flow rate can be found. This will be constant during the burn because of the external pressurization system. Since the propellant mass and average propellant mass flow rate are known, the burn time can be found: $t_b = m_p/\dot{m}_p = 70.8$ s. This long burn time (coupled with the high mass fraction of water vapor in the exhaust) will be a little rough on nozzle erosion. Thankfully, that is a real-world problem for down the road. Using the oxidizer-to-fuel ratio, the mass flow rate of the oxidizer can be found:

$$\dot{m}_O = \frac{O/F}{1+O/F}\dot{m}_p$$

$$\dot{m}_O = \frac{3.9}{1+3.9}(2.08 \text{ kg/s})$$

$$\dot{m}_O = 1.61 \text{ kg/s} \qquad (10.9)$$

Now that the amount of usable propellant is known, it is time to determine the propellant split and the regression rate information. The mass of the fuel and oxidizer can be found since the O/F ratio is known.

$$m_F = \frac{m_p}{1+O/F}$$

$$m_F = \frac{142.9}{1+3.9}$$

$$m_F = 29.2 \text{ kg} \qquad (10.10)$$

$$m_O = \frac{O/F}{1+O/F}m_p$$

$$m_O = \frac{3.9}{1+3.9} \times 142.9 \text{ kg}$$

$$m_O = 113.7 \text{ kg} \qquad (10.11)$$

Looking at Table 6.2, the propulsion engineer will note that there is no entry for HTPB/MON3. There has been some testing of this propellant combination, but published regression rates are not available at this time. Therefore the propulsion engineer makes an educated guess that the nearest propellant combination will be HTPB/N_2O and makes a note to ask the sponsor for funding to test the propellant combination as early as practical. Therefore assume: $a_0 = 1.88 \times 10^{-4}$ and $n = 0.347$ in SI units. There are plenty of ways to get fancy with this estimation, e.g., looking at other fuels with MON and changing the n exponent to be in a family with them or adjusting the value of a_o based on the difference in adiabatic flame temperature relative to the fuel surface temperature for HTPB/MON3 versus HTPB/N_2O. However, for this example, the values will be used as is. As

with all published regression rate information, it should just be used as indicative of performance and not for detailed design (recall the scaling discussion in Chapter 6, Section 6.3.8).

Assume a single, cylindrical port fuel grain. HTPB is fairly strong, so a D_f/D_i of 3 is easily achieved. The final and initial diameters can be found from Eqs. (10.12) and (10.13), respectively.

$$D_f = \left\{ \left[\frac{(2n+1)2^{2n+1}a_o}{\pi^n} \right] \times \frac{\dot{m}_O^n t_b}{1-(D_i/D_f)^{2n+1}} \right\}^{1/(2n+1)}$$

$$D_f = 0.205 \text{ m} \qquad (10.12)$$

$$D_i = \left\{ \left[\frac{(2n+1)2^{2n+1}a_o}{\pi^n} \right] \times \frac{\dot{m}_O^n t_b}{(Df/Di)^{2n+1}-1} \right\}^{1/(2n+1)}$$

$$D_i = 0.068 \text{ m} \qquad (10.13)$$

At this point, one can complete a check on the numbers already calculated to confirm that the regression rate will burn all the provided fuel in the calculated burn time. The grain was also sized using a predetermined ratio of diameters, so that is another good check: 0.205 m/0.068 m = 3. Check.

The fuel grain length can be found using Eq. (7.29) and is given below.

$$L_F = \frac{4m_F}{\pi \rho_F (D_f^2 - D_i^2)}$$

$$L_F = 1.08 \text{ m} \qquad (10.14)$$

With the ballistics completed, the design assessments presented in Chapter 7, Section 7.8 can be completed. Confirmation of the design is left as an exercise to the reader.

The quantity of usable fuel and oxidizer and the shape of the fuel grain enable the combustion chamber and oxidizer tank to be sized. Propellant residuals need to be taken into account for both the oxidizer tank and the combustion chamber. The latter will also need to account for the thickness of the insulation. Starting with the combustion chamber, assume a residual amount of fuel, say 3%, and that it is all left in the radial direction. This results in a thicker combustion chamber but does not add any length. The fuel residual, aka sliver fraction, is $m_{F,residual} = m_F \times 0.03 = 0.875$ kg, and the thickness can be calculated via Eq. (10.15). The residual fuel adds 1.4 mm to the radius of the fuel grain.

$$th_{F,residual} = \sqrt{(D_f/2)^2 + \frac{m_{F,residual}}{\rho_F \pi L_F}} - (D_f/2)$$
$$th_{F,residual} = 1.4 \text{ mm}$$
$$(10.15)$$

Now the insulation needs to be added around the fuel grain (plus residual) as well as to the fore and aft ends of the combustion chamber for this single port design. This will also increase the diameter of the motor and add appreciably to the mass.

In order to calculate the insulation thickness, choose a material from Table 9.6 in Chapter 9. Carbon cloth phenolic is used for this example, with a density of 1420 kg/m^3 and an estimated regression rate of 0.65 mm/s. The insulation along the cylindrical section of the combustion chamber will only be exposed for several seconds ($t_{ins} = 3$ s is assumed). This time is somewhat arbitrary. It is assumed that the fuel protects the combustion chamber for most of the burn, and the insulation around the cylindrical portion of the fuel grain is to protect against uneven regression (i.e., some parts of the motor case being exposed early). The thickness of the insulation can be found by multiplying the regression rate by the exposure time. For the cylindrical section: $th_{ins,cyl} = \dot{r}_{ins} \times t_{ins} = 1.9$ mm and for the post combustion chamber: $th_{ins} = \dot{r}_{ins} \times (t_b + t_{ins}) = 48$ mm. Then the insulation masses can be approximated by multiplying the surface areas of the cylindrical section and the post combustion chamber by $th_{ins,cyl}$ and th_{ins}, respectively as well as the insulation density (or

more accurately by using Eq. (7.53), and (7.54) or (7.55)). The post combustion chamber insulation mass is an overestimate, since the nozzle takes up a good deal of that space. This also assumes that the combustion chamber cannot survive any length of time with flame on it and that the ablative insulator maintains a linear erosion rate. Notice that th_{ins} was not applied to the precombustion chamber. While the precombustion chamber may actually benefit from a small amount of insulation (e.g., surface coating of the metal or a thin layer of RTV), it does not require the same level of protection as the post combustion chamber. For this example, it is assumed that the overestimate in the post combustion chamber insulation mass will more than cover whatever is required at the fore end of the motor. The propulsion engineer will note what a huge percentage of the dry mass is made up of insulation. In future design iterations, this is worth a second look. Could a different insulator or high-temperature material improve the mass? Could a heat transfer analysis be completed to show that the bare case could survive some direct heat exposure? This design process is an optimization problem that requires iteration as discussed in Chapter 7.

That was a bit of an aside into second-order design, but one now has the necessary information to size the combustion chamber: the inner diameter (ID) of the motor. The ID is the final diameter of the fuel grain plus the thickness of the residual fuel, plus the insulation thickness (Eq. (10.16)).

$$ID = D_f + 2th_{F,residual} + 2th_{ins,cyl}$$
$$ID = 205 \text{ mm} + (2 \times 1.4 \text{ mm}) + (2 \times 1.9 \text{ mm})$$
$$ID = 0.212 \text{ m}$$
$$(10.16)$$

The combustion chamber volume can now be calculated, which allows for the mass to be found by using the empirical relationship given by Eq. (7.66). Assume a composite combustion chamber ($PV/W|_{cc} = 25,000$ m) with a safety factor of 1.25.

For the combustion chamber, the volume can be found using the ID calculated above, which accounts for the residuals and the insulation, see Eq. (10.17).

$$V_{cc} = \pi(ID/2)^2 L_F + 4/3\pi(ID/2)^3$$
$$V_{cc} = \pi(0.106 \text{ m})^2 \times 1.08 \text{ m} + 4/3\pi(0.106 \text{ m})^3$$
$$V_{cc} = 0.043 \text{ m}^3$$
$$(10.17)$$

The volume and burst pressure (MEOP times the safety factor of 1.25) can now be used to find the combustion chamber mass.

$$m_{cc} = \frac{P_c k_s V_{cc}}{g_0 PV/W|_{cc}}$$
$$m_{cc} = \frac{2.07 \times 10^6 \text{ Pa} \times 1.25 \times 0.043 \text{ m}^3}{9.81 \text{ m/s}^2 \times 25,000 \text{ m}}$$
$$m_{cc} = 0.454 \text{ kg} \quad (10.18)$$

The nozzle length and mass will complete the motor design. The length can be determined (Eq. (7.5)). Assume that the nozzle is not submerged, so this entire length will be added to the back of the combustion chamber. The throat area was calculated above in Eq. (10.6), which can be used to solve for the radius, $r_{th} = (A_{th}/\pi)^{0.5}$ and then the nozzle length can be found:

$$L_{n,conical} = \frac{r_{th}(\sqrt{R}-1) + r_{th,inlet}(sec(\theta_e)-1)}{tan(\theta_e)}$$

$$L_{n,bell} = 0.80\frac{0.0219 \text{ m}(\sqrt{40}-1)+1.5\times0.0219 \text{ m}(sec(15)-1)}{tan(15)}$$

$$L_{n,bell} = 0.352 \text{ m}$$
$$(10.19)$$

The nozzle mass can be determined using an empirical relationship, see Eq. (7.56), as below:

$$m_n = 2.56 \times 10^{-5}\left[\frac{\left(m_p C^*\right)^{1.2} R^{0.3}}{\left(\frac{P_c}{10^6}\right)^{0.8} t_b^{0.6}(tan\theta_e)^{0.4}}\right]^{0.917}$$

$$m_n = 2.56$$

$$\times 10^{-5}\left[\frac{(142.9 \times 1549)^{1.2} 40^{0.3}}{2.07^{0.8}70.76^{0.6}[tan(15)]^{0.4}}\right]^{0.917}$$

$$m_n = 1.029 \text{ kg}$$
$$(10.20)$$

The first-order combustion chamber design is complete, so move on to the oxidizer tank design. Assume the hold up plus mixture ratio uncertainties are 5% for the oxidizer tank. However, since the oxidizer is a liquid, one must account for the temperature dependence of the density. The oxidizer tank is sized for the maximum possible temperature (50 °C) and MEOP. MON3 has a non-negligible vapor pressure, which can be found from Eq. (10.21) from Ref. [370]. This is used to find the density at the max temperature and pressure, using Eq. (10.22) from Ref. [369]. Both these equations need temperature in °C and give pressure in Pa and density in kg/m^3.

$$P_{vapor,O} = 3.6845 \times 6894.76 \times e^{0.0223(9T/5+32)}$$

$$P_{vapor,O}|_{sizing} = 3.859 \times 10^5 \text{ Pa}$$
$$(10.21)$$

$$\rho_O = 1693 + 0.61408T - 4.9261 \times 10^{-3}T^2$$
$$- 7.3845 + 1.6 \times 10^{-6}(P_{O,tank} - P_{vapor,O})$$
$$\rho_O|_{sizing} = 1370 \text{ kg/m}^3$$
$$(10.22)$$

At this point, a minimum ullage (say 5%) needs to be assumed for the oxidizer tank at launch. Now the total volume of the oxidizer tank can be calculated using the total mass of the oxidizer (including residual), $m_{O,total}$.

$$V_{O,tank} = \frac{m_{O,total}}{\rho_O|_{sizing}}\frac{1}{1-ullage}$$

$$V_{O,tank} = \frac{119.41 \text{ kg}}{1370 \text{ kg/m}^3}\frac{1}{1-0.05} = 0.092 \text{ m}^3$$
$$(10.23)$$

The initial conditions did not define an oxidizer tank run pressure or MEOP. However, since the combustion chamber pressure was given, the propulsion engineer can assign their own value for the margin (due to losses in the line, potential fluctuations in the combustion chamber pressure, and some additional padding). Taking this into account, a design oxidizer tank pressure of 3.8 MPa (with a MEOP of 4.0 MPa) is assumed. Assume a metal, PMD tank for the oxidizer ($PV/W|_{tank} = 10,000$ m) and a k_s of 2.

$$m_{O,tank} = \frac{P_{O,b} \Psi_O}{g_0 PV/W|_{tank}}$$

$$m_{O,tank} = \frac{4.0 \times 10^6 \text{ Pa} \times 2 \times 0.092 \text{ m}^3}{9.81 \text{ m/s}^2 \times 10,000 \text{ m}} = 7.49 \text{ kg}$$

$$(10.24)$$

The pressure in Eq. (10.24) is the burst pressure, which is equal to the MEOP times a safety factor, $P_{O,b} = P_{O,MEOP} k_s$. Remember, if a bottom-up design approach is taken for the tank and combustion chamber (instead of the PV/W scaling equation), an additional 15–30% needs to be added to the structural mass for bosses, skirts, etc. Note that taking the difference in density due to temperature (i.e., sizing for the max temperature launch conditions) results in nearly a 10% increase in tank mass.

The results of this example are summarized in Table 10.6. The propulsion engineer will have caught that the mass of the pressurization system is also included, but it was not calculated here. The steps to find these masses are determined in the next section. Note, the "other" category is a simple a percentage of the dry mass. This is not meant to be highly accurate but is kept as a placeholder until higher fidelity estimates can be made.

At this point, the major components have been sized, and the volume of the system is known. Do not forget to add some additional length for components, lines, fittings, mechanical connections, etc. Remember that the rocket

TABLE 10.6 Orbit Insertion Propulsion System Masses.

	Mass [kg]
Usable Propellant	142.90
Fuel	29.16
Oxidizer	113.73
Residual Fuel	0.87
Residual Ox	5.69
Pressurant	1.34
Insulation	9.25
Combustion Chamber	0.45
Nozzle	1.09
Oxidizer Tank	7.49
Pressurant Tank	8.13
Other (e.g., Feed System, Secondary Structure)	10.03
Total	187.24

needs to be assembled: tools such as torque wrenches and orbital welders will need to fit.

10.5.1 MON safety

Are there any safety precautions that would change the design since MON is being used here?

As with any propellant, the materials being used must be compatible. However, since MON is toxic, leakage becomes a catastrophic failure. This means three mechanical inhibits are required for any lines wetted on the ground. See Fig. 10.1 for how this safety precaution forced that design to use fill and drain valves with caps (typical fill and drain valves have two seals, and a cap adds a third). This adds about 10–50 grams to each valve [287], [286]. Enhanced safety precautions necessary during assembly, integration and test operations will also drive schedule and cost.

10.6 Pressurization selection

This example trades the mass of stored gas and pumped pressurization systems for the previous example. The stored gas masses are included in the roll-up above, and the delta for the pumped system is described below.

Stored gas

There are three main conditions that need to be assessed for the pressurization system design: sizing, BOL and EOL. The oxidizer and pressurant tank conditions calculated at each of these conditions for the stored gas system are summarized in Tables 10.7 and 10.8, at the end of this section.

As with most rocket propulsion, the devil is in the details (or the assumptions in this case). Assume that the pressurant tank is designed for a MEOP of 15 MPa (a reasonable estimate though pressurization tanks can be significantly higher pressure than this). For this stored gas pressurization system, blowdown is assumed to be nearly adiabatic. As described in Section 7.12.1 in Chapter 7, this is a conservative assumption, even for a single burn. Additionally, one would expect heaters to be used on the tank for an in-space application. Therefore a small reduction in the ratio of specific heats for He is assumed, $\gamma_p = 1.55$ (instead of 1.6 for truly adiabatic). The pressure at the launch pad is typically lower than the operating pressure and, in this case, is assumed to be 3.33×10^5 Pa. Finally, assume that the oxidizer tanks are initially at 20 °C and the ullage cools to −5 °C over the burn time. The end-of-life temperature of −5 °C is chosen here simply because it is above the freezing temperature of MON3, so by design, the ullage should be kept above this temperature. Once the detailed design phase is reached, a thermodynamic model would be necessary to confirm the drop in temperature. The system will need to be designed such that the ullage temperature does not freeze the propellant at the end of the burn (i.e. that it has some margin above the freezing point of MON). If the model predicts that the temperature will drop below this level at the end of the burn, some preheating of either the oxidizer or the helium gas (above 20 °C) will be required (or alternatively, the volume of the pressurant

tank and mass of pressurant gas would need to grow).

The volume of the oxidizer tank was calculated in Eq. (10.23) in the example above. The conditions at the beginning of life ($T = 20$ °C) can now be determined. The vapor pressure and density are again found from Eqs. (10.21) and (10.22), respectively. Then the volume of the oxidizer, $V_O|_{BOL}$, can be found by dividing the total oxidizer mass (including residuals) by this new density. Since the oxidizer tank volume is known, the ullage volume can be found from Eq. (10.25).

$$V_{ullage}|_{BOL} = V_{O,tank} - V_O|_{BOL}$$
$$V_{ullage}|_{BOL} = 0.092 \text{ m}^3 - 0.083 \text{ m}^3 = 0.009 \text{ m}^3$$
$$(10.25)$$

The partial pressure of the helium in the ullage can be calculated by subtracting the vapor pressure of the MON at BOL from the oxidizer tank pressure. It is assumed that the tank is loaded to 0.33 MPa at 20 °C in preparation for launch. It is also assumed that no gas is lost, and blowdown essentially starts from the launch pressurization; therefore the density at launch is the same as the density at BOL. The density of the helium at this pressure and BOL temperature (20 °C) can be found using NIST [22]: $\rho_{press,ullage}|_{launch} = 0.357 \text{ kg/m}^3$. There is now enough information to find the amount of helium in the ullage at launch:

$$m_{press,ullage}|_{BOL} = \rho_{press,ullage}|_{launch}$$
$$\times V_{ullage}|_{BOL}$$
$$m_{press,ullage}|_{BOL} = 0.0032 \text{ kg} \quad (10.26)$$

The density of the helium in the ullage at the end of life is then determined using NIST [22] at the oxidizer tank run pressure and −5 °C, $\rho_{press,ullage}|_{EOL} = 6.686 \text{ kg/m}^3$. The mass of helium in the ullage at the EOL can now be calculated conservatively assuming all the oxidizer is expelled:

$$m_{press,ullage}|_{EOL} = \rho_{press,ullage}|_{EOL} \mathbb{V}_{O,tank}$$

$$m_{press,ullage}|_{EOL} = 0.614 \text{ kg} \tag{10.27}$$

The EOL density of helium is calculated using polytropic blowdown, Eq. (10.28). The density of the pressurant gas at launch is defined by the maximum temperature (50 C) and MEOP (15 MPa). This density is constant until the pressurant is used (since it is just the loaded mass of pressurant divided by the tank volume) and so the pressure at BOL temperature (20 C) can be determined from NIST [22], $P_{press}|_{BOL} = 13.6$ MPa.

The oxidizer tank design pressure is 3.8 MPa. The pressurant tank pressure must be higher than this at EOL to maintain the pressure throughout the mission. Assume 1.5 MPa pressure drop between the pressurant tank(s) and oxidizer tank(s). This makes the EOL pressure, $P_{press}|_{EOL} = 5.3$ MPa. In more detailed design, this pressure drop should be verified to be less than this using the feed line analysis approach presented in Chapter 8.

The density of helium in the pressurant tank at EOL can now be calculated via Eq. (10.28).

$$\rho_{press}|_{EOL}$$
$$= \rho_{press}|_{BOL}(P_{press}|_{EOL}/P_{press}|_{BOL})^{1/\gamma_p}$$
$$\rho_{press}|_{EOL} = 11.42 \text{ kg/m}^3 \tag{10.28}$$

The total volume of the pressurant tank can be found:

$$\mathbb{V}_{press} = \frac{m_{press,ullage}|_{EOL} - m_{press,ullage}|_{BOL}}{\rho_{press}|_{BOL} - \rho_{press}|_{EOL}}$$

$$\mathbb{V}_{press} = \frac{0.614 \text{ kg} - 0.0032 \text{ kg}}{20.99 \text{ kg/m}^3 - 11.42 \text{ kg/m}^3} = 0.064 \text{ m}^3 \tag{10.29}$$

Multiplying the volume of the pressurant by the density of the helium at launch gives the mass of helium required at launch. To find the total helium necessary, the mass of pressurant in the ullage of the oxidizer tank at the beginning of life, $m_{press,ullage}|_{BOL}$, must be added to this.

$$m_{press} = m_{press}|_{launch} + m_{press,ullage}|_{BOL} m_{press}$$
$$= 1.34 \text{ kg} \tag{10.30}$$

Now, using the PV/W equation, the mass of the pressurant tank can be found. Assume a COPV tank ($PV/W|_{press,tank} = 18,000$ m) and a k_s of 1.5.

$$m_{press,tank} = \frac{P_{press,b}\mathbb{V}_{press}}{g_0 PV/W|_{press,tank}}$$

$$m_{press,tank} = \frac{15.0 \times 10^6 \text{ Pa} \times 1.5 \times 0.064 \text{ m}^3}{9.81 \text{ m/s}^2 \times 18,000 \text{ m}}$$

$$= 8.126 \text{ kg} \tag{10.31}$$

There is enough information (pressure and density) to check the temperature of the helium at the end of life to make sure it is not unreasonably cold. Using NIST [22], the temperature is $-57.25\,^\circ$C, which is not unreasonable. The components in the pressurization feed line need to be capable of operating at this temperature, and the detailed thermodynamic analysis of the ullage needs to account for this (though the temperature of the helium flowing into the ullage will be slightly warmer due to the Joule-Thompson effect).

TABLE 10.7 Pressurant Design: Conditions in the Oxidizer Tank.

	Sizing	BOL	EOL
T [°C]	50	20	−5
P_O [MPa]	4	0.33	3.8
$P_{vapor,O}$ [MPa]	0.386	0.116	
$P_{press,ullage}$ [MPa]		0.218	
ρ_O [kg/m³]	1370	1443	
$\rho_{press,ullage}$ [kg/m³]		0.357	6.686
$m_{press,ullage}$ [kg]		0.0032	0.6137
Ullage [%]	5	9.81	100

TABLE 10.8 Pressurant Design: Conditions in the Pressurant Tank.

	Sizing	BOL	EOL
T [°C]	50	20	−57.25
ρ_{press} [kg/m^3]	20.993	20.993	11.4199
P_{press} [MPa]	15	13.618	5.3

Pump

Pumped systems use the oxidizer tank at constant pressure, so they typically still have a gas pressurization system (with a regulator or bang-bang pressure control system). The benefit of pumps for liquid systems comes from the ability to run the chamber at higher pressure whilst keeping the propellant tank pressure low. Hybrid rocket propulsion systems will tend not to see the same relative mass benefit as liquid systems. This is because the combustion chamber mass of hybrid systems is a larger percentage of the total system mass (compared to liquid systems) and combustion chamber mass will increase with increasing chamber pressure. Despite this, pumped oxidizer feed lines can still have mass benefits over a purely pressure-fed pressurization system as will be shown in this example. This example trades a design using a turbopump against the pressure-fed pressurization system already calculated in the previous section. Note that electric pumps will outperform turbopumps for small systems, but the power requirements for electric pumps will lead to a first-order mass impact and so must be estimated early in the design.

The hybrid rocket design of Section 10.5 is used here. The launch pressure in the oxidizer tank is kept the same at 0.33 MPa at 20 °C. For simplicity, the combustion chamber pressure is also unchanged (even though the pumped system design will likely optimize at higher combustion chamber pressure). Since the combustion chamber pressure is held constant, the required pressure at the outlet of the turbopump is simply the oxidizer tank pressure in the previous section, namely 3.8 MPa.

The pressurant tank is designed with the same MEOP (15 MPa) and maximum temperature (50 °C). The oxidizer tank pressure is a free variable to trade in this design. The use of the turbopump allows the oxidizer tank pressure to be reduced, but the amount of reduction is limited by a lower pressure limit to avoid cavitation in the pump and buckling of the tank during launch loads. A design oxidizer tank pressure of 0.76 MPa is chosen for this example. It is assumed that the oxidizer tank MEOP is again 0.2 MPa above the design pressure, i.e. the oxidizer tank MEOP is now 0.96 MPa. The minimum oxidizer tank ullage is kept the same at 5%. The pressurization system design for this pumped system follows the approach of the previous section; the only difference here is the reduced oxidizer tank pressure. The design approach is not repeated here, see the previous section for the equations and approach used. The oxidizer and pressurant tank conditions calculated via this approach for the pumped system are provided in Table 10.9 and 10.10.

TABLE 10.9 Pressurant design: conditions in the oxidizer tank for a pumped system.

	Sizing	BOL	EOL
T [°C]	50	20	−5
P_O [MPa]	0.96	0.33	0.76
$P_{vapor,O}$ [MPa]	0.386	0.116	
$P_{press,ullage}$ [MPa]		0.218	
ρ_O [kg/m^3]	1370	1443	
$\rho_{press,ullage}$ [kg/m^3]		0.357	1.358
$m_{press,ullage}$ [kg]		0.0032	0.124
Ullage [%]	5	9.81	100

TABLE 10.10 Pressurant design: conditions in the pressurant tank for a pumped system.

	Sizing	BOL	EOL
T [°C]	50	20	−111.3
ρ_{press} [kg/m^3]	20.993	20.993	6.589
P_{press} [MPa]	15	13.62	2.26
m_{press} [kg]	0.177	0.177	0.056

The mass of helium gas in the pressurant tank required to maintain an oxidizer tank pressure of 0.76 MPa is 0.177 kg. This mass is stored in a COPV tank with an internal volume of 8.44 $\times 10^{-3}$ m^3 and a burst pressure of 22.5 MPa (burst safety factor of 1.5). The mass of the COPV tank for the pumped design is estimated via Eq. (10.24) to be 1.075 kg. Note that the EOL helium temperature is now $-111.3\,°C$. This decrease in EOL temperature is driven by the larger pressurant gas blowdown pressure ratio with the pumped design (lower EOL pressure in the pressurant tank). The EOL temperature of $-111.3\,°C$ is very low and may require that the pressurant tank and oxidizer tank be preheated. If preheating is not sufficient to keep the tank ullage temperature at EOL above $-5\,°C$, then a larger pressurant tank would also be required.

The size of the oxidizer tank is unchanged from the pressure-fed system, except that its design burst pressure is now reduced to 1.92 MPa (burst safety factor of 2.0). The mass of the oxidizer tank for the pumped design is estimated via Eq. (10.24) to be 1.796 kg.

The turbopump can be sized using the approach described in Section 7.12.3 of Chapter 7. The oxidizer mass flow rate was calculated previously to be 1.61 kg/s. The nominal oxidizer temperature is 20 °C, so the oxidizer density is 1443 kg/m^3, and the oxidizer vapor pressure is 0.12 MPa.

The increase in pressure required to be generated across the pump dictates how many stages are required, see Eq. (10.32). The required pressure rise is simply the pump outlet minus the pump inlet pressure, $\Delta P_{pump} = 3.8 - 0.76 = 3.04$ MPa. Since this pressure is less than the limit per stage for a centrifugal pump, it is reasonable to use a single-stage pump, i.e., $n_{stages} = 1$.

$$n_{stages} \geqslant \frac{\Delta P_{pump}}{\Delta P_{stage,lim}} \simeq \frac{\Delta P_{pump}}{47 \times 10^6} \quad (10.32)$$

The head pressure rise is now calculated via Eq. (10.33).

$$H_p = \frac{\Delta P_{pump}}{g_0 \rho_O} = \frac{3.04 \times 10^6}{(9.81)(1443)} = 214.8 \text{ m} \quad (10.33)$$

The Net Positive Suction Head (NPSH) is calculated via Eq. (10.34) to be 45.5 m. This value is not unreasonable as an initial estimate (see, for example, experimentally determined critical NPSH values reported in Figure 11.7 of Ref. [295]) but should be revisited in more detailed design.

$$NPSH = \frac{P_{inlet} - P_{vapor}}{g_0 \rho_O}$$
$$= \frac{0.76 \times 10^6 - 0.12 \times 10^6}{(9.81)(1443)} = 45.5 \text{ m} \quad (10.34)$$

The rotation rate of the pump, N_{pump}, can now be calculated. It is the minimum of the values calculated via Eq. (10.35) and (10.36). A reasonable suction specific speed limit for MON3 is $V_{SS} = 70$ rad m$^{0.75}$/s$^{1.5}$ [1]. As discussed in Chapter 7, the stage specific speed, N_s, in Eq. (10.36) is chosen to increase pump efficiency without requiring a large inducer diameter. For initial design a stage specific speed of $N_s = 3.0$ rad.m$^{0.75}$/s$^{1.5}$ is used [1].

$$N_{pump} = \frac{V_{SS}(NPSH)^{0.75}}{\sqrt{\frac{\dot{m}_O}{\rho_O}}} = 3.67 \times 10^4 \text{ [rad/s]} \quad (10.35)$$

$$N_{pump} = \frac{N_s \left(\frac{H_p}{n_{stages}}\right)^{0.75}}{\sqrt{\frac{\dot{m}_O}{\rho_O}}} = 5.04 \times 10^3 \text{ [rad/s]} \quad (10.36)$$

For this example, the rotation rate is determined by the limit on the stage specific speed, Eq. (10.36), and so N_s remains 3.0 rad.m$^{0.75}$/s$^{1.5}$. The rotation rate calculated of $N_{pump} = 5.04 \times$

10^3 [rad/s] is within the range of existing turbo pump capability. The pump efficiency for this stage specific speed is taken from Fig. 7.13 as $\eta_{pump} = 83.5\%$.

To determine the pump mass, the pump power and pump torque are calculated via Eqs. (10.37) and (10.38), respectively.

$$\mathcal{P}_{pump} = \frac{\Delta P_{pump} \dot{m}_O}{\rho_O \eta_{pump}} = 40.6 \text{ W} \qquad (10.37)$$

$$\tau_{pump} = \frac{\mathcal{P}_{pump}}{N_{pump}} = 0.0081 \text{ Nm} \qquad (10.38)$$

This allows the turbopump mass to be calculated via Eq. (10.39) using $\alpha_{pump} = 1.5$ and $\beta_{pump} = 0.6$.

$$m_{pump} = \alpha_{pump} \tau_{pump}^{\beta_{pump}} = 0.083 \text{ kg} \qquad (10.39)$$

It is worth pausing at this point to note that the mass of the turbopump is extremely low and is likely an artifact of extending the pump scaling equations beyond their applicable range. For the moment, this shall be ignored for the sake of completing the trade but will be discussed further momentarily as part of the next steps in the design process. The oxidizer tank, COPV tank, GHe, and pump masses can now be compared between the stored gas example and this pumped design, see Table 10.11. From this table alone (ignoring component masses and power

TABLE 10.11 A comparison of major pressurization system masses for a stored gas versus pumped pressurization system. These masses are calculated using the first-order design scaling equations of Chapter 7, but the pumped system masses are likely too low due to manufacturing/miniaturization limitations.

Mass Input	Units	Stored Gas	Pumped
Oxidizer Tank	[kg]	7.485	1.796
COPV	[kg]	8.125	1.075
GHe	[kg]	1.342	0.180
Pump	[kg]	0.000	0.083
Total	[kg]	16.952	3.134

system mass), it appears that the pumped system could save on the order of 14 kg compared to the stored gas system. This example makes the pumped design appear favorable, but in order to move forward with it, an assessment would need to be made on the overall pump size and whether the $NPSH$ is truly sufficient for this pump design and oxidizer to avoid cavitation. The calculated turbopump mass is very low and should be compared to existing pumps as some of these scaling laws will not be accurate at a small scale. In particular, miniaturization of the pump cannot be assumed to perfectly scale down to low mass due to manufacturing limitations (minimum material thicknesses, etc.) The oxidizer tank mass should also be verified before closing this trade via a bottoms-up mass estimate as discussed in Chapter 7 as at these low pressures, the oxidizer tank mass will likely be dictated by minimum wall thickness, rather than the $P\Psi$ scaling adopted here. This example was included to show both how to estimate the mass of a pumped system and also to highlight the need to critically assess the results of the scaling equations presented in Chapter 7.

10.7 Large-scale motor

The goal of this section is to look at a large-scale hybrid rocket design, such as would be required for Earth launch. The same process from the example in Section 10.5, could be applied again here. However, a number of different references have looked at this exact application. Tables 10.12 and 10.13 are adapted from an old (about a decade before liquefying fuels like paraffin were discovered), but still valuable reference. The goal was to match the performance of the Advanced Solid Rocket Motor (ASRM), which was a development meant to replace the Space Shuttle Reusable Solid Rocket Motors (RSRMs). Two different scales of hybrid rockets were traded. Two of the larger hybrid booster

TABLE 10.12   Design parameters for one (of many) design trades for Hybrid Booster Motors from Ref. [371]. Parameters varied in the study included fuel, pressurization system, number of ports, TVC (liquid injection vs flex seal). RSRM refers to Reusable Solid Rocket Motor.

Parameter	RSRM [353]	Hybrid Booster Motor [371]	
Diameter [m]	3.1	4.57	2.44
Length [m]	38.4	50.5	37.4
Number of Ports [-]	1	34	18
Number of Boosters (Shuttle)	2	2	8
Solid Fuel/Propellant	TP-H1148, PBAN, 86% solids	HTPB (40%) /Escorez (60%)	HTPB (40%) /Escorez (60%)
Liquid Oxidizer (hybrids only)		LOx	LOx
Total Impulse [Ns]	1.32×10^9	1.44×10^9	3.60×10^8
Expended Mass/Initial Mass	0.11	0.39	0.39

TABLE 10.13   Mass break down from a design trade for Hybrid Booster Motors (HBMs). Adapted Table 3-7 from Ref. [371].

Masses [kg]	4.57 m diameter HBM	2.44 m diameter HBM
Hybrid Rocket Motor	569,503	141,229
Total Solid Fuel (including residual)	138,273	33,645
Fuel Residual	8297	2019
Total Oxidizer (including residual)	355,791	90,689
Oxidizer Residual	3679	938
Subsystems (recovery, separation, interstage structure, nose cone)	17,926	5573
LOx Tank (composite with metal liner and insulation)	2703	985
LOx Pump and Hardware	2767	1588
Pump Fuel	3558	907
Pump Fuel Tank	204	56
Helium Tank	113	32
Helium	40	10
Injector	635	612
Ignition System	227	68
Motor Case	31,811 (metal with three joints)	2766 (composite)
Case Insulation	5595	1794
Flex seal TVC Nozzle	9867	2594
Total Mass Inerts	485,606	122,276

motors would be needed for a Shuttle replacement (i.e. to replace two RSRMs), while eight of the smaller hybrid boosters would have been needed per Shuttle. This report traded propellants (including some fuels that included oxygen to increase regression rate), Thrust Vector Control (TVC) system, pressurization systems, number of ports, fuel grain length and more. The selected fuel was inert and contained a large percentage of Escorez, a polycyclopentadiene reinforcing agent. The average operating pressure was 5.17 MPa for the pump-fed options. Two of the leading candidates are presented below.

As described in Chapter 7, both of these candidates outperform the solid rocket alternative. The hybrid options resulted in an increased payload mass of on the order of 10,000 kg over the shuttle RSRMs. However, this mass benefit was never realized because the designs were never matured.

Testing

11.1 Introduction

This chapter provides an introduction to testing necessary in hybrid rocket propulsion development programs. It includes characterization testing, standard practices, safety (specifically focusing on oxidizer safety), and gives some design guidelines for a ground system. A discussion of data analysis is included. Finally, an example ground system is presented, and the key features are highlighted.

Beyond the performance testing, which is the focus of this chapter, the propulsion engineer will also need to do environmental testing of the motor. Environmental tests may include thermal cycling, shock, aging, outgassing, quasi-static acceleration, radiation, as well as vibration (random and sine) for workmanship.

11.2 Fuel characterization testing

Common hybrid fuels have become more well-known (and characterized) as the field has gained popularity. However, fuel standards have not been developed to facilitate the use of data collected by multiple researchers and streamline their transition to flight use. Therefore the propulsion engineer must undertake a number of tests themselves or be forced to rely

on published data for similar fuels while understanding there is a large amount of variability across manufacturers and countries.

Solid fuel characteristics of interest for the hybrid motor designer are the fuel chemical composition, density, decomposition temperature, glass transition temperature, mechanical strength properties, coefficient of thermal expansion, heat of formation, and heat of combustion. Candidate hybrid rocket fuels, along with standard properties of these fuels, were discussed in Chapter 5. It is important to note that the properties of these fuels can and will differ based on the supplier and may even suffer from batch-to-batch variability. The authors strongly recommend that hybrid motor test campaigns include propellant characterization tests. Even without such testing, it is worthwhile for engineers to record and report as much information as possible about the propellants utilized, including vendor names and part numbers. This is discussed more in Section 11.7.1.

It can be challenging to experimentally determine a complete set of the material properties of interest for a given fuel. Some possible test options that the propulsion engineer can consider running in order to characterize their fuel are discussed here. There is often more than one way to determine a given parameter. The following sections attempt to introduce some of the techniques available; however, this is not an exhaustive list.

11.2.1 Density

Propellant density is a key parameter at all stages of rocket design, since volume estimates and propellant ballistics rely on accurate predictions of density. Thankfully, standard experimental methods are well known, see, for example, ASTM D792-13 for evaluating the density of plastics through liquid displacement. The authors recommend that density measurements are recorded by laboratories familiar with compliance to such standards, e.g., to ensure that samples are first exposed to the correct preconditioning environment for accurate measurement.

If the fuel will be processed in-house, understanding the density transition from liquid to solid states is also important. For example, shrinkage of the fuel grain during the cooling or curing process can set up residual stresses in the fuel grain.

11.2.2 Heat of combustion and heat of formation

The heat of formation directly affects the predicted performance of the propellant, see Chapter 4. The heat of combustion can be used to determine the heat of formation. The former is commonly measured via bomb calorimetry for organic materials. A known mass of fuel is placed within a sealed container of constant volume, saturated with oxygen, and ignited. The change in temperature can then be used to determine the energy of the reaction, then the change in enthalpy, and finally, by knowing the complete reaction products, the enthalpy of formation of the substance. An example of the calculation used to calculate the heat of formation from the measured heat of combustion is provided in Section 4.6.3.

11.2.3 Decomposition temperature

ThermoGravimetric Analysis (TGA) is used to measure the mass of a fuel sample as it is heated in a controlled environment. Mass, temperature, and time are recorded to provide insight into the fuel decomposition process. TGA will determine the temperature range at which the fuel sample will lose volatiles and the decomposition temperature of the material. This information is particularly useful if the designer is attempting to model the combustion physics of the rocket motor. TGA can also be used to provide insight into the thermal stability and reaction mechanisms of the fuel.

11.2.4 Glass transition temperature

Dynamic Mechanical Analysis (DMA) is useful to determine the glass transition temperature of a material. DMA is a dynamic method of characterizing the viscoelastic properties of a polymer. A sinusoidal force (stress) is applied to a sample of the material, and the strain response of the material is then measured over a range of temperatures. This method allows the storage modulus and loss modulus to be determined for viscoelastic materials. The storage modulus will rapidly decrease, and the loss modulus will reach a maximum around the glass transition temperature. The method for determining glass transition temperature from DMA results is well described in standards such as ISO 6721-11 [372]. Differential Scanning Calorimetry, described in Section 11.2.6, can also be used for this purpose.

11.2.5 Mechanical properties

The mechanical properties of the fuel must be evaluated to ensure it can survive the loads to which it will be exposed during the mission. For example, paraffin is known to be a soft and brittle material that expands and contracts significantly with temperature, and therefore launch and thermal loads can be a concern. If the mechanical properties of the fuel are not known, tensile strength can be tested by pulling samples (typically a dog bone shape), and compressive strength can be evaluated by pressing on a cube or cylinder. Test standards for many specific hybrid fuels are not available; however, commonly used materials such as plastics do have a lot of literature available on sample geometry and machine settings (e.g., ASTM). Mechanical properties should be taken at several temperatures that surpass the temperature range the materials will experience in the motor in storage, transportation, and operation.

A summary of tensile test results for paraffin wax provided in Ref. [158] gives an ultimate tensile strength of about 0.9 MPa for neat paraffin at a nominal pull rate of approximately 5 mm/min. The effects of additives on tensile strength and elongation were also reviewed in this work.

11.2.6 Thermal/environmental properties

Specific heat capacity

The specific heat capacity is the amount of heat required to raise the temperature of a material a given amount per unit mass and a useful parameter in rocket propulsion. The specific heat capacity of solid fuels can be measured using Differential Scanning Calorimetry (DSC). A known amount of heat is applied to both the material in question and reference material. The difference in the amount of heat required to heat the two samples can be used to determine the heat capacity as a function of temperature. Additionally, melting temperature and glass transition can also be determined from this data.

Thermal conductivity

The transport of heat through a material, or thermal conductivity, can be measured in a number of ways. The most common include the guarded hot plate, hot wire, and laser flash diffusivity. In the guarded hot plate method, the sample is sandwiched between a heated and a cooled (or a less heated) plate until a steady state is achieved. The heat input into the system, temperatures, and thickness of the sample are used to calculate the conductivity. In the hot wire method, as the name implies, a wire is heated and inserted into the unknown material. The temperature of the wire is then monitored over time. A modified hot wire method can be applied to solids where the wire is connected to a backing and does not penetrate the material. The authors have used a Transient Hot Bridge instrument (hot wire) for paraffin wax in the past, see, for example, ASTM D 5930-01. Finally, in laser diffusivity, a laser delivers a pulse of heat into the material, and the temperature change of the material is measured.

Adding aluminum powder increases the conductivity of the fuel, adding to the importance of characterizing the exact fuel to be used. In the case of paraffin wax, non-aluminized fuel at ambient conditions is 0.325 W/mK [373] for $C_{43}H_{88}$ or about 0.36 W/mK [157] for SP7. With the addition of 20% aluminum by mass, the thermal conductivity increases to 0.403 W/mK [373] and about 0.49 W/mK [157], respectively.

11.3 Standard test practices

11.3.1 Test operator and operating procedures

Clear operating procedures are critical to successful testing. This ensures repeatable results with a record of how each test was run for reconstruction and it maintains safety. The procedures also outline responsibilities of each person working on the team. Final calls on testing and safety should lie with the test operator and these steps should be outlined in an easily accessible portion of the procedures.

Several drafts of operating procedures are typically required. Propulsion engineers typically "grease board" the procedures before a dry run. This process gets its name from the fact that a schematic is printed and laminated and high-lighted (and erased) like a grease board (more commonly called dry erase or whiteboards to-day). Engineers go through each step with the team, highlighting where fluids are active in the system to work out all the issues with the proce-dure. Changes can be made in real-time. A dry run will work out any remaining bugs, typically related to ordering steps for process optimiza-tion. Steps that are frequently skipped should be omitted, as it may cause other, more important parts to be overlooked and can impact system safety or test success.

Operating procedures should include safety information, the assumed initial state of the sys-tem and checks, instructions for preparing the system for test, locations to record pertinent in-formation that is tracked between tests (atmo-spheric conditions, pressure levels of gas bottles, selected options, and anything else not recorded by the computer), test procedures, standard shutown/decontamination steps, system saving procedures, and emergency procedures for all potential cases (e.g., power failure, natural dis-asters).

A test operator should be identified for ev-ery test. This person is responsible for the safety of the team and makes final calls under all cir-cumstances. The test operator should be inti-mately familiar with the standard operating pro-cedures, safety concerns associated with the sys-tem, and all emergency shut-down procedures. The test team should be informed of the identi-fied operator and take all instruction from them; this is of particular importance in the event of an anomaly. All personnel interacting with the test equipment should also be familiar with the operating procedures and have received the ap-propriate training to be aware of all potential hazards and mitigation options. A two-person rule should be used whenever testing is occur-ring; that is, a minimum of two people should be present during any hazardous operations (e.g., handling oxidizers).

11.3.2 Proof testing

Proof testing is a test of workmanship and material quality and demonstrates that the pres-sure vessel will operate as designed. The pres-sure used for this test is equal to the MEOP plus a safety factor [7]. Proof testing should occur on any new feed system or any time modifications are made to an existing feed system. The system proof test should be conducted prior to person-nel being in proximity to pressurized hardware. Relief valves and pressure gauges typically need to be removed for proof testing. The propulsion engineer should cap the feed lines at the connec-tion point to such components when necessary. Feed systems will also often have different proof pressures for different sections of line. The feed system can be divided into sections with connec-tion points between segments capped to ensure that only the section of interest is proofed during a given test.

It is common practice to allow personnel around a system when the feed system is loaded with a non-hazardous gas or liquid and the pres-sure is below around 689 kPa [100 psia], even prior to conducting the proof test. Thus during proof testing, the system will often be pressur-ized to a low pressure to be checked for gross leakage. If leakage is detected, then the system should be depressurized prior to attempting to tighten any loose fittings. After the feed line is shown to not have any gross leakage at low pres-sure, then personnel should be cleared from the vicinity of the system, and it should be taken to the proof pressure. Typically the feed system is held at the proof pressure for at least 5 minutes [7].

Whenever practicable, proof testing is con-ducted as a hydro test with water to reduce the stored energy in the system. If a hydro test is to be conducted, then the system proof pres-sure is often made to be 1.5 times the Maxi-mum Expected Operating Pressure (MEOP), this is generally in line with the requirements for metallic pressurized components per Ref. [7].

See Table 9.7 of Chapter 9 for details. However, the ASME standard for pressure vessel testing (which is often applied to ground systems), Ref. [374], is slightly more conservative. It requires the proof pressure to be 1.5 times the system Maximum Allowable Working Pressure (MAWP). As discussed in Chapter 9, the MAWP is the lowest allowable working pressure of any component or line in the system whereas MEOP is the planned maximum operating pressure of the system. By design, the MEOP will always be less than or equal to the MAWP.

If the proof test is going to be conducted with a gas, rather than a liquid, then the required proof pressure is reduced to minimize the stored energy in the system. In the case of the ASME standard, the requirement reduces to 1.1 times the system MAWP [374].

11.3.3 Leak testing

System leak testing is conducted after proof testing as it generally requires personnel to be in the vicinity of the hardware and the feed lines to be returned to their final operating conditions. The approach for leak testing is typically less defined than for proof testing. Ref. [7] requires that leak checks should be conducted for a minimum of 30 minutes; however, for ground systems, this is often reduced to on the order of 10 minutes. The propulsion engineer can choose the best approach to leak test a feed system based on their specific test requirements. There are two different types of system leakage: internal leakage and external leakage. Internal leakage is the unexpected transfer of fluid within the feed line, such as across a closed isolation valve. External leakage is the leakage of fluid from within the feed line to the external environment, such as leakage through a fitting between two sections of feed line.

External leak testing

There are multiple methods to verify that there is no external leakage in a feed system.

The most rigorous (and expensive) of these is to load the system with helium and then use a helium mass spectrometer to measure the external leak rate around the feed system. This approach with the mass spectrometer gives a quantitative measurement of leakage. A simpler approach is to see if the system holds pressure across a reasonable span of time. With this method, the system is filled with pressure (typically up to the system MEOP), and the source of supply pressure is closed off. Pressure gauges or transducers are used to measure the pressure decay rate over time. The rate of leakage, if there is any measurable amount, can then be calculated based on the loss of pressure and the internal volume of the system. Perhaps the most widely used form of external leak test is a "Snoop" test. Snoop tests apply a soapy water mixture to the outside of system joints. If bubbles form around the joint, then it indicates a leak is present. There are name-brand products available for using this method on gas or liquid feed lines. Care should be taken with oxidizer feed lines to ensure that using any of these products on a loose fitting will not potentially contaminate the internal feed line.

Internal leak testing

Two common methods are often used to measure internal leakage in a feed system. As with external leakage, the most rigorous approach is to use a mass spectrometer and helium gas to quantitatively measure internal leakage across all components of interest in the feed line. To use this approach, the system must be pressurized using helium gas. The leak rate for each component is measured across one component at a time by connecting the mass spectrometer to the downstream end of the component/system. For feed lines with components in series, this method requires that such components can be held open or that test ports exist in the system in order to take the downstream measurement with the mass spectrometer. A pressure hold can also be used to indicate internal leakage. In this case, it is often helpful to first ensure that there is no external leakage (e.g., by confirming that the system holds pressure with the internal valves open and the outlet of the system capped) prior to checking internal leakage.

11.4 Safety

Combustion research and development has inherent risks. Safety precautions must be taken seriously, not only to maximize mission (or research) success but to ensure human safety. Propulsion engineers and technicians have lost limbs, reduced their lifespans or lost their lives because of unsafe testing practices. The single most important safety measure is a remote operation of all tests that involve anything remotely hazardous (including cold flows or proof tests). Major risks include highly oxidizing gases/liquids, pressure vessel/feed system safety, hot combustion gases, and, occasionally, toxic chemicals. Oxidizer safety will have its own discussion in Section 11.5. Feed system safety is discussed in Section 11.6.2. The discussion in the following sections is not exhaustive. The propulsion engineer will have to look to various sources of safety information for the propellants, the test facility or launch range, and their organization.

Propulsion engineers with intimate knowledge of the propellants, ignition system, hazards and risks are not the only people working around propulsion systems. They are installed into spacecraft or payloads are installed into them (e.g., launch vehicles). Also, simple awareness of the risks is not enough to prevent accidents. It is critical to make the system as safe as possible. This is done in many ways.

Range safety requirements necessitate multiple independent inhibits (mechanical and electrical) to ensure the system cannot be ignited prematurely [366]. The number of inhibits is dependent on the level of hazard the system imposes. Ref. [366] defines this in the context of

loss of property and life. Most propulsion systems fall into the catastrophic hazard rating, which requires three independent inhibits. This can be completed through a mechanical safe and arm device (a physical key or electrical signal is needed to complete the circuit that allows for ignition), software inhibits (a single computer can only provide a single inhibit, or it would not be independent), relays connecting the power supply to the igniter, etc. Also, the igniters are typically the last things to be installed and enabled.

Solid rocket propellant, pyrotechnics and some oxidizers (e.g., GOx) are ElectroStatic Discharge (ESD sensitive). Propulsion engineers and technicians should be grounded (and not have a cell phone in their pockets) when working with propellants. One last note on igniters: it is standard to check the integrity/resistance of a pyro igniter before installing it. However, a standard multimeter imparts enough current to potentially set off the charge. A Kelvin Bridge should be used instead.

Many ranges will require a Flight Termination System (FTS) if the rocket can be actively thrust vector controlled. The FTS is capable of destroying the rocket if it veers off course. See Section 9.12.2 for more information.

11.5 Oxidizer safety

This section focuses on safety using the most common oxidizers. This is not intended to be a complete guide but is merely meant to highlight some of the relevant safety considerations when working with strong oxidizers.

In general, all wetted surfaces must be constructed from materials that are compatible with the oxidizer being used and be maintained as clean as possible. Oxidizer compatibility is different for each oxidizer and can sometimes even change from batch to batch of material tested. Residue from cleaners should be compatible or removed. However, no matter how well as system is cleaned, contaminants can still be introduced during the assembly process. Therefore it is recommended that precautions be taken to minimize ignition in the presence of particles/contamination. Propulsion engineers are encouraged to learn from a particularly good piece of advice the authors received in graduate school. Prof. Karabeyoglu used to quote Trevor Kletz, saying the "ignition source is always free", meaning that one should always assume an ignition source can be present, even while going to great lengths to minimize such possibilities.

Appendix B contains tables to provide the propulsion engineer with an initial guide to material compatibility to the common hybrid rocket oxidizers: oxygen, nitrous oxide, hydrogen peroxide, and MON3. All materials being considered for use with a given oxidizer should be thoroughly reviewed using additional reputable literature sources prior to use. Safety Data Sheets (SDS's) should always be obtained from the selected vendor for any chemical being used in propulsion testing.

11.5.1 Oxygen safety

Most materials are flammable in pure oxygen and the flammability increases with pressure. Even "fire resistant" materials, such as Nomex, become flammable in enriched oxygen environments, making material compatibility critical in oxygen service applications. The autoignition temperature of most materials decreases with increasing pressure up to about 4 MPa, the remains are about constant [375]. Fiberglass and asbestos are the only known materials that are not flammable in pure oxygen [301]. While it is not always possible to select a material that is completely compatible with pure oxygen under any operating conditions, preferable materials and operating constraints make oxygen use feasible. For example, aluminum can be used with oxygen in low-pressure applications (typically below 13.8 MPa or 2000 psi) even though it is a good fuel.

Ref. [301] provides an excellent list of oxygen system design guidelines; highlights include: minimizing pressure in the system, ensuring adequate pressure relief, and avoiding elevated temperatures whenever possible. Do not use oxygen for secondary objectives (e.g., do not use oxygen to power pneumatic systems because it is already available). While it is rather difficult to achieve oxygen toxicity, ventilation is required to ensure materials which are not compatible are not found in oxygen-rich environments.

Ignition mechanisms should be both considered and mitigated. The two most common ignition mechanisms will be discussed here: particle impact and compression heating. See Ref. [301] for a more complete discussion of other mechanisms, including chemical reactions, electrical arcing, electrostatic discharge, external heat, friction, fresh metal exposure (oxidation), mechanical impact, thermal runaway, and resonance chambers.

The first ignition mechanism describes a particle impacting a material with sufficient velocity to ignite the particle or the material. Particles (typically contaminates) can be entrained in gases with velocities of greater than about 30 m/s [301]. The particle impact typically needs to occur at an angle between 45–90 degrees to be problematic. The occurrence of ignition can be minimized by maintaining cleanliness and designing for gradual turns.

Compression heating occurs when a gas is rapidly pressurized (on the order of one second for small diameters and several seconds for larger diameters [301]). An example of rapid pressurization is a fast actuation valve opening to allow gaseous oxygen into tubing or volume with a dead end or tight bend radius. This mechanism is typically not a concern for bulk metals, but commonly ignites nonmetallic materials (e.g., soft goods, lubricants, residual cleaning agents) by raising the gas temperature above the auto-ignition temperature for the material. Valve seats are susceptible to this mechanism as they often have nonmetallic seals at the point where the compression occurs (e.g., a dead end).

Gaseous oxygen (GOx) is slightly more dense than air, and liquid oxygen (LOx) is slightly more dense than water, so both tend to accumulate at low points. This should be taken into account when designing feed systems (e.g., placing vents and traps at these points.) Additionally, many gases and liquids are miscible or soluble in oxygen, so it can easily be contaminated [301].

Cleaning is a fundamental necessity for components used in oxygen service and will be described later in this chapter (see Section 11.6.2). It is important to be aware that having a component or section of line cleaned for oxygen service is not the same as ensuring the line is compatible with oxygen. Many companies will sell products that are cleaned to a particular specification (e.g., SC-11 [376]), even when the product itself is not suitable for use in oxygen systems. This is because SC-11 is a cleanliness standard, not a cleanliness and material compatibility standard.

Cryogenic liquids have a number of hazards associated with them, the most obvious of which is their low temperature. Direct contact with LOx can cause damage to tissue or cause frostbite. Thermal contraction is also evident in LOx tanks and piping. Eq. (7.69) in Chapter 7 illustrates this issue. It must be accounted for in the restraints/structural supports. Also, transition joints between dissimilar metals should be avoided because constant thermal cycling leads to failures [301]. A minimum flow rate for thermal conditioning (cooling the lines to prevent thermal shock, bowing, etc.) is recommended. For a nominal pipe diameter of about 2.5 cm (1 in) and high purity LOx, this minimum flow rate is 0.1 kg/s [301]. Note that the maximum flow rate of LOx for ignition hazards is substantially lower than for GOx.

Over-pressurization is a risk, since LOx is constantly boiling off into GOx. Any section of line that can be isolated and has access to a heat source (e.g., the sun) should be treated as a

pressure vessel. Cavitation should be avoided as gaseous oxygen in the line can lead to the safety hazards described above.

11.5.2 Nitrous oxide safety

Nitrous oxide is widely used as a hybrid rocket oxidizer and in the medical/dental fields as a "laughing gas"; however, even with its wide use, precautions need to be taken to ensure it is utilized safely. Nitrous oxide targets the respiratory system, the central nervous system and the reproductive system [377]. Long-term exposure to nitrous oxide can lead to infertility [378]. Breathing in nitrous oxide gas can cause dizziness, impaired judgment, and in large concentrations, can lead to unconsciousness and death from suffocation [378]. Adequate ventilation and the use of oxygen sensors are recommended when working with nitrous oxide systems to prevent personnel from being exposed to dangerous concentrations. Nitrous oxide systems should be operated remotely whenever possible, regardless of whether a "cold-flow" or a hybrid motor "hot-fire" is planned.

In addition to its use as an oxidizer in hybrid rockets, N_2O is also a monopropellant [179]. At temperatures above 650 °C (1200 F) and at pressures as low as 1 atmosphere, nitrous oxide can begin to rapidly exothermically decompose without exposure to any contamination [379]. At ambient temperatures but at elevated pressures, above approximately 1.38 MPa [200 psia], nitrous oxide decomposition can be self-sustaining with the presence of an ignition source [379]. Copper and platinum will both catalyze nitrous oxide [181]. Nitrous oxide decomposition occurs six orders of magnitude slower than hydrogen peroxide at the same temperature and pressure [181]. It should be noted that ignition of nitrous oxide occurs in the gas phase; all attempts to ignite pure liquid nitrous oxide reported in the open literature have failed [181]. However, the increased pressure decreases the energy required to ignite liquid N_2O [380]. Deflagration of nitrous oxide vapor is the primary safety concern associated with these systems. Ref. [181] provides a detailed description of the nitrous decomposition physics for those interested in learning more. Any contamination of the nitrous oxide in the form of a small amount of catalyst or fuel will greatly reduce the activation energy needed to ignite the decomposition reaction. Nitrous oxide is a very effective solvent of hydrocarbons [181] and can form an explosive mixture when contaminated with hydrocarbon fuels, which can result from nitrous exposure to some common lubricants and oils [379]. Gaseous N_2O, in particular, is very susceptible to reaction with any sort of contamination. Thus it is extremely important to carefully clean all surfaces and review all materials that will potentially be wetted with nitrous oxide. The authors advise checking multiple sources for material compatibility information and re-examining all products that are cleaned at external vendors.

The addition of diluents, such as nitrogen, helium or oxygen, to nitrous oxide, can greatly increase the ignition energy required to initiate decomposition [181]. Such diluents can either be added as a pressurant to regulate the oxidizer mass flow rate or can be used to initially increase the nitrous tank pressure, known as "supercharging," before allowing the tank to operate in blow-down mode. The Peregrine hybrid rocket, developed at NASA Ames Research Center, used helium (then later oxygen) to supercharge the nitrous tank prior to ignition [177], see Fig. 11.3.

Similar to oxygen systems, nitrous oxide ignition sources include compressive heating. Ref. [379] recommends always limiting pressurization rates to less than 0.14 MPa/s (20 psi/s), and potentially further reducing this rate under some circumstances, see Ref. [381]. Other potential ignition mechanisms for nitrous oxide include electrostatic discharge, overheating pumps, sparks, and other local heat sources.

Nitrous oxide has low electrical conductivity, which can theoretically build up sufficient static charge when flowing to initiate a decomposition reaction [379]. To reduce the likelihood of such an occurrence, nitrous oxide feed lines and tanks should therefore be constructed of conductive materials that are properly grounded [379].

Pumps running dry and then overheating have been known to initiate nitrous decomposition, see, for example, Ref. [382]. It is easy to cavitate nitrous across a pump since it is typically stored at its saturation point. Pumps are broadly used in nitrous oxide systems, they do not need to be avoided; however sufficient care should be taken to ensure that cavitation does not occur across the pump. For example, SpaceDev prechilled their pumps to avoid boiling/flashing of N_2O in the pump [379].

Nitrous oxide is typically operated along its saturation curve for blow-down systems, or slightly above it for supercharged systems where an additional gas is used to add pressure. For both of these operational scenarios, two-phase flow in the oxidizer feed lines must be considered and modeled [181]. Careful design of the injector element can ensure that the nitrous oxide fully cavitates across the injector, creates choked flow, and therefore isolates the feed line from motor feed-couple instabilities whilst vaporizing the oxidizer as it is injected into the motor. A typical cavitating venturi will allow the oxidizer to recover to a liquid, but if the downstream pressure is low enough, it could remain vaporized. See [47] for further details.

The vapor pressure and saturation conditions of nitrous oxide vary with temperature. Thus there is often a need to heat or cool nitrous to the desired test state. When heating nitrous oxide extreme care should be taken to avoid localized hot-spots. Water baths or heat exchanges with uniform wall temperature are recommended over any other type of localized heating system.

Whenever possible in ground testing, nitrous oxide tanks should be oriented in a vertical position to increase the likelihood that the nitrous oxide vapor in the tank is separated from the combustion chamber by a layer of liquid nitrous

[181]. This will greatly reduce the likelihood of a tank explosion, even in the event of a downstream nitrous decomposition reaction.

Nitrous oxide is often exposed to carbon steel during production and transportation. Therefore they should always be considered contaminated with rust (iron oxide), which is a catalyst for nitrous oxide decomposition [181]. Stainless steel filters at the inlet to nitrous systems can be used to mitigate the extent of the contamination but will not remove all of it [379].

Nitrous oxide should never be vented through the combustion chamber, as it can saturate some fuel grains, creating a detonatable solid. Test operators should also be aware that nitrous oxide is a very effective greenhouse gas. It is 300 times more effective than carbon dioxide at trapping heat [181]. While the amounts of nitrous oxide released to the atmosphere from rocket testing are negligible compared to industrial applications, it is still worth being cognizant of ramifications associated with venting this oxidizer to the atmosphere.

11.5.3 Hydrogen peroxide safety

Hydrogen peroxide is actually produced through metabolic reactions in the human body [187], and in low concentrations, it has common household uses (e.g., as a medical antiseptic). It can be toxic if swallowed and corrosive to the eyes and skin, but the fumes are not toxic. The vapors cause irritation and inflammation of the respiratory tract. It can cause temporary bleaching of the skin, chemical burns, and blindness. Ref. [184] discusses hydrogen peroxide safety in detail, see section 6.4 or Safety Data Sheets for recommended personal protective equipment, including gloves, protective clothing, eye protection, and a full face shield made of a compatible material such as polycarbonate. Eyewash stations and showers are also recommended. One should not wear cotton, wool, nylon or leather when working with hydrogen peroxide.

There are several unique safety considerations for hydrogen peroxide. Contamination in the system reduces the activation energy, facilitates decomposition and/or can cause an explosive mixture. Pure hydrogen peroxide is not flammable; however, it reacts/combusts with other materials relatively easily. For example, water, which is commonly added to hydrogen peroxide in other significant concentrations, can be a destabilizing contaminant. Detonations have been observed to propagate in liquid hydrogen peroxide at concentrations above 90–92%, but are hard to initiate below that concentration. Pure or stabilized hydrogen peroxide (>95%) is not impact-sensitive and is difficult to detonate in the liquid form. However, explosive decomposition of gaseous hydrogen peroxide is possible at concentrations above about 40% by mass [186]. This risk is greatest at low pressure and decreases with increasing pressure and typically also requires a heat source (as low as 120–140 °C) or a spark (the propulsion engineer is reminded that this ignition source should be assumed to be had for free).

The inherently low vapor pressure of hydrogen peroxide helps reduce this risk at ambient conditions. Rags or clothing that have been exposed to liquid hydrogen peroxide should be completely soaked in water prior to drying to minimize flammability. Hydrogen peroxide is not particularly susceptible to adiabatic compression.

Material compatibility is a bit challenging for hydrogen peroxide. Most of the testing was done in the 1960s. Since that time, hydrogen peroxide production has improved substantially. Several authors suggest that materials that may not have been considered satisfactory for general use with hydrogen peroxide in the past, e.g., Al 6061, may be compatible with the hydrogen peroxide produced now [188]. Counterintuitively, lower concentrations of peroxide are actually the stressing cases for material compatibility. Much of the compatibility testing for propellant applications was conducted with 90%

or higher concentrations of hydrogen peroxide. Therefore marginally compatible materials should be considered especially suspect if lower concentrations of hydrogen peroxide are used. This is because the water and H_2O_2 weakly bond together, causing the mixture to behave differently and actually allowing the H_2O_2 to decompose more easily. Specific aluminum alloys are the metal of choice for storage. Passivation, typically with nitric acid, is recommended for metals in contact with peroxide. Contamination in hydrogen peroxide can often be detected as an increased temperature in the containment vessel, [186] enabling emergency measures to be implemented.

Hydrogen peroxide decomposes more rapidly than other storable propellants, such as hydrazine [188], making it of interest as a monopropellant. Its decomposition produces heat and is therefore self-accelerating [189]. The decomposition rate is driven by Arrhenius' Law, increasing by a factor of 2.3 for every $10\,°C$ increase in temperature [24].

11.5.4 Mixed oxides of nitrogen safety

While valued for its storability and low-temperature performance, there are significant safety concerns when working with Mixed Oxides of Nitrogen. MON is toxic and reacts with water to produce nitric acid (HNO_3) and nitrous acid (HONO), both of which can destroy tissue [383]. MON will rapidly vaporize under standard atmospheric conditions. MON vapor (actually gaseous NO_2) is reddish-brown in color and heavier than air, so it will sink when vented. In the case of a spill or venting event, the vapor is clearly visible as a Big Red Cloud or a "FBRC" as it is affectionately termed. Because it is heavier than air, gravity drains should be employed in ground systems.

Protective equipment is required for handling MON and safety courses are available for dealing with such a high-hazard material. Several breaths of high concentration NO_2 vapors are enough to cause severe toxicity; thankfully, the color (red-brown) and smell of the vapor alert one to potential exposures. A liquid spill onto a person or equipment can also cause secondary contamination as the liquid off-gases.

MON is also incompatible with many materials, and therefore spills can cause substantial damage to GSE or flight hardware that is not designed for contact with it. MON leaches iron from wetted materials (e.g., steel) that can lead to substantial flow decay problems (clogging). Leaching is worsened in high temperature ($> 93\,°C$) and high H_2O cases. It is compatible with many types of aluminum, especially if they are anodized. Chromium plating was tested and provides adequate protection when using stainless steel. Teflon was found to be the best non-metal for use with NTO. "Probable compatibility" is listed for graphite, krytox and fluorolube lubricants at just above room temperature. Titanium is usually considered compatible, however, it has been reported to be sensitive to shock and is subject to stress corrosion [384]. It should be noted that most compatibility testing was done on NTO through MON-3. However, as higher NO concentration MONs (e.g., MON-25) are typically less corrosive, the lower concentration tests are considered valid [384].

11.5.5 Nitric acid safety

Nitric acid is another oxidizer requiring heightened safety considerations. Rocket propellant pioneer John D. Clark paints an exceptionally vivid picture of the potential hazard to human safety in his book, saying "RFNA attacks the skin and flesh with the avidity of a school of piranhas" [189]. Any propulsion engineer choosing to work with this oxidizer should take that to heart.

Nitric acid reacts with many organic materials and can spontaneously ignite. The reaction can be explosive in contact with certain chemicals (e.g., hydrazine) or if the ignition was delayed (e.g., a hard start) [23]. Special consideration needs to be given to using nitric acid for applications requiring recovery from the ocean. It reacts with seawater, creating toxic nitrogen oxides (see previous section) [23]. Nitric acids are also very corrosive.

Hydrofluoric acid (HF) is usually added to the fuming nitric acids to help minimize corrosion and increase storability. While the HF in IRFNA dramatically reduces corrosion with most common materials, it should be noted that it makes some already incompatible materials, even less compatible (e.g., tantalum and titanium) [189]. Aluminum is most commonly used with IRFNA, with the occasional use of some stainless steels. For RFNA, Hastelloy, chromium, gold and platinum are also compatible [384]. Compatibility tables, including temperature limits, corrosion, and swelling (soft goods), can be found in [384].

11.6 Designing a ground feed system

11.6.1 Design

The main goal of any hybrid rocket ground feed system is to deliver oxidizer from a storage tank to the combustion chamber. This should be done in the simplest manner possible, whilst also meeting the requirements needed to achieve the test goals. Complexity should only be added when needed for system capability or safety. For example, an early decision must be to evaluate whether the tank dynamics need to be replicated in test, necessitating the use of an oxidizer run tank to mimic the flight tank volume, or whether the test can be run directly from larger oxidizer tanks, such as those provided directly by the oxidizer supplier. The selection of the oxidizer itself is key, as the choice of oxidizer will dictate the requirements of the components in terms of material compatibility and the safety measures that need to be considered to work with that oxidizer.

System safety should be the first priority in feed system design. The propulsion engi-

neer must evaluate whether relief valves, check valves, vent lines, cold traps, purge lines, or remote activation capabilities are necessary for the specific design. This will be discussed in detail in the next section.

The feed line schematic should be drafted with safety in mind and then reviewed and revised to ensure it can be safely filled, operated, shut down, and decontaminated (if a toxic oxidizer is used) using a "grease board" or similar approach. Several revisions may be required to mitigate potential failure modes and ensure smooth end-to-end operation.

Test environments, e.g., the target temperature and pressure, also drive feed system design. If thermal conditioning is necessary for test success, then the expected propellant temperature range will likely become a key factor in component selection. Note that the design propellant temperature range should be used when calculating the needed propellant flow rates. Test pressure can be similarly driving in component selection as it is often desirable to test a motor at pressures representative of altitude conditions. In this case, not only do the components need to be able to survive low-pressure exposure, but feed-throughs into a vacuum chamber must be designed.

Once operating conditions and safety have been evaluated, the feed system design can proceed with the engineer considering what minimum number of valves/components are required to fill the feed lines, control the oxidizer flow, and vent/safe the system at the end of the test. The actual components can only be selected when an initial system sizing has been conducted.

After all the components are selected, it is recommended that the propulsion engineer reevaluate the expected oxidizer flow rates and pressure drops in the feed system to ensure that the system can deliver the expected range of flow rates. As a reminder, the required oxidizer flow rate to achieve a given OF ratio/thrust level in the hybrid motor can be evaluated using the design equations of Chapter 7. The feed line sizing and component selection can then be conducted using this flow rate as the design requirement. A general approach for sizing tubing for incompressible flow is given in Section 7.13.2. Tube sizing for ground systems (outer diameter and thickness) is selected in the same way as for a flight system, see Section 9.4.5 for hardware options. The feed line analysis approach is discussed in Chapter 8. Iteration on component selection and feed line sizing is not uncommon in this phase of the feed system design. Once the feed system flow rates and pressure drops are verified by calculation, then the instrumentation can be selected. Instrumentation should be selected to ensure that it will have adequate accuracy and resolution at the expected conditions. A discussion of standard hybrid rocket instrumentation is provided in Section 11.6.5.

11.6.2 Feed system safety

The propellants, operating conditions and facility capabilities should be taken into account when designing a new test set up. Specific safety precautions pertaining to common oxidizers were described previously. In this section, the propulsion engineer is introduced to hardware and system safety considerations.

A safe system is one in which the lines are at ambient conditions, and power cannot be applied to the ignition mechanism. In-ground test applications, the propulsion engineer should design the system to "fail safe". This means that if power is removed, the oxidizer will no longer flow, preventing unwanted combustion. However, the lines then need to be vented to render the system completely safe.

Remote feed system operation is strongly encouraged whenever possible. Placing distance between the operator and the hazardous activity allows the operator to learn from their mistakes and more importantly, to walk away from them. A high degree of automation is often required to enable remote operation. Automated opera-

tion has additional benefits of predictable and repeatable test conditions, which are necessary for most test campaigns. Automation is therefore encouraged but, where possible, should be coupled with a manual shut-off (typically a big red button that removes valve power from normally closed valves), the benefit of which cannot be overlooked when a computer has crashed. Both of the authors have experienced this situation firsthand. For systems with a gas purge, it is advantageous to have the power for purge systems and instrumentation separated from the power for the oxidizer lines. Then, in the event of a failure, the purge can be used to quell any unintended flames.

Relief valves are used in any situation where an overpressure could occur that could damage the downstream components. For example, the downstream components in a ground feed line typically have design pressure ratings less than the source bottle pressure, so using a suitably sized relief valve below the regulator could save the system if the regulator were to fail open. In gas systems, a standard Department of Transportation (DOT) cylinder is often used as the gas source, and a regulator is employed to achieve the desired pressure. If the regulator fails, relief valves can be used to make sure none of the other components are damaged. It is best to size the relief valve (or use multiple) to enable the system to vent the entire bottle if the regulator were to fail open. In this situation, the location to which the gas is vented is important to avoid an oxygen deficiency (or oxidizer enrichment that could enable autoignition) in the test area. Ensure that test rooms have adequate ventilation for the installed system and that any oxygen or gas monitoring devices are located in appropriate locations.

Venting is necessary to remove the pressure from the feed lines between tests. On some occasions, it is possible to vent through the main flow path (e.g., through the motor). However, this is not recommended for some oxidizers as it can create an explosive mixture for some propellant combinations. Be aware of the location of the feed line vents to ensure that personnel will not be put at risk (e.g., by being hit with high-pressure gas or having it block an exit, etc.) during venting operations.

A purge system uses an inert gas to flush oxidizer, fuel vapors, or combustion by-products from the system. The designer should also evaluate whether a purge gas is desired to rapidly extinguish combustion at the end of a test. Purge systems are often used in ground tests for added safety, but its use will make the motor shutdown transients different from flight.

Check valves are often utilized to maintain flow in a single direction. While they are typically helpful at guiding flow in the correct direction, it is possible to get some backflow across them. Check valves should not be considered an isolating device. The pressure required to seat a check valve varies widely between types of check valves. Chatter in a check valve occurs when there is not enough flow through the valve to keep it open, so it cycles between the open and closed position, potentially generating particulate and likely causing unsteady flow.

When reviewing the design of liquid feed lines, care should be taken to ensure areas of potential liquid lock are avoided. Liquid lock can occur whenever you have trapped liquid propellant without a sufficient venting mechanism. Vents can be in the form of external relief valves, or known back pressure relief designed into the installed solenoid valves. If such relief is not present, then increases in temperature can result in rapid increases in pressure (due to changes in liquid density), possibly resulting in system rupture. It is also important to look at the layout of liquid feed lines to consider the potential pooling of the oxidizer at feed line low points when venting the system, and the potential to trap gas in the system at high points when filling. Such considerations are also important when designing the system for decontamination. Cold traps remove hazardous propellants by freezing them.

All feed line dead-ends should be considered as potential locations for compression heating problems. Compression heating can be mitigated by slowly increasing the pressure in the feed line to allow local heat increases to be dissipated.

Hard-starts can occur when too much oxidizer or fuel is allowed to accumulate in the combustion chamber prior to ignition, this is of particular concern with liquid oxidizers [181], [385]. Such starts can lead to overpressurization sufficient to rupture the combustion chamber. Hard starts can be avoided by adjusting the timing of the main oxidizer valve opening and the powering of the ignition system. For liquid oxidizers, remember fire first: igniter before oxidizer valve.

There are several methods for minimizing flame propagation through a feed system in the event of an unexpected fire. High melting point materials (e.g., Monel) can be used as flame breaks, since they melt instead of burn. Flashback arrestors can also be used to stop the flame or prevent backflow, though it can be challenging to find such devices rated to typical hybrid rocket design pressures.

It should be noted that hazards presented by fuels were not given much attention here. That is not because they do not exist but because they are less common. A notable exception is the use of nano- or micro-aluminum, which is exceptionally flammable. There certainly are hazards in making the fuels, e.g., aluminum powders that were just discussed, some curatives and ingredients that should not be inhaled in making HTPB, and molten wax is surprisingly hot, to name a few. There are hazardous (e.g., toxic) fuels and additives as well. Many common hybrid fuels are valued for their safety (e.g., plastic, rubber and wax). However, more exotic materials are being used as propulsion engineers are driven to maximize performance. The Safety Data Sheet for all materials should be scrutinized before they are used. Expected combustion products should also be evaluated for safety for any new propellant combination.

11.6.3 Cleaning

Feed system cleanliness is vital to ensure test program success. Failure to carefully follow cleanliness standards during both cleaning and system assembly is, at a minimum, a cause for reliability issues (e.g., particulate contamination of valves can introduce internal leakage) and at worst, extremely dangerous (e.g., unintended ignition hazard). There are various cleaning standards available in industry to provide best practice for certain oxidizers. Examples of this for oxygen systems include ASTM G93 [386], MIL-STD-1330D [387], and CGA G-4.1 [388]. For non-welded systems, the lines and components should be cleaned at the piece-part level prior to assembly. For welded systems, it can be beneficial to clean some welded manifolds, rather than individual fittings, though components should still all be cleaned at the piece-part level. Most cleaning processes, see, for example, those used by Ref. [376] or [379], follow some or all of the following steps:

- Submersion in a suitable cleaning agent. See Ref. [389] for oxygen systems.
- Ultrasonic agitation of the part in a cleaning agent. See Ref. [390].
- Deionized water rinse. This step is often repeated at least three times to ensure the complete removal of any cleaning agent.
- Drying of the components to remove any water residue, e.g., via gentle heating or the use of a purge gas.
- Inspection of the part to ensure it is clean. This is typically just a visual inspection under a bright light and/or ultraviolet light. If the part does not pass inspection, then the entire cleaning cycle should be repeated. For flight components, particulate counts and/or Non-Volatile Residue (NVR) are often recommended in addition to a visual inspection.

- Storage of the component in a clean environment to ensure it is not recontaminated prior to final assembly.

Some systems may further require vacuum bake-out of components or even component piece parts to ensure that they do not outgas and produce contamination during operation. Note also that ideally, assembly of the feed line should be performed in a clean environment, such as a certified cleanroom. The recommended rating of the clean room depends on the level of risk associated with contamination and should be evaluated based on the needs of a particular system.

11.6.4 Feed system component selection

Component selection is discussed for both flight and ground systems in Chapter 9. The approach for ground systems is similar to flight, except that some key drivers for flight system selection, like mass, are typically not a concern for ground systems. Component selection for ground feed systems will typically be based on the system requirements for design pressure, material compatibility, flow rate/pressure drop, and potentially also response time. Mass, power, and reliability are not generally a concern for ground systems. Temperature requirements can occasionally be driving for ground systems if the oxidizer is planned to be pre-heated or pre-chilled prior to testing or if a gas is planned to be used in a blow-down configuration. Further details on component selection for each of these areas is provided in Chapter 9.

If a program is moving beyond early development tests and towards flight, then wherever possible, it is good practice to "test like you fly". For ground systems, this means that ideally, the feed system geometry and major components will mirror the flight design as much as practicable (likely with additional instrumentation in the ground test configuration). At a minimum, it is good practice to test the flight motor design with a flight representative main oxidizer valve prior to flight.

11.6.5 Instrumentation

A minimum set of measurements are typically taken for hybrid propulsion research. These include mass flow rate, either using a flow meter or venturi and differential pressure transducer, and chamber pressure. Multiple chamber pressure measurements are common for redundancy, frequency analysis, and to measure the pressure drop across the motor (fore and aft end measurements). Additional data can include strain gauges for pressure, temperature sensors, accelerometers, and microphones for acoustic measurements.

Oxidizer mass flow rate measurement

There are various methods that can be used to measure the oxidizer mass flow rate. Care must be taken to ensure that the hardware required is compatible with the oxidizer being used and that the method will be accurate for the state of the fluid (e.g., incompressible, compressible, high vapor pressure). A brief overview of various types of hardware used to measure fluid flow rate is provided in Chapter 9. In that chapter, a number of dedicated flow meters are described, such as turbine flow meters, Coriolis flow meters, ultrasonic flow meters, magnetic flow meters, and vortex flow meters. Such flow meters are often expensive but can provide very accurate steady-state measurements of oxidizer flow rate if they are able to be used with the oxidizer of choice.

A cavitating venturi can be used to give an accurate measurement of flow rate for liquid flow knowing only the upstream pressure and the fluid temperature, see Eq. (8.21) in Chapter 8. Note that this equation uses a known discharge coefficient, C_d, generally determined experimentally. In order to accurately determine C_d in a liquid system having high-pressure gas as the source of pressure, the flow rate used as

the true flow rate for calibration should be measured upstream of the venturi [321]. This is because, in this scenario, a large amount of gas will be dissolved in the pressurized liquid and come out of solution as the fluid travels through the venturi, causing errors in any downstream flow measurements [321].

A sonic venturi can also be used for gaseous flow similarly to a cavitating venturi for liquid flow, see Eq. (11.1). Chapter 8 and Ref. [322] provide more information on this approach. Again, the discharge coefficient of the venturi, C_d should be calibrated whenever practicable.

$$\dot{m}_O = \frac{\gamma C_d A_{th} P_t}{\sqrt{\gamma R T_t}} \left(\frac{\gamma + 1}{2} \right)^{-\frac{\gamma + 1}{2(\gamma - 1)}} \quad (11.1)$$

A noncritical venturi can be used to measure the mass flow of both incompressible liquids and compressible gases. In both cases, for best accuracy, the flow should be as uninterrupted as possible (including avoiding bends) for 10 tube diameters upstream of the venturi and 5 diameters downstream of the venturi.

The oxidizer mass flow rate for an incompressible fluid can be determined from differential pressure transducer measurements across a venturi with known properties, see Eq. (11.2). This equation is derived in Appendix A. Note that the conditions at location 1 refer to the inlet to the venturi, and at 2 refer to the venturi throat.

$$\dot{m} = C_d A_2 \sqrt{\frac{2\rho (P_1 - P_2)}{1 - \left(\frac{A_2}{A_1} \right)^2}} \quad (11.2)$$

The oxidizer mass flow rate for a compressible gas can be determined from differential pressure transducer measurements across a venturi along with upstream temperature and static pressure measurements, see Eq. (11.3) [323]. Again, the conditions at location 1 refer to the inlet to the venturi and at 2 refer to the venturi throat. Note that the expansibility factor, Y_{exp}, in this equation, is explicitly provided in Eq. (11.4). These relationships between the mass flow rate, upstream conditions, and the pressure change associated with a cross-sectional area change can be derived for a compressible gas as shown in Appendix A. These equations are generally considered valid if $\frac{P_2}{P_1} \geqslant 0.80$. The discharge coefficient, C_d, for a machined venturi designed in accordance with the guidelines of Ref. [323] is 0.995. The relative uncertainty of this discharge coefficient is 1% per Ref. [323]. The uncertainty of the expansibility factor, Y_{exp} is a function of the expansion area ratio, pressure drop and inlet pressure, see Ref. [323] for details.

$$\dot{m} = C_d A_2 Y_{exp} \sqrt{\frac{2(P_1 - P_2)\rho_1}{1 - \left(\frac{A_2}{A_1}\right)^2}} \qquad (11.3)$$

$$Y_{exp} = $$

$$\sqrt{\frac{\frac{\gamma}{\gamma - 1}\left(\frac{P_2}{P_1}\right)^{\frac{2}{\gamma}}\left[1 - \left(\frac{P_2}{P_1}\right)^{\frac{\gamma - 1}{\gamma}}\right]\left(1 - \left(\frac{A_2}{A_1}\right)^2\right)}{\left(1 - \frac{P_2}{P_1}\right)\left(1 - \left(\frac{A_2}{A_1}\right)^2\left(\frac{P_2}{P_1}\right)^{\frac{2}{\gamma}}\right)}}$$

$$(11.4)$$

It is challenging to accurately capture flow transients with most flow meters and with a differential measurement across a venturi. One way to get around this transient limitation is to calibrate the flow rate at given steady-state conditions, such as given upstream pressures and temperatures, and then use this calibration during transient events using the transient pressure and temperature measurements.

Fuel mass flow rate

Directly measuring the instantaneous fuel mass flow rate in hybrid rocket motors is challenging. Measurements of the real-time change in mass of the combustion chamber via simple scales are complicated by the need for accurate measurements of a typically heavy system (for a ground test motor), the need for accurate knowledge of the mass of oxidizer in the motor, and variability in any measurement from the thrust produced by the motor. As discussed in Chapter 9, there are some alternative direct measurement approaches that have been used successfully. These include ultrasonic measurements, x-ray measurements, and embedded resistors within the fuel grain. Each of these has its challenges, and so no single approach has become the standard for the hybrid rocket community.

The fuel mass flow rate is most commonly reported in the form of the fuel regression rate, $\dot{r}$. When a direct regression rate measurement is not available, the regression rate can be reasonably estimated based on other measured values. There are various such approaches used in the hybrid rocket community and not all of them are rigorous or consistent. The simplest and most common approach is to measure the change in fuel mass before and after a test, then calculate a space-time averaged regression rate from this measurement. The limitations and considerations of this approach, and other more involved approaches also recommended by the authors, are discussed in Chapter 6. Relevant equations are also provided in Chapter 6.

Thrust

Thrust stands are the best way to measure thrust and calculate the I_{sp} of the motor. They are discussed in general terms in Chapter 9 with some specific design guidelines provided here to help improve the accuracy of the measured thrust, since accurate thrust measurements are difficult to obtain. The main obstruction is the stiffness of the feed system, which must be overcome or calibrated out of the measurement. A number of design choices can be made in order to both minimize the measured thrust error and, more importantly, to make sure it is as repeatable as possible. The following list summarizes these considerations [391], [325]:

- Feed lines: lines should be anchored to both the floating side (moves to enable the thrust measurement) and ground side (fixed) of the thrust stand, orthogonal to the axis of thrust. This allows for changes in line length due to thermal effects. When the lines are small, s-bends can be used to minimize axial loading. While flexible lines are tempting because they give the appearance of not interfering with the measurement, bending them results in small but non-repeatable friction forces, which reduces the accuracy of the calibration.
- Joints: the flexibility of ball and gimbal joints can be pressure dependent, making the force non-repeatable. Loose joints introduce non-repeatable offsets (e.g., threaded joints without preload loosening over time).
- Close clearances: close clearances and obstructions should be avoided, especially in hard-to-inspect areas. Particulate contamination can be a particular challenge in these areas as well.
- Pendulum effect: the thrust stand supports can become misaligned, applying an unintended force on the stand.
- Gravity: the change in mass of a test article over time will be directly measured in a vertical configuration and needs to be well characterized/predictable. If the center of gravity (CG) moves substantially over the length of the burn, the additional error is introduced since the load will no longer be reacted through the CG of the motor.
- Thermal/pressure loads: the temperature of the hardware changes the line lengths, combustion chamber geometry, etc. Additionally, the load cell itself must be within an allowable temperature range as well as at a relatively uniform temperature to perform within calibration. The expansion of the case diameter/length due to thermal and pressure loads can be on the order of, or larger than, the load cell deflections.

- Safety: Thrust stand safety includes protecting the hardware and personnel. A factor of two on the load cell is recommended in order to survive hard starts. Shear pins show excellent repeatability and can be used to protect load cells.

Calibration with a known load is critical in order to trust data from any thrust stand. This can be done in situ or by removing the load cell from the test stand, or with a combination of both. The best results are typically achieved by first calibrating the load cells before installing them in the thrust stand and then applying a known load to the stand.

If the test facility does not have a thrust stand to give a direct measurement of thrust, it can instead be inferred based on the nozzle design, combustion efficiency, and the ideal combustion performance, see Eq. (11.5). The equivalent vacuum thrust is also calculated in a similar manner, see Eq. (11.6). Note that Eq. (11.5) assumes that the nozzle is properly sized to ensure that there is no flow separation in the nozzle during the ground test. For some motor designs, this means that a different nozzle will be used in ground test than would be planned for a flight application. In such situations, the ideal vacuum-specific impulse, $I_{sp}|_{vac}$, is different for Eq. (11.5) and Eq. (11.6), since two different nozzle area ratios would be used. Similarly, the nozzle efficiency, η_n, would likely also be different for the two equations.

$$F|_{ground} = \eta_{C^*}\eta_n \left(\dot{m}_p g_0 \, I_{sp}|_{vac} - P_a A_e \right) \quad (11.5)$$

$$F|_{vac} = \eta_{C^*}\eta_n \left(\dot{m}_p g_0 \, I_{sp}|_{vac} \right) \quad (11.6)$$

In these equations, the combustion efficiency, η_{C^*}, can be calculated via either Eq. (11.7) or (11.8). Note that the two approaches give nearly identical results as discussed in Ref. [10], where the mean absolute difference in the results was only 0.01%, indicating that the two approaches are essentially equivalent.

Ref. [131] makes a very good point that the C^* efficiency, η_{C^*}, calculated from these equations is only an indicator of combustion efficiency. If the measured C^* is only 90% of the ideal value, the propulsion engineer cannot be sure if 10% of the propellants were left unreacted or if 100% of the propellants were reacted, but other factors affecting performance were responsible for the decrease (e.g., nonlinear nozzle erosion, the inability to accurately model the fuel mass/port evolution with time, and the ablation/combustion of inerts over time).

The ideal pressure of Eq. (11.8) is calculated via Eq. (11.9). Note that in Eq. (11.9) all variables, namely γ, $T_{t,th}$, and R are evaluated at the nozzle throat and determined from CEA [74] with equilibrium combustion. The values reported as mean values, such as the mean chamber pressure, $\overline{P_c}$, are simply the values averaged over the entire burn time. The ideal characteristic velocity is calculated using CEA, Ref. [74], with the assumption of equilibrium combustion. The effect of insulation on the characteristic velocity and specific impulse is often not evaluated for simplicity but should be considered whenever practicable, and especially for long-duration burns where their contribution may be significant. The enthalpy of formation used in the calculation of the ideal C^* is a key input; where possible, it should be measured experimentally for the fuel under test, as discussed in Section 11.2.2 and in Ref. [9].

$$\eta_{C^*} = \frac{\left(\frac{P_c A_{th}}{\dot{m}_p}\right)_{meas}}{C^*|_{ideal}} \tag{11.7}$$

$$\eta_{C^*} = \frac{P_c|_{meas}}{P_c|_{ideal}} \tag{11.8}$$

$$P_c|_{ideal} = \frac{\dot{m}_p \sqrt{\gamma R T_{t,th}}}{\gamma A_{th}} \left(\frac{\gamma + 1}{2}\right)^{\frac{\gamma + 1}{2(\gamma - 1)}} \tag{11.9}$$

Chamber pressure measurements

Combustion chamber pressure (P_c) is typically the most critical measurement for determining C^* efficiency, η_{C^*}, as can be seen in both Eq. (11.7) and (11.8).

Chamber pressure measurements are most commonly taken at the fore-end of the combustion chamber because that area is more easily accessed and is a less harsh environment than the aft end of the motor. However, a substantial pressure drop is possible along the length of the combustion chamber, especially if devices to enhance mixing are used within the combustion chamber. Therefore it is more accurate to use aft end combustion chamber pressure for combustion efficiency calculations. The combustion efficiency of a motor typically changes over the duration of the burn, as do the input parameters. Eqs. (11.8) and (11.7) refer to instantaneous measurements and so they rely upon accurate knowledge of the fuel regression rate with time. This is often circumvented by using average values of all inputs for a given test.

Once test data has been collected, a Monte Carlo analysis has been shown to be a good way to determine C^*, η_{C^*} and the associated error [131]. This was shown to give a different result (by more than 1%) than using averaged values to determine the efficiency. This may seem like a small difference; however, when it comes to spaceflight applications, even a single percentage point on performance goes a long way.

Chamber pressure measurements generally need to be recorded at reasonably high frequency in order to capture transient events and any potential combustion instabilities. Nyquist sampling requires a sample rate at a minimum of twice the frequencies of interest, but sampling at frequencies of 3 to 5 times the desired maximum is often employed. Chapter 8 discusses the common combustion instabilities observed in hybrids and their typical frequency range. Consideration of these potential frequencies of interest is key to selecting the needed sampling rate for the system.

The volume between the chamber and pressure transducer should be minimized so the

TABLE 11.1 Example hybrid rocket regression rate data summary table. Note: * Calculation method for each of these variables should be reported.

Test [-]	Date [-]	Burn time* [s]	Min/avg/max oxidizer mass flux [kg/m²s]	Min/avg/max chamber pressure [MPa or bar]	Min/avg/max O/F [-]	Min/avg/max C* [m/s]	Min/avg/max thrust [N]	Min/avg/max regression rate* [m/s]	Etc.

signal is not damped. It is necessary to keep in mind that common methods for protecting instrumentation, such as using a protective barrier gel or narrow pressure transducer ports, can also attenuate data. Further, as discussed in Chapter 9, some transducers are not well suited to recording high-frequency oscillations. It can be beneficial to use different transducers on the same motor with some dedicated to quasi-steady/running mean measurements of chamber pressure and others dedicated to the measurement of higher-frequency combustion instabilities. More details on investigating and reporting chamber pressure data in the frequency domain is provided in Section 11.7.3.

11.7 Reporting test data

There has historically been very little uniformity in the way that hybrid rocket test data is reported in literature. Unfortunately, this often makes it difficult to apply test data from one set of experiments to other applications. This section discusses the recommended approach to reporting test data for the purpose of specifying motor ballistics (i.e., regression rate data), provides a quick overview of error analysis so that the propulsion engineer can provide error bounds on reported data, and describes the process for analyzing data in the Fourier domain which is often an essential tool for analyzing combustion instabilities.

11.7.1 Regression rate data reporting

A recommended approach to test data reporting is provided in Table 11.1, based on work the authors did with other members of the AIAA Hybrid Rocket Technical Committee [105]. Note that it is possible to report very different results for the same set of hybrid rocket test data by applying different data reduction techniques, see Chapter 6 for further discussion of this. As such, it is critical that any data reduction techniques that have been used be reported along with the data. Essentially, if any of the reported data is calculated, rather than read directly from an instrument, then the calculation method should be reported as either a note in the table or in the associated primary text. Most variables have some form of calculation technique associated with them, as an example, the burn time could be the nominal burn time calculated using the bisector method, the regression rate may have been calculated assuming constant C^*, and C^* may have been determined using one-dimensional equilibrium (optimistic).

Any data included in a summary table, like that of Table 11.1 should specify whether the value is instantaneous, a maximum/minimum, or an average. SI units should be used for all calculations and value reporting (e.g., regression rate should be reported in m/s). Multiple tests at similar conditions (with a minimum of two tests) should be used to show repeatability whenever possible. The general goal is to include enough information in the summary table and the supporting information to allow data from one experiment to be combined with data from another and thereby strengthen regression rate/ballistic predictions for hybrid rockets. In addition to the data of Table 11.1, the following information should also be provided whenever possible:

- The propellant combination with as much information as possible, e.g., vendors, densities, molecular weight, and initial temperature.
- Figures showing the variation of chamber pressure, thrust, mass flow rate and regression rate with time. This is particularly important if any of these vary substantially with time.
- Scale/geometry of motor. Note that if the detailed geometry cannot be provided, even a general indication of motor dimension or thrust can be helpful to a future propulsion engineer interested in using the data as it gives the ability to account for potential scaling effects.
- Unique information including injection scheme and motor internals. This information may be restricted as proprietary or export controlled; however, even broad classifications, such as specifying if the injection scheme is axial, circumferential, or swirl and whether any mixing devices or diaphragms are used in the combustion chamber can be extremely helpful.
- Instruments used to measure data and associated errors.
- Error bars on graphs and as a $\pm$ in text/tables. The definition of the error bar, e.g., if it is 3-sigma or an alternative definition, should also be specified.
- Ambient conditions at which the tests were conducted, e.g., sea level or in a vacuum chamber at altitude, at ambient initial temperature, or thermally conditioned prior to ignition.
- The motor development status, e.g., lab, breadboard/battleship/heavyweight, and flight.

Oxidizer mass flux (instead of total mass flux) has been used for the generation of nearly all available regression rate data (including the authors' work). See Table 6.2. However, a_o depends on the O/F during a test. If the mass flow rate of the oxidizer as a function of time, $\dot{m}_O(t)$, is recorded. It is possible to analyze the data using the coupled mass accumulation and regression rate equations based on G, and then a, n and possibly even m can be reported on that basis. Developments in regression rate theory and computational methods make it desirable to archive as much data as possible for later use with updated methods. A more complete discussion of the pitfalls of using a_o instead of a and a similar appeal are included in Chapter 11 of Ref. [97].

11.7.2 Error analysis

There has historically been an unfortunate habit within the hybrid rocket community of reporting data without also reporting the associated error bounds of the result(s). For some (poorly designed) experiments, this can lead to the misreporting of results and trends that, when examined more closely, are not actually statistically significant. It is therefore key that the propulsion engineer always evaluates the error bounds of any key finding. The topic of error analysis of experimental data is of broad interest to the scientific community. Here we provide a taste of the recommended approach but direct the reader to the referenced literature if more detail is required. There are two widely accepted standards on measurement uncertainty, one is the ASME Test Uncertainty Standard [392] and the other is the International Organization on Standardization (ISO) Uncertainty of Measurement Guide [393]. These two standards differ in the way that they classify error, and in their terminology, but their methodology is very similar. Either approach can be used to quantify measurement error, the end result will be similar.

Under the ASME standard, errors are categorized as either systematic or random. Systematic errors remain constant for repeated measurements at the same conditions. They are also referred to as bias errors. Systematic errors can be reduced by calibration against a suitable standard. As an example of best practice to reduce systematic error, pressure transducers should be re-calibrated annually.

Random errors vary randomly with repeated measurements at the same operating condition. As such, random errors should be characterized by repeated measurements, with reporting of the statistical estimate of the error. Each set of measurements of a variable, x_i, has a mean value, $\bar{x}_i$ and an associated uncertainty, u_i, Eq. (11.10). The uncertainty will be a function of the confidence interval chosen; e.g., for a normal distribution, it is common to specify the error as some number of standard deviations, σ. A single standard deviation ($\pm\sigma$) correspond to 68.3% confidence interval, two standard deviations ($\pm2\sigma$) corresponds to a 95.4% confidence interval, and three standard deviations ($\pm3\sigma$) correspond to 99.7% confidence interval. Given a set of n measurements of the random variable x_i, the mean, $\bar{x}_i$, can be calculated from Eq. (11.11) and the standard deviation, σ_i can be calculated from Eq. (11.12). In both of these equations $x_{i,j}$ refers to a single measurement of the variable x_i.

$$x_i = \bar{x}_i \pm u_i \qquad (11.10)$$

$$\bar{x}_i = \frac{\sum_{j=1}^{n} x_{i,j}}{n} \qquad (11.11)$$

$$\sigma_i = \sqrt{\frac{\sum_{j=1}^{n} \left(x_{i,j} - \bar{x}_i\right)^2}{n-1}} \qquad (11.12)$$

Direct measurement of a variable is not always possible. As such, error propagation must be considered for many variables of interest in hybrid rocket motors. If a value is calculated as a function of other measured values, then the error of the dependent result must take into ac-

count the error of each of the measured variables. Let y be our dependent variable that is a function of N measured variables, x_i. As such, y is a function of x_i per Eq. (11.13). The expected value of y, $\bar{y}$, is simply a function of the expected value of each measured value, Eq. (11.14). The expected error of y, u_y, can be calculated via a first-order Taylor series expansion of Eq. (11.13) around the expected value, per Eq. (11.15).

$$y = f(x_1, x_2, ..., x_N) \qquad (11.13)$$

$$\bar{y} = f(\bar{x}_1, \bar{x}_2, ..., \bar{x}_N) \qquad (11.14)$$

$$u_y = \pm \left[\sum_{i=1}^{N} \left(\left. \frac{\partial y}{\partial x_i} \right|_{x=\bar{x}} u_i \right)^2 \right]^{1/2} \qquad (11.15)$$

The best way of understanding Eq. (11.15) is to work through an example. Consider a reported space-time averaged regression rate for a hybrid fuel grain with a cylindrical port. It is common for this regression rate to be calculated based on the measured mass of the fuel grain at the end of the test (see Chapter 6) but for simplicity here, assume that the radius of the fuel grain port is directly measured before, r_i, and after, r_f, the test. The regression rate is therefore calculated via Eq. (11.16).

$$\dot{r} = \frac{r_f - r_i}{t_b} \qquad (11.16)$$

Assume that five different measurements are made of the fuel port radius, and the burn time is calculated using three different methods (see Chapter 6) as reported in Table 11.2. For this example, assume that each burn time estimate is equally valid so a mean and standard deviation can be calculated per Eqs. (11.11) and (11.12), respectively. The resulting observations of each variable are summarized in Table 11.2.

The expected regression rate is calculated from the average of each of the variables per Eq. (11.17). This gives an expected regression rate of 0.0017 m/s but not the error associated

TABLE 11.2 Example set of possible regression rate data. This data set is not from actual measurements but is intended to be somewhat representative of data that could be collected.

Variable	Units	Measurement					Mean	σ
		1	2	3	4	5		
r_i	[m]	0.0100	0.0099	0.0100	0.0097	0.0101	0.0099	0.0002
r_f	[m]	0.0250	0.0288	0.0275	0.0248	0.0285	0.0269	0.0019
t_b	[s]	10.00	10.50	9.75	-	-	10.08	0.38

with it.

$$\bar{\dot{r}} = \frac{\bar{r}_f - \bar{r}_i}{\bar{t}_b} \qquad (11.17)$$

The error in the regression rate can be calculated by substituting Eq. (11.16) into Eq. (11.15). The result of this substitution is provided in Eq. (11.18). Consider calculating the two-sigma variation of the expected value of regression rate. In this case, each $u_i = 2\sigma_i$. The various standard deviations, σ_i and means $\bar{x}_i$ from Table 11.2, are used to find that the expected two sigma variance of regression rate is 0.0004 m/s. As such the regression rate for this data set should be reported as 0.0017 ± 0.0004 m/s with a 95% confidence interval. This is quite a large error bound, but evaluation of the various terms in Eq. (11.18) shows that this error is primarily driven by the error in the measured fuel port radius at the end of the burn, which is a common issue for this type of regression rate calculation and one reason that mass is often used to determine average port radius at the end of a test. A more detailed example of error analysis applied to hybrid rocket regression rate data can be found in Ref. [205].

$$u_{\dot{r}} = \pm \left[\left(\frac{1}{\bar{t}_b} u_{r_f} \right)^2 + \left(-\frac{1}{\bar{t}_b} u_{r_i} \right)^2 \right.$$
$$\left. + \left(\frac{-(\bar{r}_f - \bar{r}_i)}{\bar{t}_b^2} u_{t_b} \right)^2 \right]^{1/2} \qquad (11.18)$$

11.7.3 Analysis of data in frequency domain

Frequency information in hybrid rocket motors is generally investigated by the application of the discrete Fourier transform to chamber pressure data. Feed system frequency data is also often used to investigate any potential feed system instability. The Fast Fourier Transform (FFT) is the most commonly used method to obtain frequency data from a raw signal, see Ref. [394] for details of this algorithm. It is generally worthwhile to have an understanding of potential combustion instability frequencies for a given test set-up prior to running any tests, see Chapter 8 for a discussion of these instabilities and their frequency ranges. As discussed in Section 11.6.5, the highest frequency of interest will drive the requirement for the minimum sampling rate of the data. It is also extremely beneficial to have these frequencies of interest in mind when analyzing test data.

Directly applying a Fourier transform to the chamber pressure data can provide meaningful insight into the frequency content of the signal. However, if the running mean chamber pressure of the system being analyzed is not approximately constant, then it can be difficult to discern low-frequency components in the pressure-time trace. The slowly changing mean chamber pressure, where slow is relative to the spectral content, introduces DC bias to the system with large amplitude in the Fourier domain. This bias, coupled with unavoidable windowing from sampling for a finite time, effectively obscures much of the important frequency content

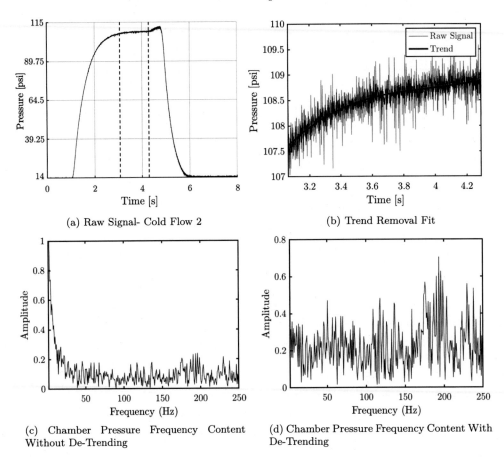

(a) Raw Signal- Cold Flow 2   (b) Trend Removal Fit

(c) Chamber Pressure Frequency Content Without De-Trending   (d) Chamber Pressure Frequency Content With De-Trending

FIGURE 11.1 Spectral analysis with and without trend removal of cold flow chamber pressure data from Ref. [85]. The normalized spectral content (FFT) presented in (c) and (d) is calculated using the data between the vertical dashed lines in (a). The spectral content of the chamber pressure for (d) is calculated after de-trending the data with a cubic fit, as shown in (b). The resolution of (c) and (d) 0.8 Hz.

of the signal, see Fig. 11.1. The most common method to remove such bias is known as trend removal [395]. Trend removal fits a least squares lower-order polynomial, typically up to third order, to the data. This 'trend' is then subtracted from the original data prior to converting it to the Fourier domain. Since low-frequency oscillations, down to approximately 5 Hz, are often observed in hybrid rockets, it can be necessary to apply trend removal in order to accurately characterize the instability. If the trend in the chamber pressure changes throughout the test,

sub-sections of pressure data can be selected for trend removal and spectral analysis. Analyzing sub-sections of the data reduces the number of data points in the sample, N, and therefore reduces the frequency resolution, Δf, of the spectral analysis, see Eq. (11.19) [396]. In this equation, f_s refers to the sampling frequency. An example of the influence of trend removal on spectral analysis results is provided in Fig. 11.1. This figure shows frequency analysis of cold flow data for a combustion visualization experiment per Ref. [85]. Analysis of the raw data at low frequency (less than around 50 Hz) is very difficult in this case without trend removal. Removal of the mean trend revealed that there were no strong low-frequency acoustic resonances in this range.

$$\Delta f = \frac{f_s}{N} \qquad (11.19)$$

The frequency content of the chamber pressure data will often change over time, and this can be due to the changing acoustics of the chamber, excitement of different higher-order instabilities, a change in the primary combustion instability, or a reduction/amplification of any instabilities. Spectrograms, also referred to as waterfall plots, are often used to investigate such changes in the frequency content of the signal over time. In order to generate a spectrogram, the frequency data needs to be separated into segments of data. This process, unfortunately, results in "windowing" of the data with a subsequent reduction in frequency resolution for each segment per Eq. (11.19) where N now becomes the number of samples in each segment. Some overlap of data between segments can be used to minimize this windowing effect. In general, though, the propulsion engineer needs to trade resolution in the time domain with resolution in the frequency domain in order to produce a spectrogram. When such trades are balanced correctly, spectrograms can be powerful tools for understanding the combustion processes within a hybrid rocket motor. Examples of spectrograms used to successfully investigate chamber pressure data are provided in Fig. 11.2 as well as Figs. 8.10, 8.12 and 8.13 of Chapter 8. Fig. 11.2 from Ref. [47] is included here to illustrate the benefit of using a spectrogram to explore the combustion behavior over the course of a burn. It is difficult to discern any clear pattern by just looking at the full set of chamber pressure data, image (a). Zooming in on a subset of this data, image (b), shows that there are clearly some excited modes, but not at which frequency they are prevalent. The spec-

trogram, image (c) gives a much clearer picture of the frequencies and relative strength of the instabilities over the course of the burn. It can be seen that there is a significant level of frequency content down in the 10–25 Hz range over the first half of the burn. There is also strong activity around 200–210 HZ, which corresponds to the first acoustic mode of this motor. This L1 acoustic mode can be seen to be particularly coherent during the second half of the test [47]. Note that trend removal was not conducted when generating the spectrogram, and hence there is an appearance of a strong signal around 0 Hz for the duration of the burn. As discussed in Ref. [47], all of the instabilities associated with this development motor were identified and resolved to produce a stable hybrid motor.

11.8 Real world example: Peregrine

Fig. 11.3 is an example of a hybrid rocket ground test Plumbing and Instrumentation Diagram (P&ID) also known as a schematic. This particular image is of an early ground test set up for the Peregrine motor at NASA Ames Research Center in Moffett Field, CA. As introduced in Chapter 1, Peregrine is a paraffin/N_2O hybrid. The tubing is color coded to depict different line sizes. Each valve type has its own symbol. While often similar, valve symbols are not standardized, so it is important to look at the legend for each schematic. This particular example was chosen because of the three main branches of this feed system: 1) run line, 2) purge and supercharge, and 3) oxidizer fill.

The main run line connects the N_2O run tank (blue) to the motor, shown at the bottom center of the image. Redundant ox valves (OV-2 and the main ox valve, OV-4) are used to isolate the oxidizer from the motor. When developing/qualifying a motor for flight, the downstream valve should be as flight-like as possible, since the ignition transient will be dependent on its performance. The upstream valve is a safety

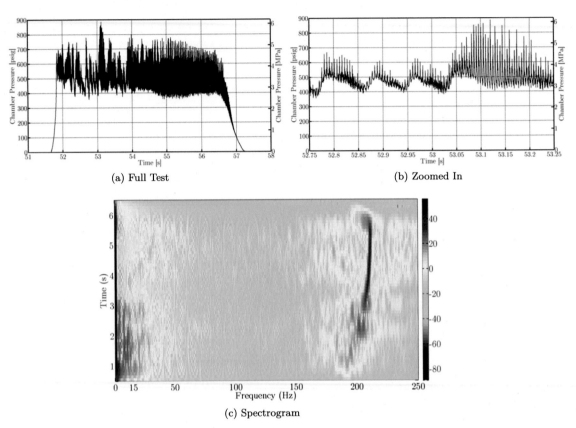

(a) Full Test

(b) Zoomed In

(c) Spectrogram

FIGURE 11.2 Chamber pressure data showing combustion instabilities from a developmental hot-fire test of the Peregrine motor, a nitrous oxide and paraffin hybrid. The figure is used with permission from Ref. [47].

precaution. A venturi (FM-1) and differential pressure transducer (DP-2) are used to measure the oxidizer flow rate in this line, a critical piece of information in determining the performance of the propulsion system. The flow rate is set by a throttle valve (OV-5) just upstream of the injector in this case. However, an orifice or cavitating venturi could also be used instead of this valve.

The rightmost part of Fig. 11.3 depicts the helium supercharge line. Note that in later iterations of the design, the authors suggest an oxygen supercharge in order to enable the oxidizer vapor to be burned. The high-pressure helium gas is regulated to the desired pressure via R-1. It is important to ensure that every part of the

line can be vented safely. In this case, because the nitrous oxide is non-toxic, and the test stand is outdoors, it can be vented directly to the atmosphere. Both a relief and vent valve is located upstream of the regulator (HV-1 and HV-2, respectively) to safe the system. A relief valve, HV-8, is also located downstream of the regulator, R-1. This relief valve should be sized to accommodate the full flow if the regulator were to fail open and is a critical safety feature. A burst disk (BD-1) is teed off of the inlet to the run tank to allow for rapid, high-flow venting in the case of an overpressure event. Shop air (called OARF pneu, for Outdoor Aerodynamic Research Facility pneumatic lines, in this exam-

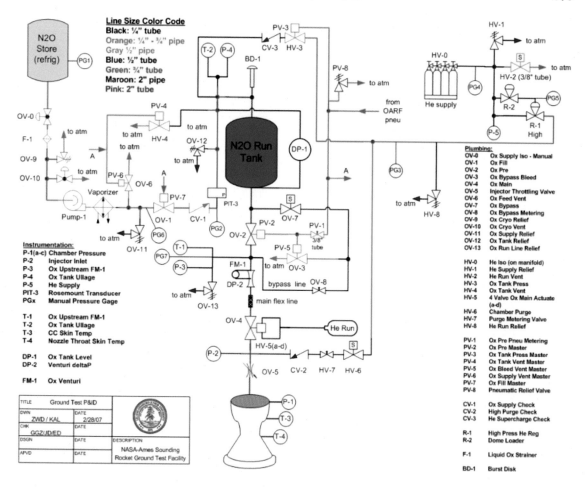

Line Size Color Code
Black: ¼" tube
Orange: ¼" - ¾" pipe
Gray ½" pipe
Blue: ½" tube
Green: ¾" tube
Maroon: 2" pipe
Pink: 2" tube

Instrumentation:
P-1(a-c) Chamber Pressure
P-2 Injector Inlet
P-3 Ox Upstream FM-1
P-4 Ox Tank Ullage
P-5 He Supply
PIT-3 Rosemount Transducer
PGx Manual Pressure Gage

T-1 Ox Upstream FM-1
T-2 Ox Tank Ullage
T-3 CC Skin Temp
T-4 Nozzle Throat Skin Temp

DP-1 Ox Tank Level
DP-2 Venturi deltaP

FM-1 Ox Venturi

Plumbing:
OV-0 Ox Supply Iso - Manual
OV-1 Ox Fill
OV-2 Ox Pre
OV-3 Ox Bypass Bleed
OV-4 Ox Main
OV-5 Injector Throttling Valve
OV-6 Ox Feed Vent
OV-7 Ox Bypass
OV-8 Ox Bypass Metering
OV-9 Ox Cryo Relief
OV-10 Ox Cryo Vent
OV-11 Ox Supply Relief
OV-12 Ox Tank Relief
OV-13 Ox Run Line Relief

HV-0 He Iso (on manifold)
HV-1 He Supply Relief
HV-2 He Run Vent
HV-3 Ox Tank Press
HV-4 Ox Tank Vent
HV-5 4 Valve Ox Main Actuate (a-d)
HV-6 Chamber Purge
HV-7 Purge Metering Valve
HV-8 He Run Relief

PV-1 Ox Pre Pneu Metering
PV-2 Ox Pre Master
PV-3 Ox Tank Press Master
PV-4 Ox Tank Vent Master
PV-5 Ox Bleed Vent Master
PV-6 Ox Supply Vent Master
PV-7 Ox Fill Master
PV-8 Pneumatic Relief Valve

CV-1 Ox Supply Check
CV-2 High Purge Check
CV-3 He Supercharge Check

R-1 High Press He Reg
R-2 Dome Loader

F-1 Liquid Ox Strainer

BD-1 Burst Disk

TITLE	Ground Test P&ID		
DWN	ZWD / KAL	DATE	2/28/07
CHK	GGZ/JD/ED	DATE	
DSGN		DATE	
APVD		DATE	DESCRIPTION
			NASA-Ames Sounding Rocket Ground Test Facility

FIGURE 11.3 Ground Testing Setup for the Peregrine Rocket in 2007 [278].

ple), typically compressed air at about 100 psi (0.7 MPa), is used to power the pneumatically actuated valves. Note that OV-4 uses a separate helium pressurization source. This allows the valve to be actuated with higher pneumatic pressure, facilitating a faster valve actuation.

The leftmost part of the diagram showing a green branch of line is used to fill the run tank. This line leverages a relief valve for safety and includes venting capability on any section of line that can be isolated. A check valve is utilized to prevent backflow into the storage tank. How-

ever, the propulsion engineer is reminded that check valves can leak and are not a foolproof way to keep fluids separate.

Manual pressure gauges are used at various points in the system. Manual gauges can be useful as a quick pressure indicator for the operator and also as a safety device to ensure lines are at safe pressures in the event of a failure in the instrumentation system.

In addition to standard testing best practices, safety considerations include proximity to a populated area and nitrous oxide hazards. Rec-

FIGURE 11.4   Hot-fire test of the Peregrine motor in 2017.

ommended safety practices for N_2O were cataloged in Section 11.5.2. Peregrine, with its test location in suburban Silicon Valley, takes several steps to ensure that failures will not cause any damage to the surrounding area. Fig. 11.4 shows the protective structure around the motor affectionately termed the "doghouse." Its function is to reduce the kinetic energy of any pieces of the composite case structure in the unlikely event of a Rapid Unplanned Disassembly (RUD). The barrier downstream of the plume can stop the nozzle in case it is ejected. The bolts securing the nozzle to the motor are notched to ensure that they will be the primary failure method during ground testing. Finally, there is a large structure filled with sand and a top barrier surrounding the N_2O run tank to absorb energy in the event of a failure.

APPENDIX

A

Derivations

A.1 Launch mission design governing equations

The governing equations for a three degree of freedom (3DOF) launch simulation, as discussed in Chapter 2, are derived here. Start with a force balance for motion in a two-dimensional plane, see Fig. A.1, and Newton's second law:

$$m\frac{d\vec{V}}{dt} = \sum \vec{F} \qquad (A.1)$$

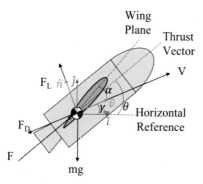

FIGURE A.1 Rocket free body diagram showing the inertial coordinate frame (blue) and body fixed co-ordinate frame (orange).

The inertial coordinate frame is defined with the orthogonal unit vectors, $\hat{i}$ and $\hat{j}$. Consider also a body fixed reference frame with unit vectors parallel to and orthogonal to the velocity

vector, call them $\hat{v}$ and $\hat{n}$, respectively. This body fixed reference frame can be both accelerating and rotating, and is therefore not inertial.

The velocity of the rocket, $\vec{V}$, can be written as:

$$\vec{V} = V\hat{v} \qquad (A.2)$$

The rate of change of the velocity vector is then:

$$\frac{d\vec{V}}{dt} = \frac{d(V\hat{v})}{dt} = \frac{dV}{dt}\hat{v} + V\frac{d\hat{v}}{dt} \qquad (A.3)$$

From consideration of Fig. A.1, the rotating frame of reference unit vectors can be expressed as:

$$\hat{v} = cos(\gamma_{fpa})\hat{i} + sin(\gamma_{fpa})\hat{j} \qquad (A.4)$$

$$\hat{n} = -sin(\gamma_{fpa})\hat{i} + cos(\gamma_{fpa})\hat{j} \qquad (A.5)$$

Recall that γ_{fpa} is the flight path angle, as defined by Fig. A.1. Differentiate Eq. (A.4) with respect to time to obtain Eq. (A.6):

$$\frac{d\hat{v}}{dt} = -\dot{\gamma}_{fpa}sin(\gamma_{fpa})\hat{i} + \dot{\gamma}_{fpa}cos(\gamma_{fpa})\hat{j} \qquad (A.6)$$

Now, decompose the force balance equation, Eq. (A.1), into the components parallel to, and perpendicular to, the instantaneous velocity vector, see Eq. (A.7) and (A.8).

$$m\frac{d\vec{V}}{dt} \cdot \hat{v} = \sum \vec{F} \cdot \hat{v} \qquad (A.7)$$

$$m\frac{d\vec{V}}{dt} \cdot \hat{n} = \sum \vec{F} \cdot \hat{n} \qquad (A.8)$$

Substitute Eqs. (A.3), (A.4), and (A.6) into the left side of Eq. (A.7) to produce Eq. (A.9). Note that the simplification in Eq. (A.9) between the second and third line relies on the fact that after the dot product is expanded, the terms associated with V cancel out, and the terms associated with $\frac{dV}{dt}$ equal 1, from the Pythagorean identity, Eq. (A.10).

$$m\frac{d\vec{V}}{dt} \cdot \hat{v} = m\left(\frac{dV}{dt}\hat{v} + V\frac{d\hat{v}}{dt}\right) \cdot \hat{v}$$

$$= m\left(\frac{dV}{dt}\left[cos(\gamma_{fpa})\hat{i} + sin(\gamma_{fpa})\hat{j}\right]\right.$$

$$+ V\left[-\dot{\gamma}_{fpa}sin(\gamma_{fpa})\hat{i}\right.$$

$$\left.+\dot{\gamma}_{fpa}cos(\gamma_{fpa})\hat{j}\right]\right)$$

$$\cdot \left(cos(\gamma_{fpa})\hat{i} + sin(\gamma_{fpa})\hat{j}\right)$$

$$= m\frac{dV}{dt}$$

$$\qquad (A.9)$$

$$sin^2(\gamma_{fpa}) + cos^2(\gamma_{fpa}) = 1 \qquad (A.10)$$

Perform equivalent substitutions of Eqs. (A.3), (A.5), and (A.6) into Eq. (A.8) to produce Eq. (A.11). Again, the simplification in Eq. (A.11) between the second and third line relies on the fact that after the dot product is expanded, the terms associated with V equal 1, and the terms associated with $\frac{dV}{dt}$ cancel out.

$$m\frac{d\vec{V}}{dt} \cdot \hat{n} = m\left(\frac{dV}{dt}\hat{v} + V\frac{d\hat{v}}{dt}\right) \cdot \hat{n}$$

$$= m\left(\frac{dV}{dt}\left[cos(\gamma_{fpa})\hat{i} + sin(\gamma_{fpa})\hat{j}\right]\right.$$

$$+ V\left[-\dot{\gamma}_{fpa}sin(\gamma_{fpa})\hat{i}\right.$$

$$\left.+\dot{\gamma}_{fpa}cos(\gamma_{fpa})\hat{j}\right]\right)$$

$$\cdot \left(-sin(\gamma_{fpa})\hat{i} + cos(\gamma_{fpa})\hat{j}\right)$$

$$= mV\dot{\gamma}_{fpa}$$

$$\qquad (A.11)$$

Substitute Eqs. (A.9) and (A.11) back into the force balance (Eqs. (A.7) and (A.8)) to obtain the governing equations for launch mission design, Eq. (A.12) and Eq. (A.13) for the direction of flight and the direction normal to flight, respectively.

$$\frac{dV}{dt} = \frac{F}{m}cos(\theta - \gamma_{fpa}) - \frac{F_D}{m} - gsin(\gamma_{fpa})$$

$$\qquad (A.12)$$

$$V\frac{d\gamma_{fpa}}{dt} = \frac{F}{m}sin(\theta - \gamma_{fpa}) + \frac{F_L}{m} - gcos(\gamma_{fpa})$$

$$\qquad (A.13)$$

A.2 Thrust

The goal is to derive an equation for the thrust from a rocket motor, F. This derivation follows the derivation approach of Ref. [97], with permission. Fig. A.2 shows a rocket secured to a thrust stand. The rocket produces thrust by expelling propellant mass from a combustion chamber through a nozzle that accelerates the flow. The rocket is held static on the test stand. Therefore the test stand applies a force to the rocket that is equal thrust produced by the rocket, F. In Fig. A.2, $\dot{m}_p$ refers to the propellant mass flow rate, A_e is the exit area of the nozzle, P_e, ρ_e, and V_e are the average pressure, density, and velocity of the gas at the nozzle exit, respectively. To analyze the rocket, a control volume is used that encompasses the full internal volume of the rocket, as shown in Fig. A.2. Some additional areas need to be defined for this derivation: A_S is the external surface of the rocket, A_C

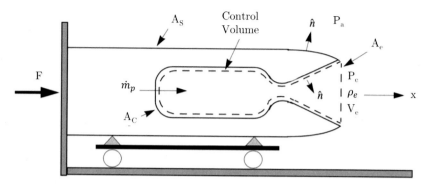

FIGURE A.2 Schematic of a rocket on a thrust stand showing the control volume (dashed lines) used to derive the thrust equation.

is the inside surface of the combustion chamber, $\hat{n}$ is the unit normal to these areas. The external atmosphere is at pressure P_a.

A force balance on the rocket in the x direction of Fig. A.2 produces Eq. (A.14). This equation is equal to 0 (the total force acting on the rocket is 0) since the rocket is at rest on the thrust stand. Note that V_{xm} is the velocity of the injected propellant in the x-direction, and so the last term in Eq. (A.14) is the force from injected propellant. P in Eq. (A.14) is the gas pressure acting at any point on the surface of the rocket. $\bar{\bar{\tau}}$ is the viscous stress tensor. $\bar{\bar{I}}$ is the identity matrix.

$$
0 = F + \int_{A_S} \left(P\bar{\bar{I}} - \bar{\bar{\tau}} \right) \cdot \hat{n} dA \Big|_x \\
+ \int_{A_C} \left(P\bar{\bar{I}} - \bar{\bar{\tau}} \right) \cdot \hat{n} dA \Big|_x + \dot{m}_p V_{xm}
$$
(A.14)

To simplify Eq. (A.14), consider the force balance on the rocket in the x direction when it is not firing. In this case, there is no thrust, or corresponding reaction force from the restraint, and the internal and external surfaces of the rocket are all at ambient pressure, P_a. In this case, the force balance is given by Eq. (A.15).

$$
0 = \int_{A_S} P_a \bar{\bar{I}} \cdot \hat{n} dA \Big|_x + \int_{A_C} P_a \bar{\bar{I}} \cdot \hat{n} dA \Big|_x
$$
(A.15)

A control volume analysis of this same scenario with the rocket inactive (fluid is all at rest) produces Eq. (A.16).

$$
0 = \int_{A_C} P_a \bar{\bar{I}} \cdot \hat{n} dA \Big|_x + \int_{A_e} P_a \bar{\bar{I}} \cdot \hat{n} dA \Big|_x
$$
(A.16)

Eq. (A.16) can be rewritten as Eq. (A.17).

$$
0 = \int_{A_C} P_a \bar{\bar{I}} \cdot \hat{n} dA \Big|_x + P_a A_e
$$
(A.17)

Substitution of Eq. (A.17) into Eq. (A.15) produces Eq. (A.18).

$$
0 = \int_{A_S} P_a \bar{\bar{I}} \cdot \hat{n} dA \Big|_x - P_a A_e
$$
(A.18)

Substitution of Eq. (A.18) into the original force balance of Eq. (A.14) produces Eq. (A.19). Note that this substitution assumes that the external surface pressure and stress distribution acting on A_S is unchanged by the rocket firing, see Eq. (A.20).

$$
0 = F + P_a A_e + \int_{A_C} \left(P\bar{\bar{I}} - \bar{\bar{\tau}} \right) \cdot \hat{n} dA \Big|_x + \dot{m}_p V_{xm}
$$
(A.19)

$$\left(\int_{A_S} \left(P\bar{\bar{I}} - \bar{\bar{\tau}} \right) \cdot \hat{n} dA \bigg|_x \right)_{during\ hot\ fire}$$

$$= \left(\int_{A_S} P_a \bar{\bar{I}} \cdot \hat{n} dA \bigg|_x \right)_{before\ hot\ fire} \quad (A.20)$$

Eq. (A.20) is not strictly true, since the plume from the rocket firing will mix with the surrounding air, setting the air around the vehicle into some motion and thereby changing the external surface pressure slightly from the atmospheric pressure. However, for rockets of reasonable size/thrust this effect is very small and so Eq. (A.19) is accurate. With the rocket firing, the momentum balance over the control volume is given by Eq. (A.21). Note that this equation uses vector notation, so that $\bar{V}$ refers to the velocity vector in the three dimensions.

$$\frac{D}{Dt} \int_V \rho \bar{V} d\mathcal{V} = \int_V \frac{\partial (\rho \bar{V})}{\partial t} d\mathcal{V}$$

$$= - \int_V \nabla \cdot \left(\rho \bar{V} \bar{V} + P\bar{\bar{I}} - \bar{\bar{\tau}} \right) d\mathcal{V} \quad (A.21)$$

In this case, the combustion is considered to be steady-state, such that the flow inside the combustion chamber is stationary, making the term on the left hand side of Eq. (A.21) equal to 0. Now, the volume integral on the right hand side of Eq. (A.21) is converted to an integral over the surface of the control volume, see Eq. (A.22).

$$0 = \int_V \nabla \cdot \left(\rho \bar{V} \bar{V} + P\bar{\bar{I}} - \bar{\bar{\tau}} \right) d\mathcal{V}$$

$$= \int_{A_C} \left(\rho \bar{V} \bar{V} + P\bar{\bar{I}} - \bar{\bar{\tau}} \right) \cdot \hat{n} dA \quad (A.22)$$

$$+ \int_{A_e} \left(\rho \bar{V} \bar{V} + P\bar{\bar{I}} - \bar{\bar{\tau}} \right) \cdot \hat{n} dA$$

Eq. (A.22) can be simplified with the knowledge that on the surface of A_C the no slip condition requires that the velocity, $\bar{V}$, is 0 except at the injection location. The terms on the surface of A_e can be written in terms of the area averaged values, P_e, ρ_e, and V_e. These two changes

produce Eq. (A.23). Note that small viscous normal forces on the surface A_e are neglected here, but the momentum of the injected propellant is included.

$$\int_{A_C} \left(P\bar{\bar{I}} - \bar{\bar{\tau}} \right) \cdot \hat{n} dA \bigg|_x + \dot{m}_p V_{xm} + \rho_e V_e^2 A_e$$

$$+ P_e A_e = 0$$

$$(A.23)$$

Eq. (A.23) can be substituted back into the original force balance, Eq. (A.19), to produce Eq. (A.24).

$$0 = F + P_a A_e - \left(\rho_e V_e^2 A_e + P_e A_e \right) \quad (A.24)$$

Rearranging Eq. (A.24) can be used to solve for the thrust of the rocket (Eq. (A.25)).

$$F = \rho_e V_e^2 A_e + P_e A_e - P_a A_e \quad (A.25)$$

One final simplification can be made to Eq. (A.25) by recalling that the total propellant mass flow rate is given by Eq. (A.26).

$$\dot{m}_p = \rho_e V_e A_e \quad (A.26)$$

Substitution of Eq. (A.26) into Eq. (A.25) produces the widely used thrust equation, Eq. (A.27).

$$F = \dot{m}_p V_e + (P_e - P_a) A_e \quad (A.27)$$

A.3 Rocket equation

The rocket equation can be derived directly from a force balance in the direction of flight. This was initially derived here in Section A.1, see Eq. (A.12). To simplify this derivation, consider a rocket where the thrust vector is perfectly aligned with the direction of flight, such that $\theta = \gamma_{fp}$ in Fig. A.1. This physically means there is no active thrust vector control or thrust misalignment. Thus, Eq. (A.12) becomes Eq. (A.28).

$$\frac{dV}{dt} = \frac{F}{m} - \frac{F_D}{m} - g sin(\gamma_{fpa}) \quad (A.28)$$

Now, assume that the performance of the rocket is constant, and recall that the thrust can be written in terms of I_{sp} per Eq. (A.29).

$$F = I_{sp}g_0\dot{m}_p \qquad (A.29)$$

Substitute Eq. (A.29) into Eq. (A.28) to obtain (A.30).

$$\frac{dV}{dt} = \frac{I_{sp}g_0\dot{m}_p}{m} - \frac{F_D}{m} - gsin(\gamma_{fpa}) \qquad (A.30)$$

Recall that the rate of change of mass of the rocket, $\dot{m}$ is the negative of the propellant mass flow rate, i.e. $\dot{m} = -\dot{m}_p$ and that $\frac{d(ln(m))}{dt} = \frac{\dot{m}}{m}$. Substituting these into Eq. (A.30) produces Eq. (A.31).

$$\frac{dV}{dt} = -I_{sp}g_0\frac{d(ln(m))}{dt} - \frac{F_D}{m} - gsin(\gamma_{fpa})$$
$$(A.31)$$

Integration of Eq. (A.31) produces (A.32).

$$\Delta V|_{achieved} = \Delta V|_{rocket} - \Delta V|_{drag} - \Delta V|_{gravity}$$
$$(A.32)$$

Where:

$$\Delta V|_{gravity} = \int_0^{t_b} gsin(\gamma_{fpa})dt \qquad (A.33)$$

$$\Delta V|_{drag} = \int_0^{t_b} \frac{F_D}{m}dt \qquad (A.34)$$

$$\Delta V|_{rocket} = I_{sp}g_0 \ln\frac{m_i}{m_f} \qquad (A.35)$$

The ΔV produced by the rocket is therefore given by Eq. (A.35), which is commonly known as the Tsiolkovsky rocket equation and is used throughout this book. (An interesting side note: while credited to Tsiolkovsky, the rocket equation was actually derived for the first time nearly a century earlier by British mathematician William Moore.) Here, m_i is the initial mass of the rocket and m_f is the mass of the rocket after the burn. Eq. (A.32) shows that the ΔV

required to be produced by the rocket must account for the ΔV needed to overcome gravity losses and drag, as well as to provide the orbital ΔV.

A.4 Water hammer pressure surge

The water hammer pressure surge equation is derived here for valve closure, though as discussed in Section 8.2.4 of Chapter 8, it is often also used for estimation of the expected water hammer during system priming. This derivation follows the approach in Ref. [252]. This derivation holds when a valve closes rapidly, i.e. when the time taken for the pressure wave to travel from the valve to the inlet and back to the valve exceeds the time taken to close the valve, $(2L/a \geqslant t_{close})$.

Consider fluid flowing at velocity V, through a tube of diameter D. When a valve in the line is closed, all of that fluid is brought to rest, producing a transient pressure surge known as water hammer. The fluid is initially traveling at velocity, V, but after the pressure surge passes through it, it is at rest. Assuming that the pressure surge moves at the speed of sound, a, then in time dt, the fluid travels a distance $L_{dt} = adt$.

The momentum of the fluid in length L_{dt} prior to the valve closing is given by Eq. (A.36). Here, A is the pipe cross sectional area at the location where momentum is halted and ρ is the density of the fluid.

$$mV = \rho A L_{dt} V = \rho A a V dt \qquad (A.36)$$

After the pressure wave passes, the momentum in L_{dt} is zero. Thus the change in momentum from the pressure wave is $\rho A a V dt$. The force that produces this change in momentum, F_{wh} is the force of the pressure wave, Eq. (A.37).

$$F_{wh} = PA \qquad (A.37)$$

Force is simply the change in momentum per unit time, Eq. (A.38).

$$F_{wh} = \frac{d(mV)}{dt} \tag{A.38}$$

Substitution of Eqs. (A.36) and (A.37) into Eq. (A.38) produces Eq. (A.39).

$$P = \rho V a \tag{A.39}$$

Eq. (A.39) is the widely used equation for predicting the water hammer pressure surge, P. Note that P in Eq. (A.39) is the peak magnitude of the transient pressure rise above the static pressure of the fluid. a is the speed of sound, or celerity, of the fluid. The expression for a can be derived by considering conservation of energy.

Consider the entire length of tube, L, in which fluid is brought to rest by the water hammer pressure wave. The water hammer pressure surge converts the initial kinetic energy in the column of fluid, E_K, to strain energy in the compressed fluid, $E_{S,f}$, and in the stretched pipe, $E_{S,p}$ [252], this is stated in Eq. (A.40).

$$E_K = E_{S,f} + E_{S,p} \tag{A.40}$$

The initial kinetic energy of the fluid moving through the tube is given by Eq. (A.41).

$$E_K = \frac{1}{2}\rho V^2 L\pi \frac{D^2}{4} \tag{A.41}$$

Eq. (A.42) gives the expression for the increased strain energy of the stretched pipe, $E_{S,p}$. Here, th_w is the wall thickness of the tubing and E is the modulus of elasticity of the tubing.

$$E_{S,p} = \frac{1}{2}\left(\frac{PD}{2th_w}\right)^2 \frac{\pi D L th_w}{E} \tag{A.42}$$

Eq. (A.43) gives the expression for the increased strain energy of the compressed fluid, $E_{S,f}$. Here, K_{prop} is the bulk modulus of elasticity of the propellant.

$$E_{S,f} = \frac{1}{2}\frac{P^2}{K_{prop}}\frac{\pi D^2 L}{4} \tag{A.43}$$

Substituting Eqs. (A.41), (A.42), and (A.43) into Eq. (A.40) and rearranging in terms of the pressure, P, gives Eq. (A.44).

$$P = \frac{V\sqrt{K_{prop}\rho}}{\sqrt{1 + \dfrac{K_{prop}D}{th_w E}}} \tag{A.44}$$

Substitute Eq. (A.39) into Eq. (A.44) and rearrange in terms of a to produce Eq. (A.45).

$$a = \frac{\sqrt{\dfrac{K_{prop}}{\rho}}}{\sqrt{1 + \dfrac{K_{prop}D}{th_w E}}} \tag{A.45}$$

At this point, recall that the hard liquid speed of sound, a_{HL}, is given by Eq. (A.46), and thus Eq. (A.45) can be rewritten in terms of a_{HL} as Eq. (A.47).

$$a_{HL} = \sqrt{\frac{K_{prop}}{\rho}} \tag{A.46}$$

$$a = \frac{a_{HL}}{\sqrt{1 + \dfrac{K_{prop}D}{th_w E}}} \tag{A.47}$$

As discussed in Chapter 8, the only difference in a_{HL} and a is that the latter accounts for the adjustment to the speed of sound as a result of elastic deformation of the tubing. Eq. (A.39) and (A.47) are together the two water hammer equations presented in Chapter 8.

A.5 Flow through a venturi

In this section, the equations used to calculate mass flow rate through a venturi (as used in Chapter 11 Section 11.6.5 for oxidizer mass flow rate measurements) are derived. The derivation is treated separately for an incompressible fluid and for a compressible gas. However, both derivations use the same terminology for the venturi: the subscript 1 denotes the conditions at the inlet to the venturi and 2 denotes conditions at the venturi throat.

A.5.1 Incompressible fluid

The derivation for the incompressible mass flow rate is relatively straightforward.

Start with Bernoulli's equation, Eq. (A.48).

$$P_1 + \frac{1}{2}\rho V_1^2 = P_2 + \frac{1}{2}\rho V_2^2 \qquad (A.48)$$

Which can be rearrange Bernoulli's Equation to get the following:

$$2(P_1 - P_2) = \rho V_2^2 - \rho V_1^2 \qquad (A.49)$$

Now recall the definition of mass flow rate, Eq. (A.50), and rearrange in terms of velocity resulting in Eq. (A.51).

$$\dot{m} = \rho V A \qquad (A.50)$$

$$V = \frac{\dot{m}}{\rho A} \qquad (A.51)$$

Substitution of Eq. (A.51) into Eq. (A.49) and recognition that conservation of mass requires that $\dot{m}_1 = \dot{m}_2$ gives Eq. (A.52).

$$2\rho(P_1 - P_2) = \frac{\dot{m}^2}{A_2^2} - \frac{\dot{m}^2}{A_1^2} = \frac{\dot{m}^2}{A_2^2}\left(1 - \left(\frac{A_2}{A_1}\right)^2\right) \qquad (A.52)$$

Now, rearrange Eq. (A.52) for mass flow rate.

$$\dot{m} = A_2 \sqrt{\frac{2\rho(P_1 - P_2)}{1 - \left(\frac{A_2}{A_1}\right)^2}} \qquad (A.53)$$

Eq. (A.53) is the ideal mass flow rate. To account for losses, use Eq. (A.54) and substitute in Eq. (A.53). Now, the mass flow rate through a real venturi can be calculated based on pressure drop via Eq. (A.55). This is the equation given as Eq. (11.2) in Chapter 11.

$$\dot{m}_{real} = C_d \dot{m}_{ideal} \qquad (A.54)$$

$$\dot{m} = C_d A_2 \sqrt{\frac{2\rho(P_1 - P_2)}{1 - \left(\frac{A_2}{A_1}\right)^2}} \qquad (A.55)$$

A.5.2 Compressible gas

For a compressible gas, assume that flow through the venturi is adiabatic and isentropic with the exception of any small losses accounted for by the discharge coefficient of the venturi, C_d, according to Eq. (A.54).

Recall again the definition of mass flow rate, Eq. (A.56).

$$\dot{m}_{ideal} = \rho V A \qquad (A.56)$$

Now, look at the conservation of energy for 1D flow across the venturi (Eq. (A.57)).

$$h_{t1} = h_{t2}. \qquad (A.57)$$

Assuming that the gas is ideal and calorically perfect, Eq. (A.57) becomes Eq. (A.58).

$$c_p T_1 + \frac{V_1^2}{2} = c_p T_2 + \frac{V_2^2}{2}. \qquad (A.58)$$

The goal is now to rewrite the specific heat at constant pressure, c_p, in terms of the ratio of specific heats, γ, and the specific gas constant, R. For an ideal gas, c_p and c_v are related by the gas constant, R, see Eq. (A.59).

$$c_v = c_p - R \qquad (A.59)$$

Now, recall the definition of γ per Eq. (A.60).

$$\gamma = \frac{c_p}{c_v} \qquad (A.60)$$

Eq. (A.59) and (A.60) together allow c_p to be written as a function of only R and γ, see Eq. (A.61).

$$c_p = \frac{R\gamma}{\gamma - 1} \qquad (A.61)$$

The ideal gas law, in the form of Eq. (A.62), is used.

$$RT = \frac{P}{\rho} \qquad (A.62)$$

It is also time to also apply the assumption of isentropic flow between the venturi inlet (1) and the venturi throat (2), see Eq. (A.63).

$$\frac{P_1}{P_2} = \left(\frac{\rho_1}{\rho_2}\right)^{\gamma}. \qquad (A.63)$$

Inserting Eqs. (A.54), (A.56), (A.61), (A.62), and (A.63), into Eq. (A.58) and rearranging, Eqs. (A.67) and (A.68) are obtained after some manipulation. This is the same form of equation as that of Eqs. (11.3) and (11.4) in Chapter 11.

$$\dot{m} = C_d A_2 \sqrt{\frac{\frac{2\gamma}{\gamma-1} P_1 \rho_1 \left[\left(\frac{P_2}{P_1}\right)^{\frac{2}{\gamma}} - \left(\frac{P_2}{P_1}\right)^{\frac{\gamma+1}{\gamma}}\right]}{1 - \left(\frac{A_2}{A_1}\right)^2 \left(\frac{P_2}{P_1}\right)^{\frac{2}{\gamma}}}} \qquad (A.64)$$

$$\dot{m} = C_d A_2 \sqrt{\frac{2(P_1 - P_2)\rho_1}{1 - \left(\frac{A_2}{A_1}\right)^2}}$$

$$\times \sqrt{\frac{\frac{\gamma}{\gamma-1}\left(\frac{P_2}{P_1}\right)^{\frac{2}{\gamma}}\left[1 - \left(\frac{P_2}{P_1}\right)^{\frac{\gamma+1}{\gamma}}\right]\left(1 - \left(\frac{A_2}{A_1}\right)^2\right)}{\left(1 - \frac{P_2}{P_1}\right)\left(1 - \left(\frac{A_2}{A_1}\right)^2 \left(\frac{P_2}{P_1}\right)^{\frac{2}{\gamma}}\right)}} \qquad (A.65)$$

$$\dot{m} = C_d A_2 \sqrt{\frac{2(P_1 - P_2)\rho_1}{1 - \left(\frac{A_2}{A_1}\right)^2}}$$

$$\times \sqrt{\frac{\frac{\gamma}{\gamma-1}\left(\frac{P_2}{P_1}\right)^{\frac{2}{\gamma}}\left[1 - \left(\frac{P_2}{P_1}\right)^{\frac{\gamma-1}{\gamma}}\right]\left(1 - \left(\frac{A_2}{A_1}\right)^2\right)}{\left(1 - \frac{P_2}{P_1}\right)\left(1 - \left(\frac{A_2}{A_1}\right)^2 \left(\frac{P_2}{P_1}\right)^{\frac{2}{\gamma}}\right)}} \qquad (A.66)$$

$$\dot{m} = C_d A_2 Y_{exp} \sqrt{\frac{2(P_1 - P_2)\rho_1}{1 - \left(\frac{A_2}{A_1}\right)^2}} \qquad (A.67)$$

$$Y_{exp} =$$

$$\sqrt{\frac{\frac{\gamma}{\gamma-1}\left(\frac{P_2}{P_1}\right)^{\frac{2}{\gamma}}\left[1 - \left(\frac{P_2}{P_1}\right)^{\frac{\gamma-1}{\gamma}}\right]\left(1 - \left(\frac{A_2}{A_1}\right)^2\right)}{\left(1 - \frac{P_2}{P_1}\right)\left(1 - \left(\frac{A_2}{A_1}\right)^2 \left(\frac{P_2}{P_1}\right)^{\frac{2}{\gamma}}\right)}} \qquad (A.68)$$

Common oxidizer material compatibility

The following information has been compiled as a general guide for the types of materials that may be compatible with oxidizers of interest in hybrid rocket propulsion. This is not an exhaustive list. The propulsion engineer is charged with ensuring the material she or he selects is appropriate (and sufficiently clean) for their application. The references cited in Tables B.1 and B.2 have significantly more detailed information that can be helpful in the design process, e.g., test data, recommendations for applications, autoignition temperatures, etc.

While a substantial amount of test data exists on material compatibility exists, most of it is quite old (e.g., tests carried out in the 1940s–1960s), and much has changed in the intervening years. For example, as discussed in Chapter 5, production of Hydrogen Peroxide has improved substantially since that time. Therefore it is likely that many of the materials listed below are more compatible with current HP [188]. It is also possible that new materials have been found for use with these oxidizers.

Several of the materials listed below are actually fuels, e.g., aluminum. They may safely be used under controlled operating conditions; however, the propulsion engineer must be aware that they are completing two of the three sides of the fire triangle. Also, the authors have first-hand experience with finding the operating conditions under which combustion with oxygen occurs for many materials that are otherwise compatible.

While not a focus of this handbook, data on the compatibility of materials with nitric acid (RFNA and IRFNA) is available in Ref. [23].

TABLE B.1 Oxidizer Compatibly with Metals.

	N_2O [397], [398], [379]	O_2 [301]	H_2O_2 [399], [188]	NTO/MON3 [23]
Aluminum	Satisfactory	Difficult to ignite, but burn at low pressures. Poor performance under friction heating. Anodization can improve resistance to particle impact. Contaminants significantly increase ignition potential.	Grades 1060, 1100, 1160, 1260, 5254, 6063, 7072, B356 have been shown to be excellent. Likely more alloys are as well, but require testing with current H_2O_2, e.g. Al 6061 was found to be good in the 1960's but may be improved now. Note: Marshall Space Flight Center has developed Al-Mg alloys particularly well suited for HP storage in support of the X-43B development.	Probably compatibility, water content must be very limited.
Brass	Acceptable. Corrosive in presence of moisture.	Satisfactory		Incompatible
Nickel	Avoid as it is a common catalyst.	Nickel 200 is suitable at high pressure (up to 69 MPa), even as a mesh e.g. filter	Poor	Incompatible
Copper	Can be satisfactory, but corrosive in presence of moisture. It is a common catalyst.	Can be suitable for use at high pressure (up to 69 MPa) depending on the alloy. Good resistance to particle impact/high velocity applications. Not suitable as wire mesh. Low ductility oxide can contaminate the system. Sintered bronze is less flammable that Monel 400. Aluminum-bronze (> 5% Al) is not recommended.	Poor	Incompatible
Ferritic Steels (e.g. Carbon steels)	Satisfactory but corrosive in presence of moisture.		Poor	Doubtful compatibility, some specific alloys are compatible
Stainless Steel	Satisfactory. Preference for 304 or 316SS	Widely used, but are flammable at relatively low pressures and susceptible to friction heating and particle impact. Wire mesh and thin walled tubes are flammable at ambient pressure. Should not be used in dynamic locations. Preferable to Al.	Good for Alloys 202, 304, 209, 310, 316, 317, 318, 321, 322	Doubtful compatibility
Titanium		Not suitable. α, $\alpha - \beta$ and β alloys show ignition at pressures as low as 7 kPa. GOx and LOx have been shown to propagate and consume all available metal.	Poor	Compatible

continued on next page

TABLE B.1 (*continued*)

	N$_2$O [397], [398], [379]	O$_2$ [301]	H$_2$O$_2$ [399], [188]	NTO/MON3 [23]
Zirconium			Excellent, improved with 2% Hafnium	Incompatible
Inconel		Suitable at high pressure (up to 69 MPa) and has excellent resistance to friction ignition (e.g. MA754). However, 718 is roughly equivalent to 300-series stainless steel.	FairPoor	Probable compatibility
Monel		Suitable at high pressure (up to 69 MPa) and with high gas velocities. K-500 for valve stems, 400-series for valve bodies. Can be used as a mesh.	Poor	Doubtful compatibility
Molybdenum			Poor	Incompatible
Hastelloy		C-22 and C-276 are much more ignition resistant than stainless steel and Inconel 718.		

TABLE B.2 Oxidizer Compatibly with Metals.

	N$_2$O [397], [400]	O$_2$ [301]	H$_2$O$_2$ [399], [188]	NTO/MON3 [23]
Polytetrafluoroethylene (PTFE), e.g. Teflon	Satisfactory	Satisfactory, but poor creep resistance. Glass filled Teflon is more susceptible to ignition than Teflon on its own.	Satisfactory for general use	Probable compatibility
Polychlorotrifluoroethylene (PCTFE), e.g. Kel-F	Satisfactory		Satisfactory for general use/repeated short term use/single short term use	Probable compatibility
Vinylidene polyfluoride (PVDF) e.g. Kynar	Acceptable but possible ignition under certain conditions			Incompatible
Polyamide (PA) e.g. Nylon	Acceptable but possible ignition under certain conditions	Satisfactory, but less favorable than fully fluorinated materials.		
Polypropylene (PP)	Acceptable but possible ignition under certain conditions	Poor		Incompatible

continued on next page

TABLE B.2 (continued)

	N_2O [397], [400]	O_2 [301]	H_2O_2 [399], [188]	NTO/MON3 [23]
Butyl (isobutene - isoprene) rubber (IIR)	Non recommended, possible ignition and significant swelling.			Incompatible
Nitrile rubber (NBR), e.g. Buna N	Non recommended, possible ignition and significant swelling.	Poor	Unsatisfactory	Incompatible
Chloroprene (CR)	Non recommended, possible ignition and significant swelling.			
Chlorofluorocarbons e.g. FKM, Viton	Non recommended, significant swelling	Satisfactory	Satisfactory for general/short term use	
Silicone (Q)	Satisfactory	Can be used if ignition sources are minimized	Satisfactory for repeated short term use/single short term use	Incompatible
Ethylene - Propylene (EPDM)	Non recommended, possible ignition and significant swelling.			
Hydrocarbon based lubricant	Non recommended, possible ignition.	Not suitable.		
Fluorocarbon based lubricant	Satisfactory	Satisfactory. These are the most commonly used lubricants for oxygen service e.g. Chlorotrifluoroethylene (CTFE), Polytetrafluoroethylene (PTFE), Fluorinated ethylene propylene (FEP), such as Fluorolube, Braycote, Krytox. Any additives must be evaluated for compatibility. For example, SiO_2 added to CTFE can lead to severe corrosion and is not recommended.	Satisfactory for repeated short term use	Probable compatibility
Mylar			Satisfactory for general use	Incompatible
Polyethylene			Satisfactory for repeated short term use	Incompatible
Vespel		Satisfactory		

C

Summary of design equations

TABLE C.1   Summary of Performance and Design Equations.

Parameter	Equation	Assumptions	Equation
Thrust	$F = \dot{m}_p V_e + A_e(P_e - P_a)$		(2.1)
	$F = \dot{m}_p C^* C_F$		(2.2)
	$F = (\dot{m}_O + \dot{m}_F) C^*_{ideal} \eta_{C^*} C_{F,ideal} \eta_n$	Real performance	(3.65), (7.43)
	$F = I_{sp} \dot{m}_p g_0 \eta_{C^*} \eta_n$	Real performance	(7.43)
Thrust Coefficient	$C_F \equiv \frac{F}{P_{t,c} A_{th}}$	Ideal Performance	(2.3)
	$C_F = \frac{\dot{m}_p V_e + (P_e - P_a) A_e}{P_{t,c} A_{th}}$		(2.4)
	$C_F = \left(\frac{P_e}{P_{t,c}}\right)\left(\frac{A_e}{A_{th}}\right)\left(\gamma M_e^2 + 1 - \frac{P_a}{P_e}\right)$	Ideal gas	(2.5)
	$C_F = \left\{ \left(\frac{2\gamma^2}{\gamma-1}\right)\left(\frac{2}{\gamma+1}\right)^{\frac{\gamma+1}{\gamma-1}} \left[1 - \left(\frac{P_e}{P_{t,c}}\right)^{\frac{\gamma-1}{\gamma}} \right] \right\}^{\frac{1}{2}}$ $+ \left(\frac{P_e}{P_{t,c}} - \frac{P_a}{P_{t,c}}\right)\frac{A_e}{A_{th}}$	Ideal gas, 1D, frozen, isentropic	(2.6), (7.9)
Characteristic Velocity	$C^* \equiv \frac{P_{t,c} A_{th}}{\dot{m}_p}$	Adiabatic, isentropic between combustion chamber and nozzle throat	(2.10)
	$C^* = \left[\frac{1}{\gamma} \left(\frac{\gamma+1}{2}\right)^{\frac{\gamma+1}{\gamma-1}} \frac{R_u T_{t,c}}{M_w} \right]^{\frac{1}{2}}$	Adiabatic, isentropic between combustion chamber and nozzle throat	(2.11)
C^* Efficiency	$\eta_{C^*} = \frac{C^*\|_{measured}}{C^*\|_{ideal}} = \frac{P_{t,c}\|_{measured}}{P_{t,c}\|_{ideal}}$		(2.12)
Effective Exhaust Velocity	$C = \frac{F}{\dot{m}_p}$		(2.13)
	$C = C_F C^*$		(2.14)
	$C = V_e \left(1 + \frac{1}{\gamma M_e^2} \left(1 - \frac{P_a}{P_e} \right) \right)$		(2.15)

continued on next page

Summary of design equations

TABLE C.1 (continued)

Parameter	Equation	Assumptions	Equation
Specific Impulse	$I_{sp} = \frac{F}{\dot{m}_p g_0} = \frac{C}{g_0}$		(2.16)
	$I_{sp} = \frac{1}{g_0} \sqrt{\frac{\gamma R_u T_e}{M_w}} \left(M_e + \frac{1}{\gamma M_e} \left(1 - \frac{P_a}{P_e} \right) \right)$		(2.17)
	$I_{sp} = \frac{1}{g_0} \sqrt{\frac{2\gamma R_u T_{t,c}}{M_w (\gamma - 1)} \left[1 - \left(\frac{P_e}{P_{t,c}} \right)^{\frac{\gamma-1}{\gamma}} \right]}$	Perfectly expanded	(2.18)
	$I_{sp,real} = I_{sp,ideal} \eta_{C^*} \eta_n$	Real performance	(7.7)
Total Impulse	$I = \int_{t_0}^{t_b} F \, dt$		(2.19)
	$I = I_{sp} m_p g_0$		(2.20)
	$I = C_F P_{t,c} A_{th} t_b$		(2.21)
Impulse Density	$I_d = \rho_p I_{sp}$		(2.22)
Delta Velocity	$\Delta V = I_{sp} g_0 \ln \frac{m_i}{m_f}$		(2.27)
	$\Delta V = I_{sp} \eta_{C^*} \eta_n g_0 \ln \frac{m_i}{m_f}$	Real performance	(7.10)
Mass Flow Rate	$\dot{m}_p = \frac{\gamma A_{th} P_{t,c}}{\sqrt{\gamma R T_{t,c}}} \left(\frac{2}{\gamma+1} \right)^{\frac{\gamma+1}{2(\gamma-1)}}$	Compressible, isentropic, choked flow	(2.8)
	$\dot{m}_p = \frac{P_c A_{th} C_d}{\eta_{C^*} C^*}$	Real performance	(7.15)
	$\dot{m}_p = \frac{F}{I_{sp} g_0 \eta_{C^*} \eta_n}$	Real performance	(7.16)
Propellant Mass and Density	$m_p = \int_{t_0}^{t_b} \dot{m}_p \, dt$		(2.9)
	$m_p = m_i \left(1 - e^{\frac{-\Delta V}{I_{sp,real} g_0}} \right)$		(7.10)
	$\rho_p = \frac{m_p}{\Psi_{total}} = \frac{\rho_O \rho_F (1 + O/F)}{\rho_O + O/F \rho_F}$		(2.23)
Fuel Mass	$m_F = \frac{m_p}{1 + O/F}$		(7.11)
Oxidizer Mass	$m_O = \frac{O/F}{1 + O/F} m_p$		(7.12)
Characteristic Length	$L^* = \Psi_C / A_{th}$		(2.24)
Nozzle Exit Mach Number	$\frac{1}{\mathcal{R}} = \frac{A_{th}}{A_e} = \left(\frac{\gamma+1}{2} \right)^{\frac{\gamma+1}{2(\gamma-1)}} \frac{M_e}{\left(1 + \frac{\gamma-1}{2} M_e^2 \right)^{\frac{\gamma+1}{2(\gamma-1)}}}$	Adiabatic, isentropic, and calorically perfect flow	(2.25)
Area Ratio	$\mathcal{R} = \frac{A_e}{A_{th}} = \frac{r_e^2}{r_{th}^2}$		(7.3)
	$\mathcal{R} = \frac{\sqrt{\gamma \left(\frac{2}{\gamma+1} \right)^{\frac{\gamma+1}{\gamma-1}}}}{\left(\frac{P_e}{P_{t,c}} \right)^{\frac{1}{\gamma}} \sqrt{\frac{2\gamma}{\gamma-1} \left[1 - \left(\frac{P_e}{P_{t,c}} \right)^{\frac{\gamma-1}{\gamma}} \right]}}$	Adiabatic, isentropic, and calorically perfect flow	(2.26)
Critical Pressure Ratio	$\frac{P_{t,c}}{P_a} \geqslant \left(\frac{\gamma+1}{2} \right)^{\frac{\gamma}{\gamma-1}}$		(2.7)
Regression Rate	$\dot{r} = a G^n x^{-m}$		(3.20), (3.45), (6.1)
	$\dot{r} = a_o G_o^n$	Dependence only on G_o	(3.48), (6.2), (7.1), (7.13)
	$\dot{r} = a \left(\frac{\dot{m}_p}{\pi r^2} \right)^n x^{-m}$	Single, cylindrical port	(3.51)
	$\bar{\bar{r}} = \frac{r_f - r_i}{t_b}$	Space-time averaged	(6.7)
Oxidizer Mass Flux	$G_o = \frac{\dot{m}_O}{\pi r^2}$		(3.49), (3.61), (7.39)

continued on next page

TABLE C.1 *(continued)*

Parameter	Equation	Assumptions	Equation
Fuel Mass Flow Rate	$\dot{m}_F = \rho_F A_b \dot{r}$		(3.62), (6.3)
	$\dot{m}_F = \frac{\dot{m}_O}{O/F}$		(6.17)
	$\dot{m}_F = 2\rho_F L_F \pi^{1-n} r^{1-2n} a_o \dot{m}_O^n$	Single, cylindrical port, dependence only on G_o	(6.5)
	$\dot{m}_F(t) = \rho_F \pi D_p L_F \dot{r}$	Single, cylindrical port	(7.40)
Oxidizer to Fuel Ratio	$O/F = \frac{\dot{m}_O}{\dot{m}_F}$		(3.63)
	$O/F = \frac{\dot{m}_O}{\dot{m}_f} = \frac{r^{2n-1} \dot{m}_O^{1-n}}{2a_o \rho_F \pi^{1-n} L_F}$	Single, cylindrical port, dependence only on G_o	(3.66), (7.42)
Chamber Pressure	$P_c = \frac{(\dot{m}_O + \dot{m}_F) C_{ideal}^* \eta_{C^*}}{A_{th} C_d}$		(3.64), (7.41)
Nozzle Throat Area	$A_{th} = \frac{\dot{m}_p C_{ideal}^* \eta_{C^*}}{P_c C_d}$		(7.2)
	$A_{th} = \frac{F}{\eta_n C_F P_c C_d}$		(7.14)
Port Diameter	$D_p(t) = \left[D_i^{2n+1} + \frac{(2n+1)(2^{2n+1})a_o}{\pi^n} \dot{m}_O^n t \right]^{\frac{1}{2n+1}}$	Single, cylindrical port, constant $\dot{m}_O$	(7.21)
	$D_f = \left\{ \left[\frac{(2n+1)2^{2n+1} a_o}{\pi^n} \right] \frac{\dot{m}_O^n t_b}{1-(D_i/D_f)^{2n+1}} \right\}^{1/(2n+1)}$	Single, cylindrical port, constant $\dot{m}_O$	(7.26)
Fuel Grain Length	$L_F = \frac{4m_F}{\pi \rho_F (D_f^2 - D_i^2)}$	Single, cylindrical port	(7.29)
Burn time	$t_b = \left[\frac{\left(D_f^{2n+1} - D_i^{2n+1} \right) \pi^n}{a_o(2n+1)2^{(2n+1)} m_O^n} \right]^{\frac{1}{1-n}}$	Single, cylindrical port, constant $\dot{m}_O$	(7.32)

References

[1] R.W. Humble, G.N. Henry, W.J. Larson, et al., Space Propulsion Analysis and Design, vol. 1, McGraw-Hill, New York, 1995.

[2] D.K. Huzel, D.H. Huang, Modern Engineering for Design of Liquid-Propellant Rocket Engines, vol. 147, American Institute of Aeronautics and Astronautics, Washington, DC, 1992.

[3] D. Balzar, T. Barksdale, D. Gilmore, F. Greene, R. Peterson, Advanced Propellant Management System for Spacecraft Propulsion Systems. Phase 1-Survey Study and Evaluation, Tech. Rep. NASA-CR-101913, 1969.

[4] J. Stark, M. Blatt, F. Bennett Jr, B. Campbell, Fluid Management Systems Technology Summaries, No. NASA CR-134748, 1974.

[5] Swagelok Company, Stainless steel seamless tubing and tube support systems fractional, metric, and imperial sizes, https://www.swagelok.com/downloads/webcatalogs/en/ms-01-181.pdf, 2018. (Accessed 30 November 2021).

[6] G.P. Sutton, O. Biblarz, Rocket Propulsion Elements, John Wiley & Sons, 2010.

[7] Standard: Space Systems - Metallic Pressure Vessels, Pressurized Structures, and Pressure Components (ANSI/AIAA S-080A-2018), Standard, American Institute of Aeronautics and Astronautics, 2018.

[8] Standard: Space Systems - Composite Overwrapped Pressure Vessels (ANSI/AIAA S-081B-2018), Standard, American Institute of Aeronautics and Astronautics, 2018.

[9] J. Rabinovitch, E.T. Jens, A.C. Karp, B. Nakazono, A. Conte, D.A. Vaughan, Characterization of Poly-MethylMethAcrylate as a Fuel for Hybrid Rocket Motors, 2018.

[10] E.T. Jens, A.C. Karp, J. Rabinovitch, A. Conte, B. Nakazono, D.A. Vaughan, Design of interplanetary hybrid CubeSat and SmallSat propulsion systems, in: 2018 Joint Propulsion Conference, 2018, p. 4668.

[11] C.S. Clair, D. Gramer, E. Rice, W. Knuth, Advanced cryogenic solid hybrid rocket engine developments - Concept and test results, 1998.

[12] M. Chiaverini, Review of Solid-Fuel Regression Rate Behavior in Classical and Nonclassical Hybrid Rocket Motors, American Institute of Aeronautics and Astronautics, 2007, pp. 37–126, 2015/07/09.

[13] T. Shimada, Hybrid Rocket will go for the next generation of space transportation, in: The Forefront of Space Science, JAXA, Institute of Space and Astronautical Science, 2012.

[14] V. Lukashov, V.V. Terekhov, V.I. Terekhov, Near-wall flows of chemical reactants: a review of the current status of the problem9, Combustion, Explosion, and Shock Waves 51 (2) (2015) 160–172.

[15] B.J. Cantwell, M.A. Karabeyoglu, D. Altman, Recent advances in hybrid propulsion, International Journal of Energetic Materials and Chemical Propulsion 9 (4) (2010).

[16] Y. Saito, T. Yokoi, H. Yasukochi, K. Soeda, T. Totani, M. Wakita, H. Nagata, Fuel regression characteristics of a novel axial-injection end-burning hybrid rocket, Journal of Propulsion and Power 34 (1) (2018) 247–259.

[17] F. Mechentel, Preliminary design of a hybrid motor for small-satellite propulsion, Ph.D. thesis, Stanford University, 2019.

[18] P. Estey, D. Altman, J. McFarlane, An evaluation of scaling effects for hybrid rocket motors, 1991.

[19] G. Story, Large-Scale Hybrid Motor Testing, American Institute of Aeronautics and Astronautics, 2007, pp. 513–552, 2015/07/09.

[20] J.M. Weiss, C.S. Guernsey, Design and Development of the MSL Descent Stage Propulsion System, 2013.

[21] R.S. Baker, A.R. Casillas, C.S. Guernsey, J.M. Weiss, Mars science laboratory descent-stage integrated propulsion subsystem: development and flight performance, Journal of Spacecraft and Rockets 51 (4) (2014) 1217–1226.

[22] E. Lemmon, M. McLinden, D. Friend, Thermophysical properties of fluid systems in NIST Chemistry WebBook, NIST Standard Reference Database Number 69, Eds. Linstrom, PJ and Mallard, WG, National Institute of Standards and Technology, Gaithersburg MD, 20899, http://webbook.nist.gov.

[23] A.C. Wright, USAF Propellant Handbooks. Nitric Acid/Nitrogen Tetroxide Oxidizers. Volume II, Report, DTIC Document, 1977.

[24] S.D. Heister, W.E. Anderson, T. Pourpoint, R.J. Cassady, Rocket Propulsion, Cambridge University Press, 2019.

[25] Northrop Grumman propulsion products catalog, https://www.northropgrumman.com/wp-content/uploads/NG-Propulsion-Products-Catalog.pdf, 2016.

[26] United States Department of Labor, Occupational Health and Safety, Nitrous oxide, https://www.

osha.gov/chemicaldata/4, 2022. (Accessed 4 February 2023).

[27] E. Dossi, J. Earnshaw, L. Ellison, G.R. dos Santos, H. Cavayed, D.J. Cleaver, Understanding and controlling the glass transitionof HTPB oligomers, Polymer Chemistry 12 (2021) 2606–2617.

[28] E. Farias, M. Redmond, A.C. Karp, R. Shotwell, F.S. Mechentel, G.T. Story, Thermal cycling for development of hybrid fuel for a notional mars ascent vehicle, in: 52nd AIAA/SAE/ASEE Joint Propulsion Conference, 2016.

[29] F. Piscitelli, G. Saccone, A. Gianvito, G. Cosentino, L. Mazzola, Characterization and manufacturing of a paraffin wax as fuel for hybrid rockets, Propulsion and Power Research 7 (3) (2018) 218–230.

[30] D. Altman, Overview and History of Hybrid Rocket Propulsion, American Institute of Aeronautics and Astronautics, 2007, pp. 1–36, 2015/07/09.

[31] R.J. Kniffen, B. McKinney, P. Estey, Hybrid Rocket Development at the American Rocket Company, AIAA, 1990.

[32] M.A. Karabeyoglu, G. Zilliac, B.J. Cantwell, S. DeZilwa, P. Castellucci, Scale-up tests of high regression rate paraffin-based hybrid rocket fuels, Journal of Propulsion and Power 20 (6) (nov 2004) 1037–1045.

[33] Z. Peterson, S. Eilers, S. Whitmore, Closed-Loop Thrust and Pressure Profile Throttling of a Nitrous-Oxide HTPB Hybrid Rocket Motor, 2012.

[34] BluShift aerospace, https://www.blushiftaerospace.com/. (Accessed 10 June 2021).

[35] P.N. Estey, J.S. McFarlane, R.K. Kniffen, J. Lichatowich, Large hybrid rocket testing results, in: AIAA Space Programs and Technologies Conference, 1993.

[36] E. Casillas, C. Shaeffer, J. Trowbridge, E. Casillas, C. Shaeffer, J. Trowbridge, Cost and performance payoffs inherent in increased fuel regression rates, 1997.

[37] H. Weyland, H. Weyland, Hybrid rocket motor fuel studies, 1997.

[38] D. Kearney, K. Joiner, M. Gnau, M. Casemore, Improvements to the Marketability of Hybrid Propulsion Technologies, 2007.

[39] M.A. Karabeyoglu, B.J. Cantwell, D. Altman, Development and testing of paraffin-based hybrid rocket fuels, in: Joint Propulsion Conferences, American Institute of Aeronautics and Astronautics, 2001.

[40] Combustion Instability and Transient Behavior, American Institute of Aeronautics and Astronautics, 2007, pp. 351–411.

[41] A. Okninski, W. Kopacz, D. Kaniewski, K. Sobczak, Hybrid rocket propulsion technology for space transportation revisited - propellant solutions and challenges, in: Progress in Hybrid Rocket Propulsion, FirePhysChem 1 (4) (2021) 260–271.

[42] E. Ewing, Operation Woodpile II, Journal of the Pacific Rocket Society 2 (2) (1947) 10–11.

[43] G. Story, J. Arves, Flight Testing of Hybrid-Powered Vehicles, American Institute of Aeronautics and Astronautics, 2007, pp. 553–591, 2015/07/09.

[44] F.B. Mead, B.R. Bornborst, Certification Tests of a Hybrid Propulsion System for the Sandpiper Target Missile, vol. AFRPL-TR-69-73, 1973.

[45] G. Story, Large-scale hybrid motor testing, in: Progress in Astronautics and Aeronautics, American Institute of Aeronautics and Astronautics, 2007, pp. 513–552.

[46] J. Arves, M. Gnau, D. Kearney, K. Joiner, Hybrid Sounding Rocket (HYSR) Program, 2003.

[47] B.S. Waxman, An Investigation of Injectors for Use with High Vapor Pressure Propellants with Applications to Hybrid Rockets, Ph.D. thesis, Stanford University, 2014.

[48] G.T. Story, A.C. Karp, B. Nakazono, G. Zilliac, B.J. Evans, G. Whittinghill, Mars Ascent Vehicle Hybrid Propulsion Effort, 2020.

[49] O. Verberne, 30kN Hybrid Engine - Green Propulsion based on Hydrogen Peroxide Technologies, Tech. Rep., NAMMO Raufoss AS, 2018.

[50] O. Verberne, A.J. Boiron, M.G. Faenza, B. Haemmerli, Development of the North Star Sounding Rocket: Getting ready for the first demonstration Launch, 2015.

[51] M. Kobald, U. Fischer, K. Tomilin, A. Petrarolo, C. Schmierer, Hybrid experimental rocket stuttgart: a low-cost technology demonstrator, Journal of Spacecraft and Rockets 55 (2) (2018) 484–500.

[52] SpaceForest Ltd, Suborbital inexpensive rocket SIR, https://spaceforest.pl/wp-content/uploads/2022/01/SpaceForest-SIR-project-brochure.pdf, 2022. (Accessed 20 December 2022).

[53] M.A. Karabeyoglu, Transient Combustion in Hybrid Rockets, Ph.D. thesis, Stanford University, 1998.

[54] L. Casalino, D. Pastrone, A Straightfotward Approach for Robust Design of Hybrid Rocket Engine Upper Stage, 2015.

[55] O. Kara, M. Karpat, M.A. Karabeyoglu, Propulsion System Design for Mars Ascent Vehicles by using the In-Situ CO_2, 2021.

[56] F. Costa, R. Contaifer, J. Albuquerque, S. Gabriel, R. Marques, Study of Paraffin/H2O2 Hybrid Rockets for Launching Nanosats, 2008.

[57] B.R. McKnight, Advanced Hybrid Rocket Motor Propulsion Unit for Cubesats, 2015.

[58] S.A. Whitmore, B.L. Chamberlain, I.W. Armstrong, S. Mathias, S.A. Fehlberg, Consumable Spacecraft Structures with Integrated, 3-D Printed ABS Thrusters, 2017.

[59] K. Veale, S. Adali, J. Pitot, C. Bemont, The structural properties of paraffin wax based hybrid rocket fuels with aluminium particles, Acta Astronautica 151 (2018) 864–873.

[60] V. Sella, A. Larkey, A. Majumder, A. Rao, Z. Abidi, N. Rasmont, A. Randeo, M. Liu, A. Moore, M.F. Lembeck, Development of a Nytrox-Paraffin Hybrid Rocket Engine, 2020.

[61] A. Ruffin, E. Paccagnella, M. Santi, F. Barato, D. Pavarin, Real-time deep throttling tests of a hydrogen peroxide hybrid rocket motor, Journal of Propulsion and Power 38 (5) (2022) 833–848.

[62] K. Ozawa, K. Kitagawa, S. Aso, T. Shimada, Hybrid rocket firing experiments at various axial–tangential oxidizer-flow-rate ratios, Journal of Propulsion and Power 35 (1) (2019) 94–108.

[63] K. Watanabe, I. Nakagawa, Study of a LOX Vaporizer for an Altering-Intensity Swirling Oxidizer flow type Hybrid Rocket, 2021.

[64] D. Lee, Y. Moon, C. Lee, Equivalence Ratio Variation and Pressure Oscillations in the Hybrid Rocket Combustion, 2017.

[65] M. Dinesh, R. Kumar, Utility of multiprotrusion as the performance enhancer in hybrid rocket motor, Journal of Propulsion and Power 35 (5) (2019) 1005–1017.

[66] C.P. Kumar, A. Kumar, Effect of diaphragms on regression rate in hybrid rocket motors, Journal of Propulsion and Power 29 (3) (2013) 559–572.

[67] L. Kamps, Y. Saito, R. Kawabata, M. Wakita, T. Totani, Y. Takahashi, H. Nagata, Method for determining nozzle-throat-erosion history in hybrid rockets, Journal of Propulsion and Power 33 (6) (2017) 1369–1377.

[68] G.D.D. Martino, S. Mungiguerra, C. Carmicino, R. Savino, Computational Fluid-dynamic Simulations of Hybrid Rocket Internal Flow Including Discharge Nozzle, 2017.

[69] T. Marquardt, J. Majdalani, Beltramian Solution for Cyclonically Driven Hybrid Rocket Engines, 2017.

[70] E. Miklaszewski, J. Dadson, D. Fox, D. Kees, L. Larson, J. Rideout, J. Stevens, T.L. Pourpoint, Hybrid Rocket Design/Build/Test Course at Purdue University, 2013.

[71] M. Kobald, C. Schmierer, U. Fischer, K. Tomilin, A. Petrarolo, A record flight of the hybrid sounding rocket HEROS 3, in: 31st International Symposium on Space Technology and Science (ISTS), June 2017.

[72] Technology readiness level (NASA), https://www.nasa.gov/directorates/heo/scan/engineering/technology/technology_readiness_level. (Accessed 29 June 2021).

[73] C. Brown, Spacecraft Propulsion, American Institute of Aeronautics and Astronautics, 1996.

[74] S. Gordon, B.J. McBride, Computer Program for Calculation of Complex Chemical Equilibrium Compositions and Applications, National Aeronautics and Space Administration, Office of Management, Scientific and Technical Information Program, 1996.

[75] R.A. Lugo, J.D. Shidner, R. Powell, S.M. Marsh, J.A. Hoffman, D.K. Litton, T.L. Schmitt, Launch Vehicle Ascent Trajectory Simulation Using the Program to Optimize Simulated Trajectories II (POST2), 2017.

[76] C. Brown, Spacecraft Mission Design, second edition, American Institute of Aeronautics and Astronautics, 1998.

[77] D.A. Vallado, Fundamentals of Astrodynamics and Applications, 4th ed., Microcosm Press, 2013.

[78] G.A. Marxman, M. Gilbert, Turbulent boundary layer combustion in the hybrid rocket, Symposium (International) on Combustion 9 (1963) 371–383.

[79] G.A. Marxman, R. Muzzy, C. Wooldridge, Fundamentals of hybrid boundary layer combustion, in: Heterogeneous Combustion Conference, in: Meeting Paper Archive, American Institute of Aeronautics and Astronautics, Dec. 1963.

[80] G.A. Marxman, Combustion in the turbulent boundary layer on a vaporizing surface, Symposium (International) on Combustion 10 (1965) 1337–1349.

[81] J.H. Perry (Ed.), Chemical Engineers' Handbook, 3rd ed., McGraw-Hill, 1950.

[82] G. Zilliac, M. Karabeyoglu, Hybrid rocket fuel regression rate data and modeling, in: 42nd AIAA/ASME/SAE/ASEE Joint Propulsion Conference & Exhibit, Joint Propulsion Conferences, American Institute of Aeronautics and Astronautics, July 2006.

[83] A.A. Chandler, An Investigation of Liquefying Hybrid Rocket Fuels with Applications to Solar System Exploration, Ph.D. thesis, Stanford University, 2012.

[84] E.T. Jens, V.A. Miller, B.J. Cantwell, Schlieren and OH* chemiluminescence imaging of combustion in a turbulent boundary layer over a solid fuel, Experiments in Fluids 57 (3) (2016) 1–16.

[85] E.T. Jens, Hybrid Rocket Combustion and Applications to Space Exploration Missions, Ph.D. thesis, Stanford University, 2015.

[86] D. Altman, R. Humble, Hybrid rocket propulsion systems, Space Propulsion Analysis and Design 379 (1995).

[87] M.J. Chiaverini, Private communication, February 2023.

[88] P. Paul, H. Mukunda, V. Jain, Regression rates in boundary layer combustion, in: Nineteenth Symposium (International) on Combustion, Symposium (International) on Combustion 19 (1) (1982) 717–729.

[89] M.J. Chiaverini, Regression Rate and Pyrolysis behavior of HTPB-based Solid Fuels in a Hybrid Rocket Motor, Ph.D. thesis, The Pennsylvania State University, 1997.

[90] S. Venkateswaran, C. Merkle, Size Scale-Up in Hybrid Rocket Motors, 1996.

[91] M.A. Karabeyoglu, D. Altman, B.J. Cantwell, Combustion of liquefying hybrid propellants: Part 1, general theory, Journal of Propulsion and Power 18 (3) (2002) 610–620.

[92] M.A. Karabeyoglu, B.J. Cantwell, Combustion of liquefying hybrid propellants: Part 2, stability of liquid films, Journal of Propulsion and Power 18 (3) (2002) 621–630.

[93] M.A. Karabeyoglu, B.J. Cantwell, J. Stevens, Evaluation of the Homologous Series of Normal Alkanes as Hybrid Rocket Fuels, American Institute of Aeronautics and Astronautics, 2005, 2014/07/09.

[94] Sigma Aldrich, Paraffin FTIR, http://www.sigmaaldrich.com/spectra/ftir/FTIR006397.PDF, 2015. (Accessed 28 November 2015).

[95] R.A. Gater, M.R. L'Ecuyer, A fundamental investigation of the phenomena that characterize liquid-film cooling, International Journal of Heat and Mass Transfer 13 (12) (1970) 1925–1939.

[96] The Candlewick Company, 160 melt point wax - 5560, http://www.candlewic.com/store/product.aspx?q=c49,p525&&title=160-Melt-Point-Wax--5560, 2015. (Accessed 29 November 2015).

[97] B.J. Cantwell, Aircraft and rocket propulsion, AA283 Course Reader, https://web.stanford.edu/~cantwell/AA283_Course_Material/AA283_Course_BOOK/AA283_Aircraft_and_Rocket_Propulsion_BOOK_Brian_J_Cantwell_July_2022.pdf, 2022. (Accessed 30 October 2022).

[98] M.A. Karabeyoglu, B.J. Cantwell, G. Zilliac, Development of scalable space-time averaged regression rate expressions for hybrid rockets, Journal of Propulsion and Power 23 (4) (2007) 737–747.

[99] B.J. Cantwell, Similarity solution of fuel mass transfer, port mass flux coupling in hybrid propulsion, Journal of Engineering Mathematics 84 (1) (2014) 19–40.

[100] A.M. Karabeyoglu, E. Toson, B.J. Evans, Effect of "O/F Shift" on Combustion Efficiency, American Institute of Aeronautics and Astronautics, 2014, 2017/12/19.

[101] C. Glaser, J. Hijlkema, J. Anthoine, Evaluation of regression rate enhancing concepts and techniques for hybrid rocket engines, Aerotecnica Missili & Spazio 101 (2022) 267–292.

[102] S. Yuasa, O. Shimada, T. Imamura, T. Tamura, K. Yamoto, A technique for improving the performance of hybrid rocket engines, 1999.

[103] N. Bellomo, F. Barato, M. Faenza, M. Lazzarin, A. Bettella, D. Pavarin, Numerical and Experimental Investigation on Vortex Injection in Hybrid Rocket Motors, 2011.

[104] T. Sakurazawa, K. Kitagawa, R. Hira, Y. Matsuo, T. Sakurai, S. Yuasa, Development of a 1500N-thrust swirling-oxidizer-flow-type hybrid rocket engine, in: Asian Joint Conference on Propulsion and Power, Gyeongju, Korea, 2008.

[105] E. Jens, A. Karp, A. Boiron, M. Faenza, HRTC Regression Rate Standards, Presented at the AIAA Propulsion and Energy 2020, New Orleans, LA, 2020.

[106] J.-Y. Lestrade, J. Messineo, J. Anthoine, A. Musker, F. Barato, Development and Test of an Innovative Hybrid Rocket Combustion Chamber, 2015.

[107] J.M. Beer, N.A. Chigier, Combustion Aerodynamics, Applied Science Publishers, 1972.

[108] E. Paccagnella, F. Barato, D. Pavarin, A.M. Karabeyoglu, Scaling of Hybrid Rocket Motors with Swirling Oxidizer Injection - Part 2, 2016.

[109] J. Majdalani, Vortex Injection Hybrid Rockets, American Institute of Aeronautics and Astronautics, 2007, pp. 247–276.

[110] T.A. Marquardt, Characterization of Swirl-Driven Hybrid Rocket Engines, Ph.D. thesis, Auburn, 2020.

[111] C. Carmicino, A.R. Sorge, Influence of a conical axial injector on hybrid rocket performance, Journal of Propulsion and Power 22 (5) (2006) 984–995.

[112] T. Viscor, L. Kamps, K. Yonekura, H. Isochi, H. Nagata, Large-scale CAMUI type hybrid rocket motor scaling, modeling, and test results, Aerospace 9 (1) (2022).

[113] T. Kato, N. Hashimoto, H. Nagata, I. Kudo, A preliminary study of end-burning hybrid rocket: Part 1 combustion stability, Journal of the Japan Society for Aeronautical and Space Sciences 49 (565) (2001) 33–39.

[114] M.A. Hitt, R.A. Frederick, Regression rate model predictions of an axial-injection end-burning hybrid motor, Journal of Propulsion and Power 34 (5) (2018) 1116–1123.

[115] Y. Saito, T. Yokoi, L. Neumann, H. Yasukochi, K. Soeda, T. Totani, M. Wakita, H. Nagata, Investigation of axial-injection end-burning hybrid rocket motor regression, Advances in Aircraft and Spacecraft Science 4 (3) (2017).

[116] J.J. Caravella, S. Heister, E. Wernimont, Characterization of fuel regression in a radial flow hybrid rocket, 1996.

[117] G. Haag, Alternative geometry hybrid rockets for spacecraft orbit transfer, Ph.D. thesis, University of Surrey, 2001.

[118] W.H. Knuth, M.J. Chiaverini, J.A. Sauer, D.J. Gramer, Solid-fuel regression rate behavior of vortex hybrid rocket engines, Journal of Propulsion and Power 18 (3) (2002) 600–609.

[119] R. Jansen, E. Teegarden, S. Gimelshein, Characterization of a Vortex-flow End-burning Hybrid Rocket Motor for Nanosatellite Applications, 2012.

[120] T. Sakurai, D. Hayashi, A Fundamental Study of a End-Burning Swirling-Flow Hybrid Rocket Engine using Low Melting Temperature Fuels, 2015.

[121] J.Y. Lestrade, J. Anthoine, A. Musker, A. Lecossais, Experimental demonstration of an end-burning swirling flow hybrid rocket engine, Aerospace Science and Technology 92 (2019) 1–8.

[122] A. Hashish, C. Paravan, A. Verga, Liquefying Fuel Combustion in a Lab-scale Vortex Flow Pancake Hybrid Rocket Engine, 2021.

[123] J. Lee, S. Rhee, J.K. Kim, H.-J. Moon, An analysis and reduction design of combustion instability generated in hybrid rocket motor, Journal of the Korean Society of Propulsion Engineers 18 (4) (08 2014) 18–25.

[124] A. Karabeyoglu, J. Stevens, B. Cantwell, Investigation of feed system coupled low frequency combustion instabilities in hybrid rockets, in: 43rd AIAA/ASME/SAE/ASEE Joint Propulsion Conference & Exhibit, 2007, p. 5366.

[125] W.E. Anderson, V. Yang, Liquid Rocket Engine Combustion Instability, American Institute of Aeronautics and Astronautics, 1995.

[126] K.K. Kuo, Principles of Combustion, 1986.

[127] K.K. Kuo, M.J. Chiaverini, Fundamentals of Hybrid Rocket Combustion and Propulsion, American Institute of Aeronautics and Astronautics, 2007.

[128] S.R. Turns, An Introduction to Combustion: Concepts and Application, McGraw-Hill, 2012.

[129] G.A. Yankura, Accurate compressible flow algorithm from straightforward approximations covering both critical and subcritical conditions, in: 27th Joint Propulsion Conference, 1991, p. 2583.

[130] M.L. Huber, E.W. Lemmon, I.H. Bell, M.O. McLinden, The NIST REFPROP database for highly accurate properties of industrially important fluids, Industrial & Engineering Chemistry Research 61 (42) (2022) 15449–15472.

[131] G. Zilliac, G.T. Story, A.C. Karp, E.T. Jens, G. Whittinghill, Combustion efficiency in single port hybrid rocket engines, in: AIAA Propulsion and Energy 2020 Forum, 2020, p. 3746.

[132] R.J. Kee, F.M. Rupley, J.A. Miller, The Chemkin Thermodynamic Data Base, Tech. Rep., Sandia National Lab. (SNL-CA), Livermore, CA (United States), 1990.

[133] D.G. Goodwin, H.K. Moffat, I. Schoegl, R.L. Speth, B.W. Weber, Cantera: an object-oriented software toolkit for chemical kinetics, thermodynamics, and transport processes, https://www.cantera.org, 2022, Version 2.6.0.

[134] National Aeronautics and Space Administration, CEARun, https://cearun.grc.nasa.gov/. (Accessed 26 December 2022).

[135] M.W. Chase Jr, C. Davies, J.R. Downey Jr, D. Frurip, R. McDonald, A. Syverud, NIST standard reference database 13, NIST JANAF thermochemical tables, in: Standard Reference Data Program, National Institute of Standards and Technology, 1998.

[136] National Institute of Standards and Technology (NIST), NIST computational chemistry comparison and benchmark database: how to get an enthalpy of formation from ab initio calculations, https://cccbdb.nist.gov/enthform2.asp, 2022.

[137] S.W. Benson, Thermochemical kinetics, in: Methods for the Estimation of Thermochemical Data and Rate Parameters, 2nd ed., New York, Wiley, 1976.

[138] G.R. Nickerson, L.D. Dang, D. Coats, Engineering and Programming Manual: Two-Dimensional Kinetic Reference Computer Program, TDK, Tech. Rep., 1985.

[139] H. Schlichting, Boundary Layer Theory, vol. 960, Springer, 1968.

[140] W. Reynolds, STANJAN: Interactive Computer Programs for Chemical Equilibrium Analysis, Mechanical Engineering Department, Stanford University, Stanford, CA, 1981.

[141] A. Ponomarenko, RPA: Design Tool for Liquid Rocket Engine Analysis, 2009.

[142] J.D. Anderson Jr, Fundamentals of Aerodynamics, Tata McGraw-Hill Education, 2010.

[143] S.W. Benson, F. Cruickshank, D. Golden, G.R. Haugen, H. O'neal, A. Rodgers, R. Shaw, R. Walsh, Additivity rules for the estimation of thermochemical properties, Chemical Reviews 69 (3) (1969) 279–324.

[144] S.A. Whitmore, Z.W. Peterson, S.D. Eilers, Comparing hydroxyl terminated polybutadiene and acrylonitrile butadiene styrene as hybrid rocket fuels, Journal of Propulsion and Power 29 (3) (2013) 582–592.

[145] S. Heister, E. Wernimont, Hydrogen Peroxide, Hydroxyl Ammonium Nitrate, and Other Storable Oxidizers, American Institute of Aeronautics and Astronautics, 2007, pp. 457–488, 2015/07/09.

[146] A. Benhidjeb-Carayon, Reactivity and Hypergolicity of Liquid and Solid Fuels with Mixed Oxides of Nitrogen, Ph.D. thesis, Purdue University, 2019.

[147] L. Simurda, G. Zilliac, Continued Testing of the High Performance Hybrid Propulsion System for Small Satellites, American Institute of Aeronautics and Astronautics, 2015, 2015/12/03.

[148] E. Doran, J. Dyer, K. Lohner, Z. Dunn, B. Cantwell, G. Zilliac, Nitrous oxide hybrid rocket motor fuel regression rate characterization, AIAA Paper 5352 (2007) 2007.

[149] F.S. Mechentel, A.M. Coates, B.J. Cantwell, Small-Scale Gaseous Oxygen Hybrid Rocket Testing for Regression Rate and Combustion Efficiency Studies, American Institute of Aeronautics and Astronautics, 2017.

[150] F.S. Mechentel, B.R. Hord, B.J. Cantwell, Optically Resolved Fuel Regression of a Clear PMMA Hybrid Rocket Motor, American Institute of Aeronautics and Astronautics, 2019.

[151] B.M. Tymrak, M. Kreiger, J.M. Pearce, Mechanical properties of components fabricated with open-source 3-D printers under realistic environmental conditions, Materials & Design 58 (2014) 242–246.

[152] S.A. Whitmore, D.P. Merkley, N.R. Inkley, Development of a Power Efficient, Restart-Capable Arc Ignitor

for Hybrid Rockets, American Institute of Aeronautics and Astronautics, 2014, 2015/03/31.

[153] S.A. Whitmore, S.L. Merkley, L. Tonc, S.D. Mathias, Survey of selected additively manufactured propellants for arc ignition of hybrid rockets, Journal of Propulsion and Power 32 (6) (2016) 1494–1504.

[154] A.A. Chandler, E.T. Jens, B.J. Cantwell, G.S. Hubbard, Visualization of the liquid layer combustion of paraffin fuel at elevated pressures, in: 63rd International Astronautical Congress, Naples, Italy, 2012.

[155] G. Story, A. Karp, B.N. Nakazono, G. Whittinghill, G.G. Zilliac, Mars Ascent Vehicle Hybrid Propulsion Developement, Nasa technical report, NASA, 2019.

[156] S. Kilic, A. Karabeyoglu, J. Stevens, B. Cantwell, Modeling the Slump Characteristics of the Hydrocarbon-Based Hybrid Rocket Fuels, 2003.

[157] E. Farias, M. Redmond, A.C. Karp, R. Shotwell, F.S. Mechentel, G.T. Story, Thermal Cycling for Development of Hybrid Fuel for a Notional Mars Ascent Vehicle, 2016.

[158] K. Veale, S. Adali, J. Pitot, M. Brooks, A review of the performance and structural considerations of paraffin wax hybrid rocket fuels with additives, Acta Astronautica 141 (2017) 196–208.

[159] J. Wang, S.J. Severtson, A. Stein, Significant and concurrent enhancement of stiffness, strength, and toughness for paraffin wax through organoclay addition, Advanced Materials 18 (12) (2006) 1585–1588.

[160] J. DeSain, B. Brady, K. Metzler, T. Curtiss, T. Albright, Tensile Tests of Paraffin Wax for Hybrid Rocket Fuel Grains, 2009.

[161] S. Ryu, S. Han, J. Kim, H. Moon, J. Kim, S.W. Ko, Tensile and compressive strength characteristics of aluminized paraffin wax fuel for various particle size and contents, Journal of the Korean Society of Propulsion Engineers 20 (5) (2016) 70–76.

[162] J.S.N. Mahottamananda, D.V. Kumar, A.K. Afreen, S. Dinesh, W. Ashiq, P.N. Kadiresh, M. Thirumurugan, Mechanical Characteristics of Ethylene Vinyl Acetate Mixed Beeswax Fuel for Hybrid Rockets, 2021, pp. 389–400.

[163] J. Nandy, P.C. Joshi, Improvement in the mechanical properties of paraffin-based propellant, in: V.H. Saran, R.K. Misra (Eds.), Advances in Systems Engineering, Springer Singapore, Singapore, 2021, pp. 91–110.

[164] R. Bisin, C. Paravan, A. Verga, L. Galfetti, Towards High-Performing Paraffin-based Fuels Exploiting the Armored Grain Concept, 2022.

[165] D. Mengu, R. Kumar, Development of EVA-SEBS based wax fuel for hybrid rocket applications, Acta Astronautica 152 (2018) 325–334.

[166] D.B. Larson, Formulation and Characterization of Paraffin-based Solid Fuels Containing Novel Additives for Use in Hybrid Rocket Motors, 2012.

[167] D. Cruise, Theoretical Computations of Equilibrium Compositions, Thermodynamic Properties, and Performance Characteristics of Propellant Systems, Report, DTIC Document, 1979.

[168] G.A. Risha, B.J. Evans, E. Boyer, K.K. Kuo, Metals, Energetic Additives, and Special Binders Used in Solid Fuels for Hybrid Rockets, 2007.

[169] G.A. Landis, Materials refining on the moon, Acta Astronautica 60 (10–11) (2007) 906–915.

[170] C. Schwandt, J.A. Hamilton, D.J. Fray, I.A. Crawford, The production of oxygen and metal from lunar regolith, in: Scientific Preparations for Lunar Exploration, Planetary and Space Science 74 (1) (2012) 49–56.

[171] M.H. Hecht, J. Hoffman, the MoxieTeam, The Mars oxygen ISRU experiment (MOXIE) on the Mars 2020 Rover, LPICo 1980 (2016) 4130.

[172] S.B. Coogan, M. Miller, Solid-State Oxygen Storage for Compact Liquid-Free Hybrid Rockets, Propulsion and Energy Forum, American Institute of Aeronautics and Astronautics, 2018.

[173] T.M. Tomsik, M.L. Meyer, Liquid Oxygen Propellant Densification Production and Performance Test Results with a Large-Scale Flight-Weight Propellant Tank for the X33 RLV, NASA/TM-2010-216247, 2010.

[174] E. Musk, Making humans a multi-planetary species, New Space 5 (2) (2017) 46–61.

[175] J.E. Zimmerman, B.S. Waxman, B.J. Cantwell, G.G. Zilliac, Comparison of nitrous oxide and carbon dioxide with applications to self-pressurizing propellant tank expulsion dynamics, in: 60th JANNAF Propulsion Meeting, 2013, pp. 1–22.

[176] J.E. Zimmerman, B.S. Waxman, B. Cantwell, G. Zilliac, Review and evaluation of models for self-pressurizing propellant tank dynamics, in: 49th AIAA/ASME/SAE/ASEE Joint Propulsion Conference, 2013, p. 4045.

[177] G. Zilliac, B.S. Waxman, M.A. Karabeyoglu, B.J. Cantwell, B. Evans, Peregrine hybrid rocket motor development, in: 50th AIAA/ASME/SAE/ASEE Joint Propulsion Conference, Propulsion and Energy Forum, American Institute of Aeronautics and Astronautics, July 2014.

[178] V. Zakirov, M. Sweeting, T. Lawrence, An update on surrey nitrous oxide catalytic decomposition research, in: Proceedings of the AIAA/USU Conference on Small Satellites, 2001.

[179] Y. Scherson, K. Lohner, B. Cantwell, T. Kenny, Small-Scale Planar Nitrous Oxide Monopropellant Thruster for "Green" Propulsion and Power Generation, 2010.

[180] K.A. Lohner, Development of a Nitrous Oxide Monopropellant Hot Gas Generator for Rocket Propellant Pressurization and Spacecraft Thruster Applications, Ph.D. thesis, Stanford University, 2012.

[181] A. Karabeyoglu, J. Dyer, J. Stevens, B. Cantwell, Modeling of N2O Decomposition Events, 2008.

[182] M.A. Karabeyoglu, Nitrous oxide and oxygen mixtures (nytrox) as oxidizers for rocket propulsion applications, Journal of Propulsion and Power 30 (3) (2014) 696–706, 2015/03/31.

[183] A.A. Chandler, B.J. Cantwell, G.S. Hubbard, M.A. Karabeyoglu, A two-stage, single port hybrid propulsion system for a Mars ascent vehicle, in: 46th AIAA/ASME/SAE/ASEE Joint Propulsion Conference & Exhibit, Joint Propulsion Conferences, American Institute of Aeronautics and Astronautics, July 2010.

[184] Chemical and Material Sciences Department Research Division, Rocketdyne, a Division of North American Aviation, Inc., Hydrogen Peroxide Handbook, AFRPL-TR-67-144, 1967.

[185] M. Faenza, A.J. Boiron, B. Haemmerli, O. Verberne, The nammo nucleus launch: Norwegian hybrid sounding rocket over 100km, in: AIAA Propulsion and Energy 2019 Forum, Joint Propulsion Conferences, American Institute of Aeronautics and Astronautics, Aug. 2019.

[186] M. Ventura, S.D. Heister, Hydrogen peroxide as an alternative oxidizer for a hybrid rocket booster, Journal of Propulsion and Power 11 (3) (1995) 562–565.

[187] M. Ventura, E. Wernimont, S. Heister, S. Yuan, Rocket grade hydrogen peroxide (RGHP) for use in propulsion and power devices - historical discussion of hazards, in: 43rd AIAA/ASME/SAE/ASEE Joint Propulsion Conference & Exhibit, Joint Propulsion Conferences, American Institute of Aeronautics and Astronautics, July 2007.

[188] M. Ventura, Long term storablity of hydrogen peroxide, in: 42nd AIAA/ASME/SAE/ASEE Joint Propulsion Conference & Exhibit, Joint Propulsion Conferences, American Institute of Aeronautics and Astronautics, July 2005.

[189] J. Clark, Ignition!: an Informal History of Liquid Rocket Propellants, Rutgers, the State University, 2017.

[190] Callery Chemical Company, Propellant Performance Data, Callery Chemical Company, 1961.

[191] B. Evans, G. Risha, N. Favorito, E. Boyer, R. Wehrman, N. Libis, K. Kuo, Instantaneous regression rate determination of a cylindrical X-ray transparent hybrid rocket motor, in: 39th AIAA/ASME/SAE/ASEE Joint Propulsion Conference and Exhibit, 2003.

[192] B. Evans, N. Favorito, K. Kuo, Study of solid fuel burning-rate enhancement behavior in an X-ray translucent hybrid rocket motor, in: 41st AIAA/ASME/SAE/ASEE Joint Propulsion Conference & Exhibit, 2005, p. 3909.

[193] M.A. Pfeil, J.D. Dennis, S.F. Son, S.D. Heister, T.L. Pourpoint, P.V. Ramachandran, Characterization of ethylenediamine bisborane as a hypergolic hybrid rocket fuel additive, Journal of Propulsion and Power 31 (1) (2015).

[194] A. Cortopassi, J.E. Boyer, Hypergolic Ignition Testing of Solid Fuel Additives with MON-3 Oxidizer, 2017.

[195] M.J. Baier, P.V. Ramachandran, S.F. Son, Characterization of the hypergolic ignition delay of ammonia borane, Journal of Propulsion and Power 35 (1) (2019).

[196] A. Benhidjeb-Carayon, M.P. Drolet, J.R. Gabl, T.L. Pourpoint, Reactivity and hypergolicity of solid fuels with mixed oxides of nitrogen, Journal of Propulsion and Power 35 (2) (2019) 466–474.

[197] D.A. Castaneda, B. Natan, Experimental Investigation of the Hydrogen Peroxide - Solid Hydrocarbon Hypergolic Ignition, 2015.

[198] N. Munjal, M. Parvatiyar, Ignition of hybrid rocket fuels with fuming nitric acid as oxidant, Journal of Spacecraft and Rockets 11 (6) (1974) 428–430.

[199] J. DeSain, T. Curtiss, K. Metzler, B. Brady, Testing Hypergolic Ignition of Paraffin Wax/LiAlH4 Mixtures, American Institute of Aeronautics and Astronautics, 2010, 2015/03/31.

[200] T.R. Sippel, S.C. Shark, M.C. Hinkelman, T.L. Pourpoint, S.F. Son, S.D. Heister, Hypergolic ignition of metal hydride-based fuels with hydrogen peroxide, in: 47th AIAA/ASME/SAE/ASEE Joint Propulsion Conference. & Exhibit, Joint Propulsion Conferences, American Institute of Aeronautics and Astronautics, March 2011.

[201] E. Wernimont, S. Heister, Reconstruction technique for reducing hybrid-rocket combustion test data, Journal of Propulsion and Power 15 (1) (1999) 128–136.

[202] R. Kumar, R. Periyapatna, Measurement of regression rate in hybrid rocket using combustion chamber pressure, Acta Astronautica 103 (2014) 226–234.

[203] B. Cantwell, Private communication, July 2022.

[204] K. Ozawa, T. Shimada, Performance of mixture-ratio-controlled hybrid rockets for nominal fuel regression, Journal of Propulsion and Power 36 (3) (2020) 400–414.

[205] R. Frederick Jr, B. Greiner, Laboratory-scale hybrid rocket motor uncertainty analysis, Journal of Propulsion and Power 12 (3) (1996) 605–611.

[206] G. Story, T. Abel, S. Claflin, O. Park, J. Arves, D. Kearney, Hybrid Propulsion Demonstration Program 250K Hybrid Motor, 2003.

[207] L. DeLuca, L. Galfetti, F. Maggi, G. Colombo, C. Paravan, A. Reina, P. Tadini, A. Sossi, E. Duranti, An Optical Time-Resolved Technique of Solid Fuels Burning for Hybrid Rocket Propulsion, American Institute of Aeronautics and Astronautics, 2011, 2015/08/22.

[208] B.E. Goldberg, J.R. Cook, Preliminary results of the NASA/industry hybrid propulsion program, 1992.

[209] Penn State High Pressure Combustion Laboratory, http://www.hpcl.psu.edu/Facilities2021-01-03.

[210] C. Wooldridge, R. Muzzy, Internal ballistic considerations in hybrid rocket design, Journal of Spacecraft and Rockets 4 (2) (1967) 255–262.

[211] P. Estey, D. Altman, J. McFarlane, An Evaluation of Scaling Effects for Hybrid Rocket Motors, 1991.

[212] A. Gany, Similarity and Scaling Effects in Hybrid Rocket Motors, American Institute of Aeronautics and Astronautics, 2007, pp. 489–511, 2015/07/09.

[213] M.J. Chiaverini, K.K. Kuo, A. Peretz, G.C. Harting, Regression-rate and heat-transfer correlations for hybrid rocket combustion, Journal of Propulsion and Power 17 (1) (2001) 99–110.

[214] S. Venkateswaran, C. Merkle, Size scale-up in hybrid rocket motors, 1996.

[215] H. Mukunda, V. Jain, P. Paul, A review of hybrid rockets: present status and future potential, Proceedings of the Indian Academy of Sciences, Engineering Sciences 2 (1979) 215–242.

[216] N. Bellomo, M. Lazzarin, F. Barato, M. Grosse, Numerical Investigation of the Effect of a Diaphragm on the Performance of a Hybrid Rocket Motor, 2010.

[217] M. Lazzarin, F. Barato, N. Bellomo, D. Pavarin, A. Bettella, M. Grosse, M. Faenza, CFD Simulation of a Hybrid Rocket Motor with Liquid Injection, 07 2011.

[218] C.D. Hill, W. Nelson, C.T. Johansen, Evaluation of a paraffin/nitrous oxide hybrid rocket motor with a passive mixing device, Journal of Propulsion and Power 38 (6) (2022) 884–892.

[219] D.D. Ordahl, D. Altman, Hybrid Propellant Combustion Characteristics and Engine Design, 1962.

[220] P. George, S. Krishnan, P. Varkey, M. Ravindran, L. Ramachandran, Fuel regression rate in hydroxyl-terminated-polybutadiene/gaseous-oxygen hybrid rocket motors, Journal of Propulsion and Power 17 (1) (2001) 35–42.

[221] M.K. Hudson, A.M. Wright, C. Luchini, P.C. Wynne, S. Rooke, Guanidinium azo-tetrazolate (GAT) as a high performance hybrid rocket fuel additive, Journal of Pyrotechnics 19 (2004) 37–42.

[222] B. Greiner, R. Frederick Jr, Results of labscale hybrid rocket motor investigation, in: 28th Joint Propulsion Conference and Exhibit, 1992, p. 3301.

[223] A.M. Karabeyoglu, Lecture 10 - hybrid rocket propulsion design issues, in: AA284a Advanced Rocket Propulsion Course, Stanford University, 2011.

[224] R. Muzzy, Applied hybrid combustion theory, in: AIAA/SAE Eighth Joint Propulsion Specialist Conference, New Orleans, LA, Paper, 1972.

[225] E. Doran, J. Dyer, K. Lohner, Z. Dunn, B. Cantwell, G. Zilliac, Nitrous Oxide Hybrid Rocket Motor Fuel Regression Rate Characterization, 2007.

[226] B.J. Evans, A.M. Karabeyoglu, Development and Testing of SP7 Fuel for Mars Ascent Vehicle Application, 2017.

[227] Y. Yun, J. Huh, Y. Kim, S. Heo, H. Kim, S. Kwon, Scale-up validation of hydrogen peroxide/high-density polyethylene hybrid rocket with multiport solid fuel, Journal of Spacecraft and Rockets 58 (2) (2021) 552–565.

[228] X. Li, H. Tian, N. Yu, G. Cai, Experimental investigation of fuel regression rate in a HTPB based lab-scale hybrid rocket motor, Acta Astronautica 105 (1) (2014) 95–100.

[229] B. Vignesh, R. Kumar, Effect of multi-location swirl injection on the performance of hybrid rocket motor, Acta Astronautica 176 (2020) 111–123.

[230] A. Pons Lorente, N. Yu, B. Zhao, Testing and evaluation of a double-tube hybrid rocket motor, in: 51st AIAA/SAE/ASEE Joint Propulsion Conference, 2015.

[231] E. Rice, D. Gramer, C.S. Clair, M. Chiaverini, Mars ISRU CO/O2 rocket engine development and testing, in: Seventh International Workshop on Microgravity Combustion and Chemically Reacting Systems, vol. 2, 2003, p. 101.

[232] H. Nagata, S. Hagiwara, N. Wakita, T. Totani, T. Uematsu, Optimal fuel grain design method for CAMUI type hybrid rocket, in: 47th AIAA/ASME/SAE/ASEE Joint Propulsion Conference & Exhibit, 2011, p. 6105.

[233] R.A. Frederick Jr, J.J. Whitehead, L.R. Knox, M.D. Moser, Regression rates study of mixed hybrid propellants, Journal of Propulsion and Power 23 (1) (2007) 175–180.

[234] Z. Wang, X. Lin, F. Li, X. Yu, Combustion performance of a novel hybrid rocket fuel grain with a nested helical structure, Aerospace Science and Technology 97 (2020) 105613.

[235] M. Grosse, Effect of a diaphragm on performance and fuel regression of a laboratory scale hybrid rocket motor using nitrous oxide and paraffin, in: 45th AIAA/ASME/SAE/ASEE Joint Propulsion Conference & Exhibit, 2009.

[236] R. Kumar, P. Ramakrishna, Enhancement of hybrid fuel regression rate using a bluff body, Journal of Propulsion and Power 30 (4) (2014) 909–916.

[237] F.D. Quadros, P.T. Lacava, Swirl injection of gaseous oxygen in a lab-scale paraffin hybrid rocket motor, Journal of Propulsion and Power 35 (5) (2019) 896–905.

[238] E. Jens, B.J. Cantwell, G.S. Hubbard, Hybrid rocket propulsion systems for outer planet exploration missions, Acta Astronautica 128 (2016) 119–130.

[239] S.A. Whitmore, S.D. Walker, D.P. Merkley, M. Sobbi, High regression rate hybrid rocket fuel grains with helical port structures, Journal of Propulsion and Power 31 (6) (2015) 1727–1738.

[240] M. Kahraman, I. Ozkol, M.A. Karabeyoglu, Regression rate enhancement of hybrid rockets by introducing novel distributed tube injector, Journal of Propulsion and Power 38 (2) (2022) 200–211.

[241] K. Araki, Y. Hirata, S. Oyama, K. Ohe, S. Aso, Y. Tani, T. Shimada, A study on performance improvement of paraffin fueled hybrid rocket engines with multi-section swirl injection method, in: 49th AIAA/ASME/SAE/ASEE Joint PropulsionConference, 2013, p. 3634.

[242] Y. Nobuhara, L.T. Kamps, H. Nagata, Fuel regression characteristics of CAMUI type hybrid rocket using nitrous oxide, in: AIAA Propulsion and Energy 2021 Forum, 2021, p. 3521.

[243] H. Sakashi, Y. Saburo, H. Kousuke, S. Takashi, Effectiveness of concave-convex surface grain for hybrid rocket combustion, in: 48th AIAA/ASME/SAE/ASEE Joint Propulsion Conference & Exhibit, 2012.

[244] M. Kumar, P. Joshi, Regression rate study of cylindrical stepped fuel grain of hybrid rocket, Materials Today: Proceedings 4 (8) (2017) 8208–8218.

[245] R. Bisin, A. Verga, D. Bruschi, C. Paravan, Strategies for paraffin-based fuels reinforcement: 3d printing and blending with polymers, in: AIAA Propulsion and Energy 2021 Forum, 2021, p. 3502.

[246] N. Bellomo, F. Barato, M. Faenza, M. Lazzarin, A. Bettella, D. Pavarin, Numerical and experimental investigation of unidirectional vortex injection in hybrid rocket engines, Journal of Propulsion and Power 29 (5) (2013) 1097–1113.

[247] K.-H. Moon, H.-C. Kim, S.-J. Lee, W.-J. Choi, J.-P. Lee, H.-J. Moon, H.-G. Sung, J.-K. Kim, A study on combustion characteristic of the hybrid combustor using non-combustible diaphragm, in: Proceedings of the Korean Society of Propulsion Engineers Conference, The Korean Society of Propulsion Engineers, 2011, pp. 258–262.

[248] H. Tian, X. Sun, Y. Guo, P. Wang, Combustion characteristics of hybrid rocket motor with segmented grain, Aerospace Science and Technology 46 (2015) 537–547.

[249] M. Bouziane, A. Bertoldi, P. Milova, P. Hendrick, M. Lefebvre, Performance comparison of oxidizer injectors in a 1-kN paraffin-fueled hybrid rocket motor, Aerospace Science and Technology 89 (2019) 392–406.

[250] J.R. Wertz, W.J. Larson, D. Kirkpatrick, D. Klungle, Space Mission Analysis and Design, vol. 8, Springer, 1999.

[251] G.G. Haselden, Cryogenic Fundamentals, Academic Press, London, 1971.

[252] E. Ring, Rocket Propellant and Pressurization Systems, Prentice-Hall, 1964.

[253] A.A. Chandler, B.J. Cantwell, G.S. Hubbard, Hybrid propulsion for solar system exploration, in: 47th AIAA/ASME/SAE/ASEE Joint Propulsion Conference & Exhibit, Joint Propulsion Conferences, American Institute of Aeronautics and Astronautics, July 2011.

[254] A.J. Boiron, B.J. Cantwell, Hybrid rocket propulsion and in-situ propellant production for future Mars missions, in: Joint Propulsion Conferences, American Institute of Aeronautics and Astronautics, 2013.

[255] E.T. Jens, B.J. Cantwell, G.S. Hubbard, B. Nakazono, Hybrid CubeSat propulsion system with application to a Mars aerocapture demonstration mission, in: 65th International Astronautical Congress, Toronto, Canada, September 2014, Proceedings, International Astronautical Federation, Toronto, Canada, 2014.

[256] J.D. Anderson Jr, Modern Compressible Flow with Historic Perspective, McGraw-Hill Education, 2021.

[257] G.V.R. Rao, Exhaust nozzle contour for optimum thrust, Journal of Jet Propulsion 28 (6) (1958) 377–382.

[258] E. Toson, A.M. Karabeyoglu, Design and Optimization of Hybrid Propulsion Systems for In-Space Application, Propulsion and Energy Forum, American Institute of Aeronautics and Astronautics, 2015.

[259] Report of the Propulsion and Energetics Panel Working Group 25, Structural Assessment of Solid Propellant Grains, Tech. Rep., Advisory Group for Aerospace Research and Development, 1997.

[260] G. Zilliac, B. Waxman, E. Doran, J. Dyer, A. Karabeyoglu, B. Cantwell, Peregrine Hybrid Rocket Motor Ground Test Results, 2012.

[261] NASA Spaceflight Human-System Standard, vol 2: Human Factors, Habitabilty and Environmental Health. NASA Technical Standard NASA-STD-3001, Rev. B, Standard, National Aeronautics and Space Administration, 2019.

[262] F. Barato, M. Grosse, A. Bettella, Hybrid rocket fuel residuals - an overlooked topic, 2014.

[263] J.T. Harding, R.H. Tuffias, R.B. Kaplan, High Temperature Oxidation Resistant Coatings, Afrpl tr-84-036, Prepared for the Air Force Rocket Propulsion Laboratory, Pacoima, CA, June 1984.

[264] B. Kahraman, H. Karakas, B.N. Eren, A. Karabeyoglu, Erosion Rate Investigation of Various Nozzle Materials in Hybrid Rocket Motors, 2020.

[265] D. Bianchi, F. Nasuti, Numerical analysis of nozzle material thermochemical erosion in hybrid rocket engines, Journal of Propulsion and Power 29 (3) (2013) 547–558.

[266] Toray Composite Materials America, Inc., Toray T1000G intermediate modulus carbon fiber, https://www.toraycma.com/wp-content/uploads/T1000G-Technical-Data-Sheet-1.pdf.pdf, 2021. (Accessed 28 December 2021).

[267] Northrop Grumman, Pressurant tanks data sheets – sorted by volume, https://www.northropgrumman.

com/space/pressurant-tanks-data-sheets-sorted-by-volume/, 2023. (Accessed 29 January 2023).

[268] Arde Aerojet Rocketdyne, Product line: COPV, https://www.ardeinc.com/copv.html, 2022. (Accessed 11 February 2020).

[269] Northrop Grumman, Diaphragm tanks data sheets – sorted by volume, https://www.northropgrumman.com/space/diaphragm-tanks-data-sheets-sorted-by-volume/, 2023. (Accessed 29 January 2023).

[270] Northrop Grumman, PMD tanks data sheets – sorted by part number, https://www.northropgrumman.com/space/pmd-tanks-data-sheets-sorted-by-part-number/, 2022. (Accessed 11 February 2020).

[271] E.W. Lemmon, M.L. Huber, M.O. McLinden, NIST Reference Fluid Thermodynamic and Transport Properties, REFPROP, 2002.

[272] G.W. Howell, T.M. Weathers, Aerospace Fluid Component Designers' Handbook. Volume I, Revision D, Tech. Rep., TRW SYSTEMS GROUP REDONDO BEACH CA, 1970.

[273] A. Benhidjeb-Carayon, E. Jens, A. Casillas, Rapid Blowdown Analysis and Application to Mars Lander Design, in preparation, 2023.

[274] J.E. Zimmerman, Self Pressurizing Propellant Tank Dynamics, Ph.D. thesis, Stanford University, 2015.

[275] G. Zilliac, M.A. Karabeyoglu, Modeling of Propellant Tank Pressurization, 2005.

[276] E. Paccagnella, F. Barato, R. Gelain, D. Pavarin, CFD Simulations of Self-pressurized Nitrous Oxide Hybrid Rocket Motors, 2018.

[277] K.M. Broughton, D.R. Williams, M.J. Brooks, J. Pitot, Development of the Phoenix-1B Mk II 35 km Apogee Hybrid Rocket, 2018.

[278] J. Dyer, E. Doran, Z. Dunn, K. Lohner, C. Bayart, A. Sadhawani, G. Zilliac, B. Cantwell, A. Karabeyoglu, Design and Development of a 100 km Nitrous Oxide/Paraffin Hybrid Rocket Vehicle, vol. 5362, 2007, p. 2007.

[279] J.E. Zimmerman, B.J. Cantwell, G. Zilliac, Parametric Visualization Study of Self-Pressurizing Propellant Tank Dynamics, 2015.

[280] I.J. Karassik, J.P. Messina, P. Cooper, C.C. Heald, et al., Pump Handbook, vol. 3, McGraw-Hill, New York, 2001.

[281] E. Jens, A.C. Karp, B. Nakazono, D.B. Eldred, M.E. DeVost, D. Vaughan, Design of a hybrid CubeSat orbit insertion motor, in: 52nd AIAA/SAE/ASEE Joint Propulsion Conference, 2016, p. 4961.

[282] E.T. Jens, A.C. Karp, B. Nakazono, K.T. Williams, J. Rabinovitch, D. Dyrda, F.S. Mechentel, Low Pressure Ignition Testing of a Hybrid SmallSat Motor, 2019.

[283] VACCO Industries, VACCO thruster valves, https://www.vacco.com/images/uploads/pdfs/liquid_propellant_thrusters.pdf, 2002. (Accessed 8 January 2023).

[284] Moog Space and Defense, Bipropellant thruster valves, https://www.moog.com/content/dam/moog/literature/sdg/space/propulsion/moog-bipropellant-thruster-valves-datasheet.pdf, 2020. (Accessed 8 January 2023).

[285] Moog Space and Defense, Monopropellant thruster valves, https://www.moog.com/content/dam/moog/literature/sdg/space/propulsion/moog-monopropellant-thruster-valves-datasheet.pdf, 2020. (Accessed 8 January 2023).

[286] Moog Space and Defense, Fill and drain service valves, https://www.moog.com/content/dam/moog/literature/sdg/space/propulsion/moog-manual-fill-and-drain-valves-datasheet.pdf, 2022. (Accessed 8 January 2023).

[287] Ariane Group, Space propulsion valves, https://www.space-propulsion.com/brochures/valves/space-propulsion-valves.pdf, 2023. (Accessed 8 January 2023).

[288] Stellar Technologies, Pressure transducer, https://www.stellartech.com/media/ST1300.pdf, 2021. (Accessed 8 January 2023).

[289] Tavis Corporation, Pressure transducer, https://pressure-transducers.taviscorp.com/compare/all-categories/pressure-transducers-for-satellite-propulsion, 2023. (Accessed 8 January 2023).

[290] VACCO Industries, VACCO check valves, https://www.vacco.com/images/uploads/pdfs/check_valves.pdf, 2004. (Accessed 8 January 2023).

[291] VACCO Industries, VACCO filtration, https://www.vacco.com/images/uploads/pdfs/VACCO_Filtration_Catalog_042121_FINAL_with_bookmarks_web.pdf, 2017. (Accessed 8 January 2023).

[292] VACCO Industries, VACCO latch valves, https://www.vacco.com/images/uploads/pdfs/latch_valves_high_pressure.pdf, 2004. (Accessed 8 January 2023).

[293] VACCO Industries, VACCO latch valves, https://www.vacco.com/images/uploads/pdfs/latch_valves_low_pressure.pdf, 2004. (Accessed 8 January 2023).

[294] Standard: Mass Properties Control for Space Systems (ANSI/AIAA S-120A-2015(2019)), 2015.

[295] F.M. White, Fluid Mechanics, 2011.

[296] P.J. Pritchard, J.W. Mitchell, Fox and McDonald's Introduction to Fluid Mechanics, John Wiley & Sons, 2011.

[297] D. Zigrang, N. Sylvester, A review of explicit friction factor equations, Journal of Energy Resources Technology 107 (2) (1985) 280–283.

[298] R.P. Benedict, Fundamentals of Pipe Flow, John Wiley & Sons, 1980.

[299] H. Ito, Pressure losses in smooth pipe bends, Journal of Basic Engineering (1960).

[300] D.C. Rennels, H.M. Hudson, Pipe Flow: A Practical and Comprehensive Guide, John Wiley & Sons, 2012.

[301] H.D. Beeson, S.R. Smith, W.F. Stewart, Safe Use of Oxygen and Oxygen Systems: Handbook for Design, Operation, and Maintenance, ASTM International, 2007.

[302] R.P. Benedict, N. Carlucci, S. Swetz, Flow losses in abrupt enlargements and contractions, Journal of Engineering for Power 88 (1966) 73–81.

[303] D. Guthrie, R. Wolf, Non-acoustic combustion instability in hybrid rocket motors, AIAA Paper 2916 (1991) 11.

[304] D.T. Harrje, F.H. Reardon, Liquid propellant rocket combustion instability. NASA SP-194, NASA Special Publication 194 (1972).

[305] B.S. Waxman, J.E. Zimmerman, B. Cantwell, G. Zilliac, Effects of injector design on combustion stability in hybrid rockets using self-pressurizing oxidizers, in: 50th AIAA/ASME/SAE/ASEE Joint Propulsion Conference, 2014, p. 3868.

[306] M.S. Natanzon, Combustion Instability, American Institute of Aeronautics and Astronautics, 2008.

[307] F.E. Culick, Acoustic oscillations in solid propellant rocket chambers, Astronautica Acta 12 (2) (1966) 113–126.

[308] F. Culick, V. Yang, Prediction of the Stability of Unsteady Motions in Solid-Propellant Rocket Motors, American Institute of Aeronautics and Astronautics, 1992.

[309] F. Culick, Combustion Instabilities in Solid Propellant Rocket Motors, Tech. Rep., California Institute of Technology, 2004.

[310] S.-J. Rhee, J.-P. Lee, H.-J. Moon, H.-G. Sung, J.-G. Kim, Hybrid rocket instability I, in: Proceedings of the Korean Society of Propulsion Engineers Conference, The Korean Society of Propulsion Engineers, 2012, pp. 81–85.

[311] J. Leea, S. Rheeb, J. Kimc, H. Moond, O. Shynkarenkoe, D. Simonef, T. Moritag, Combustion Instability for Hybrid Rocket Motors with a Diaphragm, 2019.

[312] E.T. Jens, A.C. Karp, V.A. Miller, G.S. Hubbard, B.J. Cantwell, Experimental visualization of hybrid combustion: results at elevated pressures, Journal of Propulsion and Power 36 (1) (2020) 33–46.

[313] M. Jones, T. Abel, D. Weeks, M. Jones, T. Abel, D. Weeks, Subscale hybrid rocket motor testing at the marshall space flight center in support of the hybrid propulsion demonstration program (HPDP), in: 33rd Joint Propulsion Conference and Exhibit, 1997, p. 2800.

[314] T.J. Barber, Final Cassini propulsion system in-flight characterization, in: 2018 Joint Propulsion Conference, 2018, p. 4546.

[315] C. Diaz, S. McKim, Development and Testing of the Europa Mission's Venturi Flow Meter, 2017.

[316] M. Hagopian, A. Dibbern, Failure of Pyrotechnic Operated Valves with Dual Initiators, NASA Engineering and Safety Center Technical Bulletin 09-01, NESC, 2009.

[317] Peter Paul Electronics, Solenoid Valves 101 - Whitepaper, https://peterpaul.com/pdf/Solenoid-Valves-101-Whitepaper.pdf, 2012. (Accessed 22 August 2023).

[318] Mission Evaluation Team, Apollo 15 Mission Report: Chapter 14, Tech. Rep., NASA, 1971.

[319] Swagelok Company, Proportional relief valves, https://www.swagelok.com/downloads/webcatalogs/en/ms-01-141.pdf, 2021. (Accessed 21 December 2021).

[320] NuSpace, Data sheets, https://keyengco.com/data-sheets/, 2022. (Accessed 11 February 2020).

[321] L. Randall, Rocket applications of the cavitating venturi, Journal of the American Rocket Society 22 (1) (1952) 28–38.

[322] J.D. Wright, A.N. Johnson, et al., ASME MFC-7–2016: Measurement of Gas Flow by Means of Critical Flow Venturis and Critical Flow Nozzles, 2016.

[323] Z.D. Husain, R.J. DeBoom, et al., ASME MFC-3M–2004: Measurement of Fluid Flow in Pipes Using Orifice, Nozzle, and Venturi, 2004.

[324] A. Ibrahim, Y. Ryu, M. Saidpour, Stress analysis of thin-walled pressure vessels, Modern Mechanical Engineering 05 (01) (2015) 9.

[325] D.P. Ankeney, C.E. Woods, DesignCriteria for Large Accurate Solid-Propellant Static-Thrust Stands, Tech. Rep. NOTS Technical Publication 3240, NAVWEPS Report 8353, Department of the Navy, Test Department, 1963.

[326] R.W. Miller, F.Z. Lee, et al., ASME MFC-3M–2004: Measurement of Fluid Flow in Pipes Using Orifice, Nozzle, and Venturi, 1986.

[327] Z.D. Husain, R.J. DeBoom, et al., ASME MFC-11–2014: Measurement of Fluid Flow by Means of Coriolis Mass Flowmeters, 2014.

[328] ASME, ASME MFC-5.1–2018: Measurement of Liquid Flow in Closed Conduits Using Transit-Time Ultrasonic Flowmeters, 2018.

[329] ASME, ASME MFC-5.3–2013: MFC- 5.3 Measurement of Liquid Flow in Closed Conduits Using Doppler Ultrasonic Flowmeters, 2013.

[330] F. Dijkstra, P. Korting, R.V.D. Berg, Ultrasonic regression rate measurement in solid fuel ramjets, 1963.

[331] J.-Y. Lestrade, J. Anthoine, O. Verberne, A.J. Boiron, G. Khimeche, C. Figus, Experimental demonstration of the vacuum specific impulse of a hybrid rocket engine, Journal of Spacecraft and Rockets 54 (1) (2017) 101–108.

[332] T. Boardman, L. Porter, F. Brasfield, T. Abel, An Ultrasonic Fuel Regression Rate Measurement Technique for Mixture Ratio Control of a Hybrid Motor, 1995.

[333] Orbital Technologies Corp. (ORBITEC), Embedded Sensors for Measuring Surface Regression, Tech. Rep., NASA, 2006.

[334] T. Boardman, D. Brinton, R. Carpenter, T. Zoladz, An experimental investigation of pressure oscillations and their suppression in subscale hybrid rocket motors, 1995.

[335] C. Carmicino, Acoustics, vortex shedding, and low-frequency dynamics interaction in an unstable hybrid rocket, Journal of Propulsion and Power 25 (6) (2009) 1322–1335.

[336] S.D. Zilwa, G. Zilliac, A. Karabeyoglu, M. Reinath, Combustion Oscillations in High Regression Rate Hybrid Rockets, 2003.

[337] J. Dyer, G. Zilliac, A. Sadhwani, A. Karabeyoglu, B. Cantwell, Modeling Feed System Flow Physics for Self-Pressurizing Propellants, 2012.

[338] W.H. Nurick, Orifice cavitation and its effect on spray mixing, Journal of Fluids Engineering (1976).

[339] Federal Aviation Administration, Metallic Materials Properties Development and Standardization (MMPDS-04), Battelle Memorial Institute, 2008.

[340] L. Jeffus, Welding: Principles and Applications, Cengage Learning, 2012.

[341] L.J. Martini, Practical Seal Design, CRC Press, 1984.

[342] G. Zilliac, Lecture 9 - seals, in: AA284b Advanced Rocket Propulsion Course, Stanford University, 2013.

[343] Parker Hannifin Corporation, Parker O-ring handbook ORD 5700, https://www.parker.com/Literature/O-Ring, 2018. (Accessed 26 April 2019).

[344] efunda, Design guidelines for radial seals, https://www.efunda.com/designstandards/oring/design_guidelines.cfm, 2022. (Accessed 22 January 2022).

[345] D. Tsang, B. Marsden, S. Fok, G. Hall, Graphite thermal expansion relationship for different temperature ranges, Carbon 43 (14) (2005) 2902–2906.

[346] P. Narsai, Nozzle Erosion in Hybrid Rocket Motors, Stanford University, 2016.

[347] D. Bianchi, M.T. Migliorino, M. Rotondi, L. Kamps, H. Nagata, Numerical analysis of nozzle erosion in hybrid rockets and comparison with experiments, Journal of Propulsion and Power (2021) 1–22.

[348] Kennametal, Pyrolytic graphite, technical data, https://s7d2.scene7.com/is/content/Kennametal/B-14-04141_KMT_Pyrolytic_Graphite_Datasheetpdf-1, 2014. (Accessed 3 February 2022).

[349] R. Bunker, A. Prince, Hybrid rocket motor nozzle material predictions and results, 1992.

[350] Matweb, Matweb materials database, www.matweb.com, 2022. (Accessed 3 February 2022).

[351] Laminated Plastics, Technical Data Sheet: Paper Phenolic, https://laminatedplastics.com/paperphenolic.pdf, 2022. (Accessed 3 February 2022).

[352] R.D. Carnahan, Mechanical Testing of Silica Phenolic Composites at Elevated Temperatures, Samso-tr-68-450, Prepared for the Air Force Systems Command, Los Angeles Air Force Station, Los Angeles, CA, October 1968.

[353] Northrup Grumman Innovation Systems, Propulsion products catalog, https://www.northropgrumman.com/Capabilities/PropulsionSystems/Documents/NGIS_MotorCatalog.pdf, 2018. (Accessed 26 April 2019).

[354] C. Hohmann, J. Bill Tipton, M. Dutton, Propellant for the NASA Standard Initiator, 2000.

[355] D.K. Huzel, D.H. Huang, Design of Liquid Propellant Rocket Engines, Office of Technology Utilization, National Aeronautics and Space Admistration, 1967.

[356] S.-G. Jang, J.-m. Hwang, S.-H. Baek, Analysis on shock wave and sensitivity of explosives in through-bulkhead initiator, Journal of the Korean Society of Propulsion Engineers 21 (08) (2017) 36–43.

[357] M.G. Faenza, A.J. Borion, B. Haemerli, S. Lennart, T. Vesterås, O. Verberne, Getting ready for space: Nammo's development of a 30kN hybrid rocket based technology demonstrator, in: 7th European Conference for Aeronautics and Space Sciences (EUCASS), 2017.

[358] E.S. Jung, S. Kwon, Autoignitable and restartable hybrid rockets using catalytic decomposition of an oxidizer, Journal of Propulsion and Power 30 (2) (2014) 514–518.

[359] Y.D. Scherson, Energy Recovery from Waste Nitrogen through N_2O Decomposition: the Coupled Aerobic-Anoxic Nitrous Decompostion Operation (CANDO), Ph.D. thesis, Stanford University, 2012.

[360] E.J. Wernimont, S.D. Heister, Combustion experiments in hydrogen peroxide/polyethylene hybrid rocket with catalytic ignition, Journal of Propulsion and Power 16 (2) (2000) 318–326.

[361] L. Simurda, G. Zilliac, High Performance Hybrid Propulsion System for Small Satellites, 2013.

[362] D.M. Dyrda, V. Korneyeva, B.J. Cantwell, Diode laser ignition mechanism for hybrid propulsion systems, Journal of Propulsion and Power 36 (6) (2020) 901–911.

[363] S.A. O'Briant, S.B. Gupta, S.S. Vasu, Review: laser ignition for aerospace propulsion, Propulsion and Power Research 5 (1) (2016) 1–21.

[364] D.M. Dyrda, A study of laser ignition for hybrid propulsion systems, Ph.D. thesis, Stanford University, 2020.

[365] Procedures for Performing a Failure Mode, Effects and Criticality Analysis, 1980.

[366] Air Force Space Command, Range safety user requirements manual volume 3 – launch vehicles, payloads, and ground support systems requirements, in: Manual 91-710, vol. 3, 2019.

[367] Ensign-Bickford Aerospace & Defense, Safe & arm device, electro mechanical, https://www.ebad.com/safe-arm-device-electro-mechanical/, 2020. (Accessed 29 December 2022).

[368] A.C. Karp, B. Nakazono, J. Benito Manrique, R. Shotwell, D. Vaughan, G.T. Story, A Hybrid Mars Ascent Vehicle Concept for Low Temperature Storage and Operation, 2016, p. 4962.

[369] M.J. Mueller, Density Fit for MON Oxidizer Blends Including Accuracy, 2015.

[370] Liquid Propulsion Technical Committee, Special Project: Fire, Explosion, Compatibility, and Safety Hazards of Nitrogen Tetroxide (AIAA SP-086-2001), American Institute of Aeronautics and Astronautics, 2001.

[371] G.E. Jensen, J. Keilbach, A. Holzman, R. Parsley, S.O. Leisch, J. Humphrey, Hybrid Propulsion Technology Program, Conceptual Design Package, Contract no nas 8-37778, Prepared for Geroge C. Marshall Space Flight Center, San Jose, CA, October 1989.

[372] Plastics — Determination of dynamic mechanical properties — Part 11: Glass transition temperature, Standard, International Organization for Standardization, Geneva, CH, 2019.

[373] Y. Murakami, Y. Matsumoto, R. Tanaka, K. Takahashi, I. Tanabe, M.K. Umasaki, Thermal behavior of aluminized wax-based solid fuel, Transactions of JSASS, Aerospace Technology Japan 16 (3) (2018) 280–284.

[374] ASME, ASME B1.3–2020: Process Piping, 2020.

[375] Anaesthetic and respiratory equipment — Compatibility with oxygen, Standard, International Organization for Standardization, June 2010.

[376] Swagelok, Special cleaning and packaging (SC-11), https://www.swagelok.com/downloads/webcatalogs/en/ms-06-63.pdf, 2017. (Accessed 1 April 2022).

[377] NIOSH Pocket Guide to Chemical Hazards, [Cincinnati, Ohio]: U.S. Dept. of Health and Human Services, Public Health Service, Centers for Disease Control and Prevention, National Institute for Occupational Safety and Health; Washington, DC: For sale by the U.S. G.P.O., Supt. of Docs., [2007], 2007.

[378] National Institute for Occupational Safety and Health, Workplace safety and health topics: nitrous oxide, https://www.cdc.gov/niosh/topics/nitrousoxide/default.html, 2018. (Accessed 1 December 2019).

[379] J. Campbell, F. Macklin, Z. Thicksten, Handling considerations of nitrous oxide in hybrid rocket motor testing, in: 44th AIAA/ASME/SAE/ASEE Joint Propulsion Conference & Exhibit, 2008.

[380] Scaled Composites, LLC, Scaled composites safety guidelines for N_2O, https://www.ibb.ch/publication/N2O/N2OSafetyGuidelines.pdf, 2009. (Accessed 20 December 2022).

[381] G. Rhodes, Investigation of Decomposition Characteristics of Gaseous and Liquid Nitrous Oxide, Tech. Rep., AIR FORCE WEAPONS LAB KIRTLAND AFB NM, 1974.

[382] K. Munke, Nitrous oxide trailer rupture, July 2, 2001, in: Report at CGA Seminar "Safety and Reliability of Industrial Gases, Equipment and Facilities", October 15 - 17, 2001, St. Louis, Missouri, 2001.

[383] B. Nufer, Hypergolic Propellants: the Handling Hazards and Lessons from Use, Nasa technical report, NASA, 2010.

[384] A.C. Wright, Martin Marietta Corporation, USAF Propellant Handbooks Nitric Acid/Nitrogen Textroxide Oxidizers, vol. II, AFRPL-TR-76-76, 1977.

[385] Associated Press, Liquid O_2 leak may have caused Butte rocket explosion, https://billingsgazette.com/news/state-and-regional/montana/liquid-o2-leak-may-have-caused-butte-rocket-explosion/article_74e61ba8-669d-11e0-a9f9-001cc4c002e0.html, 2011. (Accessed 26 December 2022).

[386] ASTM G93/G93M-19: Standard Guide for Cleanliness Levels and Cleaning Methods for Materials and Equipment Used in Oxygen-Enriched Environments, ASTM International, 2019.

[387] Precision Cleaning and Testing of Shipboard Oxygen, Helium, Helium-Oxygen, and Nitrogen Systems, 2007.

[388] CGA G-4.1-2018: Cleaning Equipment for Oxygen Service - Seventh Edition, Compressed Gas Association, Inc., 2018.

[389] ASTM G127-15: Standard Guide for the Selection of Cleaning Agents for Oxygen-Enriched Systems, ASTM International, 2015.

[390] ASTM G131-96(2016)e1: Standard Practice for Cleaning of Materials and Components by Ultrasonic Techniques, ASTM International, 2016.

[391] R. Runyan, J. Rynd Jr, J. Seely, Thrust stand design principles, 1992.

[392] PTC, ASME Test Uncertainty, PTC 19.1—2018 (R2018), ASME, New York, NY, USA, 1998.

[393] International Organization for Standardization, Uncertainty of Measurement-Part 3: Guide to the Expression of Uncertainty in Measurement (GUM: 1995), ISO, 2008.

[394] J.S. Walker, Fast Fourier Transforms, CRC Press, 2017.

[395] J.S. Bendat, A.G. Piersol, Data analysis: trend removal, in: Random Data: Analysis and Measurement Procedures, fourth edition, 2010, pp. 361–362.

[396] R.S. Figliola, D.E. Beasley, Theory and Design for Mechanical Measurements, John Wiley and Sons, Inc., 2006.

[397] Air Liquide, Nitrous oxide compatibility, https://encyclopedia.airliquide.com/nitrous-oxide#safety-compatibility.

[398] Gas cylinders — Compatibility of cylinder and valve materials with gas contents — Part 1: Metallic materials, Standard, International Organization for Standardization, Geneva, CH, 2020.

[399] W.K. Boyd, W.E. Berry, E. White, Compatibility of Materials with Rocket Propellants and Oxidizers, Defense Metals Information Center, Battelle Memorial Institute, 1965.

[400] Gas cylinders — Compatibility of cylinder and valve materials with gas contents — Part 2: Non-metallic materials, Standard, International Organization for Standardization, Geneva, CH, 2021.

Index

Printed in the United States
by Baker & Taylor Publisher Services